Applied Lubrication, Insulation, and Cooling Fluids for Electrical Power Systems

Applied Lubrication, Insulation, and Cooling Fluids for Electrical Power Systems provides evidence-based strategies for selecting, applying, and monitoring lubricants, coolants, and insulating fluids in power systems. It addresses the demands of turbines, wind drivetrains, nuclear actuators, and electrified systems, emphasizing fluid reliability as a critical factor in uptime, efficiency, and safety.

Condition monitoring is presented as a core reliability function, integrating American Society for Testing and Materials (ASTM) and International Organization for Standardization (ISO) test methods with operational data to detect early degradation modes like varnish, additive depletion, and electrical discharge erosion. The book explores viscosity stability, oxidation resistance, dielectric integrity, and thermal conductivity, with detailed coverage of both mineral-based and synthetic chemistries.

This book equips professionals to reduce failures, prevent outages, and optimize hybrid and electrified assets, making it invaluable for practicing original equipment manufacturers, maintenance, reliability, and mechanical engineers, condition monitoring specialists, and industrial consultants.

Michael D. Holloway is the President of 5th Order Industry, USA, which provides competency development and training programs for many industries and offers standard, customized, and certification preparation courses. Mr. Holloway holds a Bachelor of Science in Chemistry and a Bachelor of Arts in Philosophy from Salve Regina University, USA, as well as a Master of Science in Polymer Engineering from the University of Massachusetts, Lowell, USA. He has over 38 years of experience in the industry. His career began with research and development roles at Olin Chemical and WR Grace. He then moved into product management for Rohm & Haas, followed by technical marketing and application engineering for GE Plastics. Later, he focused on business development and management for NCH Corporation, ALS Global, and SGS Switzerland. Mr. Holloway has developed 18 online learning management systems, authored or edited 15 technical books, and holds 16 professional certifications, including certified lubrication specialist (CLS) and machinery lubrication engineer (MLE).

Applied Lubrication, Insulation, and Cooling Fluids for Electrical Power Systems

Michael D. Holloway

CRC Press
Taylor & Francis Group
Boca Raton London New York

CRC Press is an imprint of the
Taylor & Francis Group, an informa business

Designed cover image: John Van Ness

First edition published 2027
by CRC Press
2385 NW Executive Center Drive, Suite 320, Boca Raton FL 33431

and by CRC Press
4 Park Square, Milton Park, Abingdon, Oxon, OX14 4RN

CRC Press is an imprint of Taylor & Francis Group, LLC

ISBN: 978-1-041-30572-9 (hbk)
ISBN: 978-1-041-30575-0 (pbk)
ISBN: 978-1-003-77520-1 (ebk)

DOI: 10.1201/9781003775201

Typeset in Times
by Newgen Publishing UK

Dedication

This work is dedicated to my brother Mark Holloway who was instrumental in helping enact increased safety and disposal protocols for the Millstone Nuclear Power facility. The world should have many more like him.

Contents

SECTION V Failure Forensic Case Studies

SECTION VI Emerging Trends and Future Systems

Foreword by Greg Livingstone

Electric power is one of those things we rarely stop to think about until it is not there. We flip a switch, start a machine, cool a building, power a hospital, charge a phone, and move on with our day. But behind that quiet reliability is a world of engineering discipline, technical judgment, and constant attention to details that most people never see.

Some of the most important of those details are fluids.

That may sound like an odd statement to anyone outside our field, but those of us who have spent our careers around rotating equipment and reliability know better. Lubricants, insulating fluids, and cooling fluids are not side notes in power generation. They are essential to how equipment performs, how long it lasts, and how reliably it serves the people and industries that depend on it.

That is why this book matters, and why Mike is the right person to write it.

I have known Mike for more than a decade, and over that time I have come to appreciate something that sets him apart. He is deeply knowledgeable, yes, but knowledge alone is not what makes him unusual. What makes Mike unique is the way he thinks. I do not know anyone more thoughtful about the science of lubrication. He is endlessly curious, always asking one more question, always trying to understand not just what works, but why it works. And beyond that, he is always trying to connect the dots – to see the larger system, the broader implication, the big picture.

That quality is rare.

A lot of people can collect information. A much smaller number can turn information into understanding. Mike does that exceptionally well. He has built a body of work that reflects not only technical depth, but genuine intellectual discipline. He has written more than perhaps anyone else in the lubricant industry, and he has done it in a way that helps people think more clearly about our space. Mike elevates the conversation around lubrication.

What I have always admired most, though, is that his curiosity is not limited to the narrow boundaries of work. It is simply part of who he is.

I remember walking through a souk in Dubai with Mike. It would have been easy to stroll through, glance at the displays, and keep moving. Mike did the opposite. He stopped to talk with the vendors selling spices, perfumes, gold jewelry – whatever was in front of him. He wanted to know where things came from, how they were made, what they meant, and who the people were behind them. He was not making small talk. He was genuinely interested. He wanted to understand the culture, the history, and the people connected to what he was seeing.

That memory has stayed with me because it captures something essential about him. Mike moves through the world with real curiosity. He pays attention. He wants to learn. He wants to understand. And the lubricant industry is fortunate that he has brought that same mindset to our world. This book is a reflection of that.

Applied Lubrication, Insulation, and Cooling Fluids for Electrical Power Systems is not just a technical manual, though it is certainly technically rich. It is a serious and thoughtful treatment of a subject that does not always get the attention it deserves.

Power generation is the backbone of modern life. It supports health, safety, economic growth, industrial productivity, and the everyday comfort most people take for granted. None of that happens without reliable systems. And reliable systems do not happen by accident.

They depend on sound engineering decisions, disciplined operating practices, and a clear understanding of the materials that make performance possible. Fluids are a major part of that story.

What Mike has done in this book is bring together the knowledge that matters in one place. He helps the reader understand how these critical fluids function, how they should be selected, how they degrade, how they should be monitored, and how they influence the performance and reliability of power generation equipment. He gives the topic the seriousness it deserves.

That resonates with me personally. I have spent my career studying lubrication for rotating equipment, and I read this book with a mix of admiration and envy. Admiration because it is excellent work. Envy because more than once I caught myself thinking, *I wish I had written this book.*

I mean that sincerely.

For reliability engineers, lubrication engineers, plant operators, managers, and anyone interested in the science behind machine performance, this book will be valuable. But more than that, it will help reinforce an important truth: the details matter. In our field, the small things are often not small at all. A fluid choice, a contamination issue, a thermal problem, a missed warning sign in an analysis report. These can become the difference between dependable operation and costly failure.

Mike understands that. He also understands how to explain it in a way that is grounded, practical, and useful.

Mike has given the industry something meaningful here. He has created a book that captures a critical body of knowledge for the power generation space and presents it with both rigor and clarity. Readers will benefit from that for years to come.

I am pleased to introduce this work, and even more pleased to see Mike's thinking brought to life in these pages.

This is an important book. And it could not have been written by a more fitting author.

Greg Livingstone, March 2026

Introduction

Lubricants in power generation must meet the combined mechanical, thermal, and chemical demands of each system. Equipment reliability depends on selecting formulations matched to load, speed, temperature, and environmental exposure. The wrong choice or poor monitoring leads to wear, varnish, insulation breakdown, or thermal runaway.

Thermal plants, nuclear stations, wind turbines, hydroelectric units, gas turbines, and electrified systems each present different lubricant service conditions. In steam turbines, high thermal stability and oxidation resistance are critical to avoid sludge and deposit formation in bearings and control systems. Gas turbines require lubricants with high film strength, resistance to thermal degradation, and rapid water separation to prevent bearing distress. Hydroelectric units demand long-drain interval turbine oils with strong demulsibility and rust inhibition.

Wind turbine gearboxes operate under high torque at fluctuating speeds and low temperatures; synthetic gear oils with high micropitting resistance, extreme-pressure performance, and oxidation stability are standard. In nuclear power, lubricants must meet radiation stability requirements and comply with nuclear-grade specifications to avoid chemical degradation and contamination.

Electrification brings additional fluid demands. Electric motors, generators, and transformers require dielectric strength, thermal conductivity, and compatibility with winding insulation. EV drivetrains use integrated e-motor and gearbox fluids with electrical insulation, copper corrosion protection, and thermal stability under high-voltage conditions. Cooling fluids in battery and inverter systems must prevent galvanic corrosion, maintain low conductivity, and remain stable under thermal cycling.

Condition monitoring in these environments must combine physical, chemical, and electrical property testing. ISO and ASTM methods for viscosity, oxidation stability, particle count, dielectric breakdown, and additive depletion form the baseline. Failure investigations require correlation of lubricant condition with operating data to identify root causes such as thermal stress, contamination ingress, additive depletion, or electrical discharge erosion.

Reliability programs in power generation demand lubricant selection, application, and monitoring strategies tailored to each generation method. This requires knowledge of system loads, contamination sources, and component metallurgy, combined with a program of oil analysis, filtration, and fluid replacement based on condition rather than fixed intervals.

ACHIEVEMENT PATHWAY FOR PROSPECTIVE GOALS

Lubricant selection and management determine the reliability of rotating equipment across power generation assets. Failures linked to fluid breakdown, additive depletion, or contamination are preventable when lubrication engineering is treated as a core reliability discipline rather than a consumable purchase. Power plants operate under diverse mechanical and environmental stresses. Steam and gas turbines run at high thermal loads with tight bearing tolerances. Hydroelectric systems experience slow-speed, high-load contact with risk of water ingress. Wind turbines combine fluctuating torque cycles with remote service challenges. Nuclear and electrified platforms impose additional requirements for dielectric strength, radiation resistance, and system compatibility.

A structured program is required to align lubricant chemistry with asset duty, validate condition in service, integrate with electrical and thermal performance requirements, and continually refine specifications through root cause feedback. When this framework is applied consistently, lubricant-related failures are reduced, unplanned outages are avoided, and operational costs decline without compromising environmental compliance.

TAILORED LUBRICANT APPLICATION STRATEGIES

This work provides the technical framework for matching lubricant chemistries to the operational envelope of each power generation technology. By defining the mechanical, thermal, and chemical demands unique to steam turbines, gas turbines, hydroelectric gear systems, wind turbine drivetrains, nuclear plant rotating equipment, and electrified systems, it ensures that selection is evidence-based and not driven by generic supplier claims. The result is a marked reduction in lubricant-induced failures and unplanned outages.

INTEGRATED CONDITION MONITORING AND DIAGNOSTICS

The proposed approach embeds oil analysis and dielectric testing into the maintenance routine, using ASTM/ISO standardized test suites for viscosity, oxidation stability, water contamination, particulate load, dielectric breakdown, and additive depletion. These results are correlated with operational and environmental data to detect early-stage degradation mechanisms. This enables predictive interventions before component distress escalates to catastrophic failure, thereby extending asset life and reducing lifecycle costs.

CROSS-DISCIPLINARY RELIABILITY ALIGNMENT

Lubrication requirements are integrated with electrical, thermal, and mechanical system considerations. For electrified systems, this includes dielectric compatibility with insulation, prevention of stray current corrosion, and heat transfer stability in cooling circuits. The methodology aligns lubrication engineering with electrical reliability and thermal management disciplines, enabling a unified maintenance strategy for hybrid and fully electric generation assets.

FAILURE INVESTIGATION FEEDBACK LOOP

Root cause analysis of lubricant-related failures feeds directly back into the lubricant selection, application, and monitoring program. By maintaining a closed-loop feedback system linking lab analysis, field observations, and component teardown data, the program continually refines specifications and operating practices. This iterative process raises the standard of reliability and minimizes the recurrence of known failure modes.

OPERATIONAL SUSTAINABILITY AND COST CONTROL

Condition-based lubricant change intervals, enhanced contamination control, and high-performance fluid selection reduce both lubricant consumption and waste. This aligns with environmental compliance requirements while reducing the total cost of ownership. The approach also supports the integration of renewable generation assets by ensuring lubricants meet extended-life and low-maintenance demands typical of remote or offshore installations.

BEST USE OF THIS WORK

Apply this framework as both a reference and an operational playbook. Use the fluid selection and compatibility guidelines during design and procurement stages to ensure correct specification before installation. Implement the condition monitoring protocols as part of routine maintenance to establish performance baselines and track degradation trends. Integrate the predictive diagnostic methods into existing computerized maintenance management systems or reliability programs to shift toward planned interventions and minimize forced outages. Reference the lifecycle extension strategies when evaluating reconditioning, filtration upgrades, or alternative fluid chemistry. For compliance and sustainability initiatives, use the standards alignment and environmental sections to validate practices against regulatory requirements and corporate environmental, social, and governance goals. Treated as an active, working document, this material should inform technical training, guide field inspections, and support continuous improvement across all power generation fluid systems.

Section I

Foundations of Fluid Functionality in Power Systems

1 The Role of Fluids in the Power Landscape

Fluids in power generation and electrified systems are functional components, not passive consumables. Their role is divided into three primary categories: lubrication, insulation, and cooling. Lubricants reduce mechanical friction and wear in rotating equipment such as turbine bearings, gearbox meshes, and motor shaft journals. Insulating fluids maintain dielectric separation between conductive elements, preventing arcing and electrical discharge while providing partial thermal management. Cooling fluids transfer heat away from critical components, maintaining operational stability and preventing thermal runaway.

These functions must meet cross-disciplinary demands. Lubricants in generator bearings must tolerate high temperature and resist oxidation while coexisting with insulation systems that require non-conductive, moisture-free environments. Transformer oils must combine dielectric integrity with heat dissipation properties. Coolants in power electronics must manage localized thermal spikes without degrading surrounding seals, wiring, or coatings.

Fluid demands vary widely across the power infrastructure spectrum. Steam and gas turbines require high thermal stability turbine oils to manage heat and resist deposit formation under continuous load. Nuclear actuator assemblies may use specialty greases and hydraulic fluids compatible with radiation exposure and extreme temperature gradients. Renewable assets like wind turbines place high stress on gear oils and greases due to variable load cycles and environmental exposure. Electrified transport and stationary storage systems rely on dielectric coolants or immersion fluids that manage heat and prevent electrical bridging between high-voltage components.

Fluid performance is directly linked to equipment reliability. Viscosity loss, oxidation, contamination, or additive depletion can initiate mechanical failure, electrical breakdown, or thermal instability. Condition monitoring of these fluids is a core reliability practice. Safety also depends on correct fluid specifications, as flash point, toxicity, and decomposition products must align with the operational environment and regulatory requirements. Sustainability considerations influence fluid selection through biodegradability, life extension potential, and reduced environmental impact during production, service, and disposal.

DOI: 10.1201/9781003775201-2

FUNCTIONAL CATEGORIES: LUBRICATION, INSULATION, COOLING

Lubrication

Applied to reduce friction between moving surfaces, prevent adhesive wear, and control surface fatigue (see Table 1.1). In power generation, lubricants support bearings, gears, and hydraulic actuators in turbines, generators, pumps, and auxiliary drives. Performance requirements include maintaining viscosity under thermal load, resisting oxidation, controlling varnish formation, and providing corrosion inhibition. Formulations vary from mineral-based turbine oils to synthetic esters and PAOs, each selected for thermal stability, load-carrying capacity, and compatibility with seals and metallurgy.

Insulation

Provides dielectric separation between energized conductors, preventing short circuits and electrical discharge. Insulating fluids such as mineral transformer oils, synthetic esters, and fluoroketones also remove heat from windings and cores. Critical properties include dielectric strength, moisture tolerance, oxidation stability, and gas absorption profile for dissolved gas analysis. Gaseous insulators such as SF_6 and CO_2 blends serve in switchgear, while solid impregnated materials and insulating greases protect terminations and contacts.

Cooling

Removes heat from electrical windings, bearings, reaction zones, and other components under thermal load. Cooling media include water-based systems, dielectric liquids, phase-change materials, and gases such as hydrogen in large generators. Effective cooling depends on thermal conductivity, specific heat capacity, flow rate, and compatibility with metals, elastomers, and coatings in the circuit. Inadequate cooling accelerates insulation aging, increases lubricant oxidation rate, and can initiate component distortion or seizure.

CROSS-DISCIPLINARY DEMANDS: ELECTRICAL, THERMAL, MECHANICAL

In power generation and high-duty industrial systems, functional fluids are subjected to simultaneous electrical, thermal, and mechanical stresses. No single property can be treated in isolation. A fluid that meets dielectric requirements may fail under thermal oxidation, while one that resists shear may degrade electrically under transient surges. Reliability depends on balancing these demands across the operating envelope.

The electrical dimension centers on dielectric integrity: high breakdown voltage, low conductivity, and resistance to moisture or particulate contamination. These

TABLE 1.1
Functional Categories in Power Generation Systems

Category	Primary Function	Typical Media	Critical Properties	Applications in Power Assets	Failure Implications
Lubrication	Reduce friction, control wear, prevent surface fatigue	Mineral turbine oils, synthetic esters, PAOs	Viscosity stability under load, oxidation resistance, varnish control, corrosion inhibition	Bearings, gears, hydraulic actuators in turbines, pumps, and auxiliary drives	Increased wear, varnish-induced sticking, bearing seizure, unplanned outages
Insulation	Provide dielectric separation, prevent electrical discharge	Transformer oils, synthetic esters, fluoroketones, SF_6, CO_2 blends, insulating greases	Dielectric strength, moisture tolerance, oxidation stability, gas absorption profile (DGA suitability)	Transformers, switchgear, generator windings, terminations, contacts	Short circuits, partial discharge, insulation breakdown, catastrophic electrical failure
Cooling	Remove heat from components under load	Water-based fluids, dielectric liquids, hydrogen gas, phase-change materials	Thermal conductivity, heat capacity, flow stability, material compatibility	Electrical windings, generator cores, bearings, reactor zones	Accelerated insulation aging, lubricant oxidation, distortion of components, thermal runaway

requirements are critical in transformers, switchgear, and power electronics where fluids flow near energized conductors.

The thermal dimension focuses on heat transport. Fluids must remove energy from reaction zones and winding cores without chemical breakdown. Properties such as thermal conductivity, heat capacity, and viscosity under load determine cooling efficiency. Long-term oxidative stability governs service life in turbines, dielectric coolants, and solar thermal loops.

The mechanical dimension emphasizes film strength, shear stability, and wear control. Bearings, gears, and hydraulic circuits require lubricants with defined viscosity and additive performance under variable load and vibration. Mechanical reliability also includes compatibility with seals and elastomers, preventing fluid-induced degradation of contact materials.

Taken together, these cross-disciplinary demands define the operating envelope of lubricants, coolants, and insulating fluids. Each property must be validated against the electrical, thermal, and mechanical stresses present in the system to prevent early degradation and unplanned failure.

Electrical

Fluids must maintain dielectric properties under operating voltage and transient surges. This includes high breakdown voltage, low electrical conductivity, and stability in the presence of partial discharges or corona. In transformers and switchgear, contamination by moisture or conductive particles reduces insulation integrity. In power electronics, dielectric coolants must prevent tracking and arc propagation while circulating close to energized components.

Thermal

Fluids must transfer heat efficiently from high-load zones to heat exchangers or radiators without degradation. Thermal conductivity, specific heat capacity, and viscosity at operating temperature define cooling performance. Oxidative stability under sustained high temperature is critical for turbine oils and synthetic dielectric coolants. For nuclear, solar thermal, and concentrated photovoltaic systems, the fluid must remain stable across wide temperature ranges and resist thermal decomposition into corrosive or deposit-forming byproducts.

Mechanical

Lubricants and hydraulic fluids must maintain film strength under load, resist shear thinning, and protect against wear in high-contact stress regions. Gearboxes, journal bearings, and pump rotors demand load-carrying capacity defined by base oil viscosity and additive chemistry. Fluids must also tolerate vibration, pressure fluctuations, and aeration without cavitation damage or foaming that disrupts film integrity. Mechanical stability also applies to seals and gaskets, fluids must not cause swelling, shrinkage, or embrittlement in contact materials.

FROM STEAM TURBINES TO BATTERY PACKS: EVOLVING FLUID NEEDS

Steam turbines operate with mineral or synthetic turbine oils engineered for high thermal stability, oxidation resistance, and deposit control in continuous-duty environments. These systems prioritize lubrication film integrity under high load and sustained temperature, along with varnish control to prevent servo valve sticking and bearing failure.

Gas turbines require similar base oil stability but face higher combustion gas ingress, necessitating strong antioxidant and detergent chemistry. Combined-cycle units integrate both steam and gas turbine lubrication demands, often using shared or closely related oil specifications.

Wind turbine drivetrains rely on gear oils with high micropitting resistance, water tolerance, and additive systems that protect against fretting corrosion from oscillatory loads. Bearings in nacelle-mounted generators demand greases that resist washout and maintain consistency under variable speeds and temperatures.

Hydroelectric units require lubricants and hydraulic fluids with exceptional water separation and corrosion resistance, often with environmental acceptability certifications for use near waterways.

In nuclear plants, actuator and control rod drive lubricants must maintain stability under radiation exposure and temperature extremes, while primary and secondary coolants manage heat removal without introducing corrosive species.

For power electronics, including inverters and converters, dielectric coolants must manage localized thermal hotspots while preventing electrical tracking between components. Transformers depend on insulating fluids with combined dielectric and thermal performance.

Battery packs, especially in electric grid storage and EV applications, employ immersion coolants or liquid thermal management fluids with high dielectric strength, low viscosity, and strong oxidation resistance to prevent cell-to-cell shorting and control heat during charge-discharge cycles. These fluids also require chemical stability in contact with cell casing materials, busbars, and polymer separators.

OVERVIEW OF FLUID INTERACTIONS WITH SYSTEM RELIABILITY, SAFETY, AND SUSTAINABILITY

Fluids are active reliability components in power generation and industrial systems. Their chemical and physical condition determines how well machinery, electrical systems, and thermal circuits perform under load. Reliability depends on viscosity retention, dielectric strength, and thermal stability, each of which can degrade through oxidation, contamination, or chemical breakdown. A lubricant that loses viscosity under shear exposes bearings to fatigue; an insulating fluid contaminated with moisture compromises dielectric separation; a coolant fouled with deposits reduces heat transfer and accelerates overheating.

Safety is tied directly to fluid properties and their degradation pathways. Low flash point fluids, unstable under heat or arcing conditions, increase fire and explosion risk.

In electrical systems, failure of dielectric fluids can propagate arcs or short circuits, leading to catastrophic events. Decomposition of working fluids may release corrosive agents or hazardous gases, raising exposure risks for operators in confined spaces or high-voltage environments.

Sustainability considerations shape fluid selection, operation, and disposal practices. Long-life synthetics reduce waste volume, while biodegradable fluids mitigate ecological damage in the event of leaks. Low global warming potential alternatives are replacing fluorinated and persistent compounds in insulating and cooling applications. Reclamation and re-refining programs extend fluid utility and align with regulatory requirements for emissions and waste management.

Together, these factors illustrate how fluids are not passive consumables but integrated elements of system design that influence performance, risk, and environmental impact across the operating lifecycle (see Table 1.2).

TABLE 1.2
Fluid Interactions with Reliability, Safety, and Sustainability

Dimension	Failure Modes	Risks	Mitigation Strategies
Reliability	Viscosity loss, oxidation, varnish formation, additive depletion, water ingress, particulate contamination, coolant fouling	Bearing fatigue, gear pitting, servo valve sticking, insulation breakdown, overheating, shortened service intervals, unplanned outages	Routine oil analysis (ASTM/ISO), dielectric strength testing, filtration upgrades, water removal systems, correct fluid specification, condition-based replacement
Safety	Low flash point, fluid volatility, dielectric breakdown, arc propagation, decomposition into corrosive or toxic byproducts, hydraulic leaks onto hot surfaces	Fire, explosion, transformer rupture, toxic gas release, electrical faults, operator injury, collateral equipment damage	High flash point or fire-resistant fluids (HFDU, HFDR, HFDC), dielectric testing under load, arc suppression fluids, gas monitoring, leak detection, material compatibility validation
Sustainability	Frequent changeouts, non-biodegradable spills, fluids with high GWP, improper disposal or venting, lack of reclamation programs	High waste volume, ecological persistence, regulatory non-compliance, increased lifecycle emissions	Use of biodegradable and EALs, long-life synthetics, re-refining/recycling programs, low-GWP dielectric and refrigerant fluids, fluid monitoring to extend service life

Reliability

Fluid condition directly affects mechanical integrity, electrical insulation performance, and thermal stability. Lubricant viscosity loss or oxidation can cause bearing wear, gear pitting, and servo valve sticking. Moisture ingress into insulating fluids lowers dielectric strength and accelerates winding degradation. Coolant fouling reduces heat transfer, leading to component overheating and shortened service intervals. Contaminant control, additive health, and correct fluid specification are essential to preventing unplanned outages.

Safety

Fluid selection influences fire risk, toxicity, and equipment failure modes. Low flash point or thermally unstable fluids increase ignition potential in fault conditions. Electrical arc suppression and insulation integrity rely on the fluid's dielectric stability under load. Chemical decomposition products can create hazardous gases or corrosive agents. In confined or high-voltage environments, improper fluid choice or degradation can cause catastrophic failure such as transformer explosions, coolant leaks onto energized conductors, or hydraulic fires.

Sustainability

Sourcing, in-service management, and end-of-life handling determine environmental impact. Biodegradable lubricants and environmentally acceptable lubricants (EALs) reduce persistence in case of leaks. Long-life synthetic fluids lower changeout frequency, reducing waste volume. Fluids with low global warming potential (GWP) and no persistent fluorinated compounds limit climate and ecological impact. Effective reclamation, re-refining, and recycling programs extend fluid utility while meeting regulatory compliance for disposal and emissions.

2 Overview of Power Generation and Electrified Equipment

Power generation and electrified industrial systems depend on the performance of working fluids for lubrication, insulation, and thermal management. Each asset type imposes a distinct combination of mechanical, thermal, and electrical stresses on the fluids in service. Coal, oil, and natural gas plants operate turbines, engines, and auxiliary drives that demand turbine oils and engine lubricants with high resistance to oxidation, viscosity breakdown, and contamination. Nuclear systems add requirements for radiation stability, coolant chemistry control, and long-life lubricants compatible with specialized seals and drive mechanisms. Renewable generation assets such as wind and hydroelectric plants operate in environments where moisture, variable load profiles, and extended service intervals place equal emphasis on water tolerance, antiwear protection, and environmental acceptability of the fluids. Solar thermal and geothermal systems impose wide temperature ranges and exposure to corrosive working media, requiring fluids with strong oxidative and thermal stability.

Distributed assets including diesel and biogas generators, microgrids, and hybrid configurations introduce further complexity with frequent cycling, diverse fuels, and integrated energy storage. In these applications, lubricants must manage soot, fuel dilution, and acid formation while coolants provide effective temperature control under variable duty. Across motors, gearboxes, transformers, switchgear, inverters, and batteries, reliability depends on fluids that maintain film integrity, dielectric strength, and heat transfer performance under long service durations. Electrified equipment in mining, rail, heating, ventilation, and air conditioning (HVAC), and process industries imposes additional challenges of vibration, shock load, and contaminant exposure that require careful matching of fluid chemistry to component duty and environment.

This overview establishes the cross-asset importance of fluid specification and condition monitoring. Reliability, safety, and sustainability outcomes are directly tied to how well these fluids retain their functional properties under demanding service conditions.

DOI: 10.1201/9781003775201-3

THERMAL: COAL, OIL, NATURAL GAS (TURBINE + ENGINE)

Coal-fired power plants use steam turbines driven by boilers that convert water to high-pressure steam. Lubrication requirements focus on turbine bearings, gearboxes, and feedwater pump drives. Turbine oils must maintain viscosity stability, resist oxidation, and control varnish under continuous high-temperature operation. Steam purity is critical to prevent deposit formation on turbine blades.

Oil-fired units operate on similar steam turbine principles but require additional fuel-handling systems with pumps and valves lubricated by mineral or synthetic oils. Combustion residue management is less severe than coal but still demands fluid cleanliness to protect control systems and reduce acid formation.

Natural gas power plants operate in two primary modes: simple-cycle gas turbines and combined-cycle plants that integrate gas and steam turbines. Gas turbines require turbine oils with superior oxidation resistance and deposit control due to higher operating temperatures and potential for fuel gas ingress into the lubrication system. Auxiliary systems such as starting motors, reduction gears, and generator bearings demand compatible lubricants.

Reciprocating engine-based plants, whether oil- or gas-fueled, use engine oils meeting American Petroleum Institute (API) or original equipment manufacturer (OEM)-specific classifications. These fluids must control wear, neutralize combustion acids, and resist nitration. Cooling systems employ water-glycol or other heat transfer fluids to manage high cylinder and head temperatures. Fuel contamination control is essential to maintaining oil integrity in these systems.

NUCLEAR: PWR, BWR, HTGR, MSR, SMR

Pressurized water reactors (PWR) use high-pressure light water as both coolant and neutron moderator. Reactor coolant chemistry control is critical to prevent corrosion of primary loop materials and to maintain radiation shielding performance. Lubricants in control rod drives, coolant pumps, and auxiliary turbines must tolerate radiation exposure, high temperatures, and potential contamination with borated water.

Boiling water reactors (BWR) generate steam directly in the reactor vessel, exposing turbine and balance-of-plant systems to trace radioactivity. Turbine oils must resist oxidation and maintain cleanliness under potential radiation exposure. Cooling and lubrication systems must prevent cross-contamination between radioactive and non-radioactive circuits.

High temperature gas-cooled reactors (HTGR) use helium as the primary coolant. Gas purity is essential to avoid graphite core degradation and component wear. Helium circulators and drive systems require lubricants and seals that resist permeation and thermal breakdown at elevated temperatures.

Molten salt reactors (MSR) operate with liquid fluoride or chloride salts as coolant and, in some designs, fuel carrier. Salt chemistry control prevents corrosion of piping and heat exchangers. Lubricants in pumps and control mechanisms must be compatible with high ambient temperatures and corrosive vapors from the salt circuit.

Small modular reactors (SMR) may employ PWR, BWR, or advanced coolant designs at reduced scale. Compact layouts demand fluids with high thermal performance, radiation resistance, and long service life. Lubricants, coolants, and insulating fluids in SMR auxiliary systems must be selected for minimal maintenance intervals and resistance to degradation in isolated, potentially remote installations.

RENEWABLE: WIND, HYDROELECTRIC, SOLAR (PV AND THERMAL), GEOTHERMAL

Wind turbines require gear oils with high load-carrying capacity, micropitting resistance, and water tolerance for the main gearbox. Generator and yaw/pitch bearings use greases with stability under variable loads, temperature fluctuations, and exposure to moisture. Hydraulic fluids for pitch control systems must resist oxidation and prevent varnish formation while maintaining low-temperature pumpability.

Hydroelectric plants operate in high-moisture environments, demanding lubricants and hydraulic fluids with excellent water separation, corrosion inhibition, and compatibility with environmentally acceptable lubricant (EAL) formulations to meet waterway protection regulations. Turbine wicket gate bearings and Kaplan hub oil systems require fluids with strong antiwear protection and resistance to emulsification.

Solar photovoltaic (PV) systems have minimal lubrication needs but may employ dielectric coolants in inverters and transformers. Concentrated solar thermal systems use heat transfer fluids such as synthetic aromatic oils or molten salts. These must retain thermal stability at sustained high temperatures, resist oxidation, and avoid fouling in receiver tubes and storage tanks.

Geothermal power systems handle fluids that can be high in dissolved solids, gases, and corrosive species. Turbine and pump lubricants must resist contamination from steam or brine leaks. Geothermal brine heat exchangers and circulation pumps use material-compatible lubricants and coolants capable of withstanding mineral scaling, corrosion, and thermal cycling without rapid degradation.

DIESEL AND BIOGAS GENERATORS

Diesel generator engines require lubricants meeting API CK-4, FA-4, or OEM-specific standards to manage soot loading, fuel dilution, and acid formation from sulfur compounds. Oils must maintain viscosity under load, neutralize combustion acids, and resist oxidation and nitration in standby or continuous-duty service. High total base number (TBN) formulations are preferred where fuel sulfur levels are higher. Cooling systems use water-glycol or extended-life coolants to control head and cylinder temperatures, with additive packages that protect against cavitation erosion in wet liners.

Biogas-fueled generators operate under higher risk of oil contamination from acidic and corrosive combustion byproducts such as hydrogen sulfide and siloxanes. Oils require strong oxidation stability, nitration resistance, and ash control to prevent pre-ignition and valve deposits. Frequent oil analysis is necessary to monitor acid number increase, silicon contamination, and additive depletion. Coolants must be compatible with engine metallurgy and maintain inhibitor effectiveness against the corrosive compounds present in biogas combustion condensate.

MICROGRIDS AND DISTRIBUTED GENERATION

Microgrids combine multiple generation assets, diesel or gas engines, microturbines, photovoltaic arrays, wind turbines, fuel cells, and battery energy storage, operating either in grid-connected or islanded mode. Distributed generation systems are typically located close to the load, which reduces transmission losses but increases equipment diversity and the range of fluid requirements.

Reciprocating engines in microgrids use lubricants matched to fuel type, duty cycle, and emissions control technology. Gas engines require low-ash formulations to protect catalytic converters, while diesel units demand oils with strong soot dispersion and acid neutralization. Microturbines use synthetic turbine oils with high oxidative stability and deposit control.

Battery energy storage systems in microgrids may use immersion cooling or indirect liquid cooling with dielectric fluids to control cell temperature and prevent electrical bridging. Power electronics such as inverters and converters require cooling fluids with high dielectric strength, thermal stability, and compatibility with polymeric and metallic components.

Hybrid configurations increase the importance of fluid compatibility across subsystems. Hydraulic fluids, lubricants, insulating fluids, and coolants must be selected to operate reliably in environments with variable load profiles, frequent start-stop cycles, and rapid thermal transients. Condition monitoring is essential to detect degradation early and maintain system availability in these often unmanned or remotely managed installations.

MOTORS, GEARBOXES, TRANSFORMERS, SWITCHGEAR, INVERTERS, BATTERIES

Motors

Industrial and utility-scale electric motors rely on grease-lubricated or oil-lubricated bearings. Grease selection depends on bearing size, operating speed, load, and ambient temperature. Polyurea, lithium complex, and calcium sulfonate thickened greases are common, with base oils matched to temperature and speed conditions. Lubricants must resist channeling, bleeding, and contamination ingress while maintaining film integrity under vibration and misalignment.

Gearboxes

Gear reducers in wind turbines, hydro plants, and auxiliary drives use mineral or synthetic gear oils formulated for antiwear and extreme pressure performance. Micropitting resistance is critical in high-torque, high-speed gear meshes. Additive systems must be compatible with bearing metallurgy and seals while resisting oxidative thickening and deposit formation over long drain intervals.

Transformers

Liquid-filled transformers use mineral, synthetic ester, or natural ester insulating fluids. These provide dielectric separation, heat transfer, and oxidation resistance. Key properties include breakdown voltage, moisture tolerance, and resistance to sludge formation. Solid insulation life is directly tied to fluid conditions. Monitoring for dissolved gases, acidity, and moisture is central to preventive maintenance.

Switchgear

Medium- and high-voltage switchgear may be insulated with SF_6 gas, SF_6 alternatives such as Novec blends, or solid/dry-type insulation with auxiliary insulating greases. Fluids and gases must maintain dielectric strength and arc-quenching properties over service life while avoiding moisture and particulate contamination.

Inverters

Inverters for renewable integration and storage systems generate concentrated heat in power electronics modules. Liquid dielectric coolants or water-glycol mixtures control component temperature and prevent thermal cycling damage. Coolant chemistry must prevent corrosion of copper busbars, aluminum heatsinks, and polymer seals while maintaining electrical isolation.

Batteries

Stationary and mobile high-capacity batteries use immersion-cooled or indirect liquid-cooled designs to maintain uniform temperature across cells. Coolants require high dielectric strength, oxidation stability, and chemical compatibility with cell casings, busbars, and sealing materials. Effective thermal management slows capacity fade, reduces safety risks, and stabilizes performance under rapid charge-discharge cycles.

ELECTRIFIED MINING, RAIL, HVAC, PROCESS INDUSTRY EQUIPMENT

Electrified Mining Equipment

Battery-electric haul trucks, loaders, and drills use liquid- or air-cooled traction motors, inverters, and battery packs. Coolants must be dielectric, thermally stable, and compatible with high-vibration environments. Gearboxes in wheel drives and swing mechanisms require synthetic gear oils with high load-carrying capacity, micropitting resistance, and water tolerance due to frequent exposure to moisture and dust. Hydraulic systems for implements use fire-resistant or biodegradable fluids where environmental regulations apply.

Electrified Rail

Traction motors, power converters, and onboard transformers demand fluids that combine electrical insulation with heat transfer capability. Bearings in traction motors use greases designed for high-speed, shock-load, and wide temperature operation. Gearboxes on powered axles use EP-rated synthetic gear oils with strong oxidative stability for extended service intervals. Pantograph mechanisms and auxiliary hydraulic systems require low-temperature pumpable fluids with antiwear protection.

HVAC Systems

Large industrial and commercial HVAC systems use refrigerants, compressor oils, and chilled-water loops. For variable-speed electric compressors, lubricants must provide dielectric safety, miscibility with refrigerants, and stability against moisture ingress. Secondary cooling loops may use water-glycol blends or brines with corrosion inhibitors matched to system metallurgy. Motor bearings in fans and pumps require greases with low bleed rates and compatibility with elevated operating temperatures in confined spaces.

Process Industry Equipment

Electrified pumps, mixers, extruders, and conveyors in manufacturing facilities require motor bearing lubrication with high oxidation resistance and resistance to washout where water is present. Gearboxes in continuous-duty production lines use synthetic oils with high film strength and resistance to thermal degradation. In high-contamination environments, sealing materials must be compatible with the lubricant and coolant chemistry to avoid premature leakage or ingress. Power electronics controlling these systems may use liquid dielectric coolants to prevent overheating and maintain voltage isolation.

Section II

Functional Fluid Families

3 Overview of Lubricants in Power and Electrified Systems

Turbine oils are critical lubricants that ensure the reliability and efficiency of steam, gas, and wind turbine systems. Unlike general-purpose lubricants, they must operate under continuous high-load conditions, often in environments characterized by heat, moisture, and air entrainment. The choice of base oil, whether mineral (Group I–III), polyalphaolefin (PAO) synthetics, or esters, has a direct influence on performance factors such as oxidation resistance, solvency, volatility, and environmental compatibility. Additive systems then tailor these base stocks to withstand the unique stresses of each turbine type, from the high humidity of steam plants to the intense thermal cycling of gas turbines and the low-temperature, high-torque demands of wind energy applications.

By understanding the distinctions between base oil groups and the role of additives, operators and engineers can better align lubricant selection with the specific mechanical, thermal, and environmental challenges of their turbines. The following sections provide a detailed overview of turbine oil formulations, highlighting the strengths and limitations of each base stock category and the additive chemistries that ensure long-term performance and asset protection.

TURBINE OILS (STEAM, GAS, WIND): GROUP I–III, PAO, ESTERS

Steam turbine oils are designed for long service life under steady high-temperature and high-humidity conditions. They must resist oxidation, shed water, and control deposit formation. **Group I mineral oils**, refined by solvent extraction, contain higher levels of sulfur, aromatics, and unsaturated molecules. These impurities make them more prone to oxidation and shorter-lived in severe service, but they offer natural solvency, which helps dissolve additives and manage deposits. **Group II mineral oils**, refined by hydroprocessing, have lower sulfur and aromatic content, giving them better oxidation resistance, lower volatility, and improved stability. They are now the dominant choice for modern steam turbines. Additive packages typically include antioxidants, rust inhibitors, and antifoam agents. In electrohydraulic control (EHC) systems, high fluid cleanliness is required to prevent servo valve sticking.

Gas turbine oils are subjected to higher thermal stress and possible contamination by combustion gases. They require base oils with very high thermal stability and

DOI: 10.1201/9781003775201-5

resistance to deposit formation. **Group II and Group III mineral oils** are common, with Group III being more highly refined through hydrocracking to yield a very pure base oil with excellent oxidation stability and very low volatility. **Polyalphaolefin (PAO) synthetics**, produced by chemically assembling uniform hydrocarbon molecules, provide even greater oxidation resistance, low volatility, and excellent low-temperature performance. PAOs extend oil life, maintain viscosity control under stress, and resist deposit formation. Deposit control additives are critical to reduce coking in bearing and seal drain regions.

Wind turbine lubrication differs in its mechanical demands but shares requirements for oxidation stability and cleanliness. Gearboxes operate at low speed and high torque, while yaw and pitch systems may use turbine-grade hydraulic oils. Here, **ester-based synthetics** are widely used, including diesters and polyol esters. Esters are chemically synthesized from acids and alcohols, giving them a high viscosity index, excellent oxidative stability, and natural detergency. Many are also biodegradable, making them suitable for wind farms located in environmentally sensitive areas. In cold climates, ester synthetics and PAOs outperform mineral oils because they maintain fluidity at very low temperatures, reducing startup wear and ensuring full-film lubrication during slow-speed, high-torque operation.

ADDITIVES USED IN TURBINE OILS

Turbine oils used in steam, gas, and wind turbine applications are formulated to provide long-term oxidation stability, deposit control, and corrosion protection under conditions of continuous operation, high temperatures, and in some cases, exposure to moisture or air entrainment. In mineral oil-based Group I–III formulations, the additive system is typically built around a synergistic blend of phenolic and aminic antioxidants, which work together to neutralize free radicals and slow the propagation of oxidative chain reactions. Group I oils, with higher natural solvency, can dissolve and suspend oxidation products more effectively but require higher antioxidant loading to offset their lower oxidative stability compared to Group II and III base stocks. Group II and III oils, with lower sulfur and aromatic content, have inherently better oxidation resistance but poorer solvency, which can lead to varnish deposition if the antioxidant system is depleted; here, the additive balance often includes dispersant-like deposit control agents to maintain cleanliness. PAO-based turbine oils offer superior thermal stability, very low volatility, and excellent low-temperature performance, but also have low natural solvency, so they rely more heavily on ashless dispersants, deposit control additives, and carefully chosen antioxidants to prevent lacquer and varnish formation. Ester-based turbine oils, often used in high-performance or environmentally sensitive applications, bring inherent lubricity, detergency, and excellent solvency, allowing for cleaner operation, but their hydrolytic stability must be protected by the use of metal deactivators and moisture-tolerant additive chemistries.

Across all base oil types, additional key additives include overbased or neutral calcium sulfonates or other ashless rust inhibitors to protect ferrous surfaces, metal deactivators such as benzotriazole derivatives to prevent copper-catalyzed oxidation, and silicone or organic polymer-based defoamants to control foam

TABLE 3.1
Turbine Oil Quality Control Tests

Acceptance (A):	In-Service (IS):
Viscosity and VI – ASTM D445, ASTM D2270	RULER – ASTM D6971 (remaining antioxidant)
RPVOT – ASTM D2272 (oxidation stability)	MPC / Varnish Potential – ASTM D7843
TOST Life – ASTM D943 (steam turbines)	Air Release – ASTM D3427
Foam Test – ASTM D892	Particle Count – ISO 4406
Demulsibility – ASTM D1401	Ferrography / Wear Metals – ASTM D5185
Rust Test – ASTM D665 A/B	Foam Tendency – ASTM D892
Copper Strip Corrosion – ASTM D130	Demulsibility – ASTM D1401 (steam turbines)
Acid Number (TAN) – ASTM D974	PQ Index – Ferrous density for wind gearbox
Water Content – ASTM D6304 (Karl Fischer)	

formation without impairing air release. In water-prone environments, especially for steam and wind turbines, demulsifiers are critical to allow rapid separation of free water in reservoir systems. The overall formulation strategy balances the intrinsic strengths and weaknesses of the base oil type with an additive system tuned to the oxidative, thermal, and environmental stresses of the turbine application, while ensuring compatibility with seals, coatings, and other materials in the lubrication circuit.

The following table lists the tests run for quality control and assurance acceptance (see Tables 3.1 and 3.2) and in-service condition monitoring followed by functional additives used in turbine oils:

GEAR OILS AND SYNTHETIC INDUSTRIAL LUBRICANTS

Gear oils in power generation and electrified systems are selected for load-carrying capacity, micropitting resistance, and thermal stability under continuous operation. Mineral-based formulations use Group I or Group II base stocks with extreme pressure (EP) additives, typically sulfur-phosphorus chemistry, to protect gear tooth surfaces under high contact stress. Synthetic gear oils, primarily PAO- or polyalkylene glycol (PAG)-based, provide extended service life, improved oxidative stability, and wider operating temperature ranges.

Wind turbine main gearboxes require synthetic gear oils with high viscosity index, resistance to scuffing, and ability to retain film thickness under variable loads and slow-speed, high-torque conditions. Micropitting resistance is critical to prevent surface fatigue on case-hardened gear teeth. Water tolerance and demulsibility are important for gearboxes exposed to condensation or environmental moisture.

In hydroelectric plants, gear oils must resist water contamination and emulsification while maintaining corrosion protection for bronze and ferrous components.

TABLE 3.2
Functional Additives in Turbine Oils (Steam, Gas, Wind)

Additive Function	Typical Additive Chemistry	Typical Concentration Range (% wt)	Key Laboratory Tests (ASTM / ISO / OEM)	Notes
Oxidation Inhibitors	Hindered phenols, aromatic amines, phenyl-α-naphthylamine	0.1–1.0	ASTM D2272 (RPVOT), ASTM D943 (TOST), ASTM D6186 (PDSC)	Gas turbines and wind gearboxes require higher oxidation stability; esters have inherent oxidative stability but still benefit from additives.
Rust and Corrosion Inhibitors	Calcium sulfonates, alkenyl succinic acids, amine phosphates	0.05–0.3	ASTM D665 (A/B), ASTM D130 (copper strip)	Steam turbines at higher water exposure risk require robust rust inhibitors.
Antiwear (AW) Agents	Ashless dithiophosphates, phosphate esters, estolides	0.05–0.5	ASTM D4172 (4-ball wear), ASTM D5183 (seizure load)	Wind turbine gearboxes often use higher AW treat rates; steam turbines generally have minimal AW to avoid deposit formation.
EP Additives (Wind Gearbox)	Sulfur-phosphorus chemistries, borate esters	0.5–2.0	ASTM D2783 (EP), FZG (ISO 14635-1)	Not used in most steam/gas turbine oils unless gearbox coupling lubrication is needed.
Antifoam Agents	Polysiloxanes, polyacrylates	10–100 ppm	ASTM D892, ISO 6247	Prevents aeration in reservoirs; wind gearboxes require foam control under splash lubrication.
Demulsifiers	EO/PO block copolymers, polyalkylene glycols	0.01–0.1	ASTM D1401 (demulsibility), ISO 6614	Steam turbines require strong demulsibility for water removal; some gas turbine OEMs accept slower demulsibility to enhance air release.
Viscosity Modifiers	Polymethacrylates, olefin copolymers (shear stable)	0.1–1.0	ASTM D445 (viscosity), ASTM D2270 (VI)	Used in PAO and ester turbine oils for wide temperature range applications; wind gearboxes often use ISO VG 320–460 with VI improvers.
Pour Point Depressants	Polymethacrylates, alkylated naphthalenes	0.02–0.1	ASTM D97, ASTM D5950	Important for wind turbines in cold climates; not as critical for baseload steam/gas units.
Detergents / Dispersants *(Wind gearboxes)*	Overbased sulfonates, succinimides	0.05–0.3	ASTM D2896 (TBN), ASTM D4739 (TBN)	Keeps gearboxes clean under mixed-duty cycles; not typical in pure turbine oils.
Seal Compatibility Additives	Ester cobase fluids, aromatic hydrocarbons	0.1–0.5	ASTM D471 (rubber immersion), ISO 1817	Maintains elastomer swell and prevents leaks.

PAG-based lubricants may be used where high efficiency and low traction coefficients are desired, provided compatibility with seals and paints is confirmed.

Synthetic industrial lubricants extend beyond gearboxes to include circulating oils for bearings, compressors, and hydraulic systems in auxiliary equipment. These fluids are formulated to resist oxidation, maintain viscosity under shear, and minimize deposit formation. Selection is based on load profile, operating environment, and compatibility with component metallurgy and elastomers.

ADDITIVES USED IN GEAR OILS FOR POWER SYSTEMS

Gear oils for power systems, whether used in industrial gearboxes, wind turbine main gear drives, or high-load enclosed gear units are formulated to manage extreme contact pressures, resist micropitting, and maintain film integrity under conditions of sliding, rolling, and mixed lubrication. The additive system is built around EP agents, most commonly sulfur–phosphorus (S–P) chemistries in mineral and synthetic formulations, which react with metal surfaces under high load and temperature to form protective tribofilms that prevent adhesive wear and scuffing. In applications where yellow metal compatibility is critical, such as worm gears, lower-activity S–P additives or alternative ashless EP chemistries are selected to reduce the risk of corrosive attack. Antiwear agents, typically zinc dialkyldithiophosphate (ZDDP) in certain gear oils or ashless phosphates in others, provide a secondary line of defense under moderate load conditions and help control oxidative degradation.

To maintain cleanliness and prevent rust and corrosion, neutral or overbased calcium sulfonates, carboxylates, or other ashless rust inhibitors are included. Oxidation inhibitors are often a balanced combination of aminic and phenolic antioxidants, extend oil life by interrupting free radical oxidation pathways, especially important in high-speed or elevated-temperature service. Foam control is handled with silicone or organic polymer defoamants in very low concentrations (typically below 0.005%), preventing aeration that can lead to microdieseling and accelerated oil oxidation. Demulsifiers are used in gear oils where water ingress is a concern, allowing rapid separation of free water from the oil phase, a critical factor in wind turbine and hydroelectric plant gearboxes.

For synthetics such as PAO-based gear oils additives may be optimized for seal compatibility and to offset the lower natural solvency of the base stock, which can otherwise promote varnish formation; PAG gear oils may require entirely different additive systems due to base additive miscibility constraints. In some high-performance or environmentally sensitive power system applications, ashless gear oil packages are used with phosphate esters or borate-based EP agents to meet biodegradability requirements while retaining load-carrying capacity. Overall, gear oil additive chemistry in power systems is a balancing act between achieving high load-carrying capacity, maintaining oxidative stability, preventing micropitting, and ensuring compatibility with the metallurgy, seals, and operational environment of the gearbox.

The following table (see Table 3.3) lists the tests run for quality control and assurance acceptance and in-service condition monitoring followed by functional additives used in gear oils (see Table 3.4).

TABLE 3.3
Gear Oil Quality Control Tests

Acceptance (A):	In-Service (IS):
Viscosity and VI – ASTM D445, ASTM D2270	Remaining Antioxidant – ASTM D6971 (RULER)
EP and AW Performance – ASTM D2783, FZG ASTM D4172	Varnish Potential – ASTM D7843 (MPC)
Foam Test – ASTM D892	Particle Count – ISO 4406
Demulsibility – ASTM D1401	Wear Metals – ASTM D5185 (ICP), ferrous density (PQ Index)
Rust Test – ASTM D665 A/B	Water Content – ASTM D6304 (Karl Fischer)
Copper Strip Corrosion – ASTM D130	Foam Tendency – ASTM D892
Pour Point – ASTM D97	FZG Retest *(for R&D or severe cases)*
TAN / TBN – ASTM D974, ASTM D2896	In-Service (IS):

DIESEL AND GAS ENGINE OILS (API CK-4, FA-4, ASH CONTROL, NITRATION RESISTANCE)

Diesel engine oils meeting API CK-4 specifications are designed for high-temperature oxidation stability, shear stability, and wear protection in both modern and legacy engines. They maintain viscosity under load, disperse soot, and neutralize acids from fuel combustion. Higher total base number (TBN) formulations are used where fuel sulfur content is greater. Detergent and dispersant chemistry prevents deposit buildup in piston ring zones and keeps contaminants suspended for filtration.

API FA-4 oils target newer, high-efficiency diesel engines with lower high-temperature high-shear (HTHS) viscosity for improved fuel economy while maintaining wear control. They use Group II, Group III, or synthetic base oils to ensure oxidation resistance and volatility control at reduced viscosities. These oils are not backward-compatible with all engines and require original equipment manufacturer (OEM) verification before use.

Gas engine oils for natural gas or biogas applications must control nitration, oxidation, and deposit formation from gaseous combustion byproducts. Ash control is critical to avoid pre-ignition and exhaust valve recession, with low-ash formulations specified to protect catalyst systems. Oils for biogas-fueled engines require additional acid-neutralizing capacity to counteract corrosive compounds such as hydrogen sulfide and siloxanes. Regular oil analysis for nitration, oxidation, and base number depletion is essential for both diesel and gas engine service to prevent accelerated wear and maintain operational reliability.

ADDITIVES USED IN ENGINE OILS FOR POWER SYSTEMS

Diesel and gas engine oils used in stationary power generation particularly those meeting API CK-4, FA-4, and specialized ash-controlled or nitration-resistant

TABLE 3.4
Functional Additives in Gear Oils and Synthetic Industrial Lubricants (Power Generation)

Additive Function	Typical Additive Chemistry	Typical Concentration Range (% wt.)	Key Laboratory Tests (ASTM / ISO / OEM)	Notes
EP Additives	Sulfur–phosphorus compounds, borate esters, synthetic polyglycols (EP grade)	2–6	ASTM D2783 (EP), FZG (ISO 14635-1)	Critical for heavily loaded gears; load stage rating is a key acceptance spec.
Antiwear Agents	Ashless dithiophosphates, phosphate esters	0.1–0.5	ASTM D4172 (4-ball wear), ASTM D5183 (seizure load)	Reduces wear during boundary lubrication; wind gearboxes often combine AW and EP chemistries.
Rust & Corrosion Inhibitors	Calcium sulfonates, alkenyl succinic acids, triazoles	0.05–0.3	ASTM D665 (A/B), ASTM D130 (copper strip)	Protects ferrous and non-ferrous metal surfaces; important for gearboxes in humid environments.
Oxidation Inhibitors	Hindered Phenols, aromatic amines	0.2 – 1.0	ASTM D2272 (RPVOT), ASTM D943 (TOST), ASTM D6971 (RULER)	Extends lubricant life; depletion leads to sludge and varnish formation.
Antifoam Agents	Polysiloxanes, polyacrylates	10 – 100 ppm	ASTM D892, ISO 6247	Prevents foam in splash-lubricated systems; essential for oil circulation systems with high agitation.
Demulsifiers	EO/PO block copolymers, polyalkylene glycols	0.01–0.1	ASTM D1401, ISO 6614	Facilitates water separation; water contamination accelerates additive depletion and corrosion.
Viscosity Modifiers	Polymethacrylates, olefin copolymers (shear-stable)	0.1–1.0	ASTM D445 (viscosity), ASTM D2270 (VI)	Used in multi-grade gear oils for temperature stability; shear stability is critical in wind applications.

(*Continued*)

TABLE 3.4 (Continued)
Functional Additives in Gear Oils and Synthetic Industrial Lubricants (Power Generation)

Additive Function	Typical Additive Chemistry	Typical Concentration Range (% wt.)	Key Laboratory Tests (ASTM / ISO / OEM)	Notes
Pour Point Depressants	Polymethacrylates, alkylated naphthalenes	0.02–0.1	ASTM D97, ASTM D5950	Enables operation in cold climates; wind turbine gear oils require very low pour points.
Friction Modifiers	Organic molybdenum compounds, fatty acid esters	0.05–0.5	ASTM D5183, SRV friction tests (DIN 51834)	Improves efficiency and reduces operating temperatures.
Detergents/ Dispersants	Overbased sulfonates, succinimides	0.05–0.3	ASTM D4739 (TBN), ASTM D2896 (TBN)	Keeps systems clean under high-load, intermittent operation; used in industrial gear oils for contamination-prone environments.
Seal Compatibility Additives	Ester cobase fluids, aromatic hydrocarbons	0.1–0.5	ASTM D471 (rubber immersion), ISO 1817	Prevents seal shrinkage or excessive swell, maintaining oil retention.
Antimicrobials *(water-containing synthetics)*	Isothiazolinones, bronopol	0.01–0.05	ASTM E2275 (microbial count)	Prevents microbial fouling in water-glycol synthetics; critical for some hydraulic gear drives.

specifications are formulated to handle sustained high-load operation, elevated sump temperatures, and fuel or combustion byproduct contamination over extended drain intervals. The additive package begins with a robust detergent system, typically overbased calcium and/or magnesium sulfonates, phenates, or salicylates, which neutralize acidic combustion byproducts, control deposits, and maintain TBN. In gas engines, especially those burning natural gas, landfill gas, or biogas, ash control becomes critical; the detergent treat rate and type are adjusted to minimize sulfated ash (often ≤ 0.6% by weight) to prevent pre-ignition, spark plug fouling, and ash deposits in the combustion chamber.

Dispersants usually polybutenyl succinimides or borated variants are included at high treat rates (often 4–8% of the finished oil) to keep soot, oxidation products, and nitration-derived polymers finely suspended, preventing sludge and ring belt deposits. Antiwear protection is primarily supplied by zinc dialkyldithiophosphate (ZDDP), which also serves as an antioxidant; phosphorus limits (e.g., ≤ 800–1000 ppm for CK-4/FA-4) are enforced to protect aftertreatment devices. Supplemental ashless antiwear agents (e.g., phosphites, thiophosphates) may be used in low-sulfated ash, phosphorus, and sulfur formulations to maintain wear protection without exceeding phosphorus caps.

Oxidation and nitration resistance is enhanced with a carefully balanced aminic/phenolic antioxidant system, sometimes supported by molybdenum dithiocarbamates (MoDTC) that provide both antiwear and friction modification benefits. In high-nitration environments typical of lean-burn gas engines, nitration inhibitors and antioxidant loadings are elevated to slow the formation of nitro-compounds that lead to oil thickening and deposit formation. Rust and corrosion inhibitors, often amine succinates or overbased sulfonates, protect ferrous and non-ferrous surfaces, especially during standby periods. Foam control is provided with silicone or organic defoamants at trace levels to prevent aeration and cavitation in high-flow lubrication systems.

Viscosity index improvers (OCP, PMA, or star polymers) are shear-stable grades in multi-grade oils, ensuring viscosity retention under severe hydrodynamic shear common in large stationary engines. For gaseous-fueled units, the base oil and additive balance may also include copper deactivators such as benzotriazole derivatives to prevent catalyst-driven oxidation from fuel contaminants. Across CK-4, FA-4, and ash-controlled/nitration-resistant formulations, the goal is to maintain deposit control, neutralize acids, minimize wear, and ensure oxidative/nitrative stability while meeting the specific ash, phosphorus, and sulfated ash limits dictated by engine design and emission control systems in power generation service.

The following table (see Table 3.5) lists the tests run for quality control and assurance acceptance and in-service condition monitoring followed by functional additives (see Table 3.6) used in engine oils for power generation.

GREASES FOR MOTORS, BEARINGS, AND ACTUATORS (POLYUREA, CALCIUM SULFONATE, PFPE)

Greases play a critical role in power generation equipment, where bearings, actuators, and motors are often subjected to high loads, variable speeds, elevated temperatures,

TABLE 3.5
Engine Oil Quality Control Tests

Acceptance (A):	In-Service (IS):
Viscosity Grade and HTHS – ASTM D445, ASTM D4683	TBN Retention – ASTM D4739
TBN / Sulfated Ash – ASTM D2896, ASTM D874	TAN Increase – ASTM D664
Phosphorus/Zinc Content – ASTM D5185	Oxidation/Nitration/Soot – ASTM E2412 (FTIR), ASTM D5967 (soot)
Foam – ASTM D892	Wear Metals – ASTM D5185
Pour Point – ASTM D97	Insolubles – ASTM D7899 or D893
HTHS Viscosity – ASTM D4683 (FA-4)	Viscosity Shift – ASTM D445 (fuel dilution/ shear)
	Coolant Contamination – ASTM D6130 (glycol), sodium/potassium via ICP
	MPC / Varnish Potential – ASTM D7843 (gas engines)

and environmental challenges such as water ingress or chemical exposure. Unlike circulating oils, greases must maintain their structural integrity and protective properties over long relubrication intervals or in many cases, for the entire service life of the component. The performance of grease is defined by its base oil, thickener chemistry, and carefully selected additive system, each of which must be matched to the unique mechanical and environmental demands of the application.

In power generation, three grease families dominate specialized use cases: polyurea greases, known for their long service life in sealed electric motor bearings; calcium sulfonate complex greases, valued for their inherent load-carrying capacity and corrosion resistance in oscillating or water-prone applications; and perfluoropolyether (PFPE) greases, reserved for extreme-duty environments where chemical inertness and non-flammability are mandatory. Each system has distinct advantages and limitations, requiring tailored additive packages that enhance wear protection, oxidation resistance, and environmental durability without compromising compatibility with sensitive materials or operating conditions.

The following sections examine the role of these grease types in power generation systems, along with the additive chemistries that extend their effectiveness and reliability in mission-critical service.

Polyurea Greases

Common in electric motor bearings due to oxidation stability, low bleed, and long service life in sealed-for-life applications. Compatible with most motor winding insulation materials and suitable for high-speed operation without excessive channeling. Good resistance to moisture ingress and mechanical shear makes them effective

TABLE 3.6
Functional Additives in Diesel and Gas Engine Oils (Power Generation)

Additive Function	Typical Additive Chemistry	Typical Concentration Range (% wt)	Key Laboratory Tests (ASTM / ISO / OEM)	Notes
Detergents (Base Reserve / Ash Control)	Overbased calcium sulfonates, magnesium sulfonates, phenates	Ca: 0.1–0.3, Mg: 0.05–0.15 (as % element); TBN 7–15 (diesel), 3–5 (gas engines)	ASTM D2896 (TBN), ASTM D4739 (TBN), ASTM D874 (sulfated ash)	Controls deposit and neutralize acids; low-ash formulations (<0.6% SA) used in gas engines to protect exhaust catalysts and avoid pre-ignition.
Dispersants	Polyisobutylene succinimides, borated dispersants	2–6	ASTM D5185 (insolubles via ICP), ASTM D7899 (pentane/hexane insolubles)	Keeps soot and oxidation products in suspension; depletion leads to sludge and filter plugging.
Antiwear Agents	Zinc dialkyldithiophosphates (ZDDP), ashless phosphorus esters	P: 0.08–0.12 (diesel), ≤0.08 (gas engines)	ASTM D4172 (4-ball wear), ASTM D5185 (P/Zn levels)	Reduces cam/follower wear; gas engines may limit ZDDP to protect catalysts.
Antioxidants	Hindered phenols, aromatic amines, phenyl-α-naphthylamine	0.2–1.0	ASTM D6971 (RULER), ASTM D2272 (RPVOT)	Slows oxidation/nitration; key for high-temp stationary gas engines.
Nitration Control Additives	Aminic antioxidants, phenolic antioxidants, dispersants	Incorporated within antioxidant/dispersant package	ASTM E2412 FTIR (nitration index)	Controls varnish/oxidation from NOx exposure in gas engines; depletion shows as rapid nitration rise in FTIR.
Friction Modifiers	Organic molybdenum (MoDTC, MoDTP), fatty esters	0.05–0.2	ASTM D5183, SRV (DIN 51834)	Improves fuel economy and load efficiency; limited in gas engines.

(continued)

TABLE 3.6 (Continued)
Functional Additives in Diesel and Gas Engine Oils (Power Generation)

Additive Function	Typical Additive Chemistry	Typical Concentration Range (% wt)	Key Laboratory Tests (ASTM / ISO / OEM)	Notes
Corrosion Inhibitors	Benzotriazole derivatives, amine phosphates	0.05–0.2	ASTM D130 (copper strip), ASTM D665 (A/B)	Protects bearings and cooler components from acid/coolant contamination.
Foam Inhibitors	Polysiloxanes, polyacrylates	5–50 ppm	ASTM D892 (foam), ISO 6247	Low dose prevents air entrainment in sump.
Pour Point Depressants	Polymethacrylates, alkylated naphthalenes	0.02–0.1	ASTM D97, ASTM D5950	Maintains pumpability in cold starts.
Viscosity Modifiers	Shear-stable polymethacrylates, olefin copolymers	0.5–1.5 (multi-grades)	ASTM D445 (viscosity), ASTM D4684 (CCS), ASTM D5293 (MRV)	Maintains viscosity across load/ temperature; FA-4 oils have lower HTHS (2.9–3.2 cP).

in industrial motors and generators operating under variable load and temperature conditions.

Calcium Sulfonate Complex Greases

Used in bearings and actuators exposed to high loads, shock loading, or water contamination. Offer natural corrosion protection, high load-carrying capacity, and mechanical stability. Perform well in oscillating motion applications such as wind turbine pitch and yaw bearings or hydroelectric wicket gate bearings. Withstand water washout and maintain structural integrity in wet or corrosive environments.

PFPE Greases

Applied in high-value or extreme environments where chemical inertness, nonflammability, and thermal stability are critical. Suitable for nuclear actuator assemblies, cleanroom robotics, and aerospace-grade electromechanical actuators. Maintain lubricating properties in the presence of radiation, aggressive chemicals, or temperatures exceeding the limits of conventional synthetic greases. Chemically stable with most plastics, elastomers, and metals, making them suitable for mixed-material assemblies in mission-critical systems.

ADDITIVES USED IN GREASE FOR POWER GENERATION

Greases used in motors, bearings, and actuators in power generation whether based on polyurea, calcium sulfonate complex, or PFPE chemistries are formulated to provide mechanical stability, wear protection, and environmental resistance under continuous or intermittent service, often in high-temperature, high-load, or contamination-prone conditions. The base oil type (mineral, PAO, ester, or PFPE) and thickener system largely define inherent performance characteristics, but the additive system is tailored to the application and the limitations imposed by the thickener and operating environment.

Polyurea greases, widely used in electric motor bearings for their oxidative stability, low bleed, and long life, often rely on ashless antioxidant systems because ZDDP and metal-containing EP additives can shorten electric motor life by increasing electrical conductivity or corroding copper windings. Phenolic and aminic antioxidants are common for high-temperature resistance, while rust inhibitors such as alkenyl succinic acids or amine phosphates protect bearing steel surfaces during idle or humidity exposure. Molybdenum disulfide (MoS_2) or other solid lubricants are generally avoided in precision electric motor greases to prevent increased noise or electrical contact issues, but may be used in actuator or slow-speed bearing greases.

Calcium sulfonate complex greases inherently contain overbased calcium sulfonate thickener, which provides both thickening structure and built-in rust/corrosion inhibition along with EP and antiwear benefits through the formation of protective tribofilms. These greases often require minimal supplemental EP additives but may incorporate sulfur–phosphorus chemistries or organomoly compounds in heavy-load or oscillating bearing applications common in hydroelectric or wind turbine pitch systems. Additional antioxidants (aminic/phenolic) extend life under sustained high temperatures, and polymer tackifiers may be added for adhesion in exposed or oscillating actuator linkages.

PFPE greases, used in high-temperature, chemically aggressive, or oxygen-rich environments (such as turbine starter actuators, nuclear plant instrumentation bearings, or cleanroom controls), are typically additive-lean due to PFPE's exceptional oxidative and chemical stability. When additives are used, they are fluorinated or otherwise compatible chemistries such as PTFE solids for antiwear, fluorinated antioxidants, and perfluorinated corrosion inhibitors to ensure no incompatibility or degradation of the PFPE structure. PFPE greases are inherently non-flammable and resistant to most solvents and reactive gases, making them critical for extreme duty, but their additive palette is much narrower and highly specialized.

Across all three grease types, defoamants are not typically relevant, but water resistance is critical; in calcium sulfonate and polyurea systems, hydrophobic base oils and surface-active rust inhibitors are used to repel water ingress, while in PFPE systems, moisture has minimal chemical effect but can still cause corrosion, so compatible inhibitors are essential. The final additive strategy is a balance of oxidative stability, corrosion control, antiwear protection, and environmental resistance, tuned to the demands of the specific motor, bearing, or actuator service in the power generation environment.

The following tables list the tests (see Tables 3.7 and 3.8) run for quality control and assurance acceptance and in-service condition monitoring followed by functional additives used in greases for power generation.

FIRE-RESISTANT FLUIDS (HFDU, HFDR, HFC, HFAE)

In modern power generation, hydraulic fluids must not only transmit power efficiently but also protect against one of the greatest risks in high-temperature, high-pressure

TABLE 3.7
Grease Quality Control Tests

Acceptance (A):	In-Service (IS):
Worked Penetration – ASTM D217	FTIR (oxidation, contamination, thickener ID)
Dropping Point – ASTM D2265	Ferrous Debris – FDM or PQ Index
EP Load and Wear – ASTM D2596, ASTM D2266	Wear Metals – ICP (ASTM D5185 after solvent extraction)
Water Washout – ASTM D1264	Water Content – Karl Fischer (ASTM D6304 after extraction)
Oil Separation – ASTM D6184	Consistency Changes – ASTM D217 (penetration shift)
Oxidation Stability – ASTM D942	Contamination – Particle Count (optical microscopy)
	Purge Analysis – Visual/odor for burnt or degraded base oil

TABLE 3.8
Greases for Motors, Bearings, and Actuators (Polyurea, Calcium Sulfonate, PFPE)

Additive Function	Typical Additive Chemistry	Typical Concentration Range (% wt)	Common in Polyurea?	Common in Ca-Sulfonate?	Common in PFPE?	Key Laboratory Tests (ASTM / ISO / OEM)	Field/Lab Relevance
Thickener System	Polyurea compounds; Overbased Ca sulfonate; PTFE for PFPE	5–20 (thickener)	✓	✓	PTFE in PFPE	ASTM D2265 (dropping point), FTIR (chemical ID)	Defines high-temp stability, water resistance.
Base Oil	Mineral (Grp I–III) or synthetic (PAO, ester); PFPE fluorinated oils	70–90	✓	✓	PFPE base	ASTM D445 (viscosity), ASTM D2596 (EP), GC for base oil ID	Determines low-temp torque, oxidation life.
EP Additives	Sulfur–phosphorus, MoS_2, borate esters	0.5–5.0	Limited	✓	Rare (PFPE often EP-free)	ASTM D2596 (4-ball EP), ASTM D2783	Loss = reduced load-carrying capacity.
AW Agents	ZDDP, ashless phosphates, moly compounds	0.1–1.0	✓	✓	Rare	ASTM D2266 (4-ball wear), ASTM D4172	Wear scar ↑ = AW depletion or contamination.
Rust and Corrosion Inhibitors	Calcium sulfonate (dual role), benzotriazole	0.05–0.5	✓	✓	✓	ASTM D1743 (rust), ASTM D5969	Fail = water ingress or pH drift.
Oxidation Inhibitors	Aromatic amines, hindered phenols	0.1–1.0	✓	✓	Not required for PFPE	ASTM D942 (oxidation stability), ASTM D5483 (PDSC)	RULER or PDSC drop = shorter relube interval.

(continued)

TABLE 3.8 (Continued)
Greases for Motors, Bearings, and Actuators (Polyurea, Calcium Sulfonate, PFPE)

Additive Function	Typical Additive Chemistry	Typical Concentration Range (% wt)	Common in Polyurea?	Common in Ca-Sulfonate?	Common in PFPE?	Key Laboratory Tests (ASTM / ISO / OEM)	Field/Lab Relevance
Solid Lubricants	MoS_2, graphite, PTFE	0–5	Optional	Optional	PTFE inherent	ASTM D5707 (wear), FTIR (PTFE ID)	For boundary lubrication in oscillating/ slow-speed.
Tackifiers	Polyisobutylene, polybutenes	0.1–2.0	✓	✓	Rare	Visual tack test, cone penetration	Retains grease in vertical/ horizontal shafts.
Antifoam Agents	Polysiloxanes	<0.1	✓	✓	✓	ASTM D892	Prevents grease aeration in churning.
Water Resistance Agents	Hydrophobic thickener chemistry, waxes	Formulation dependent	✓	✓	PFPE inherently hydrophobic	ASTM D1264 (water washout), ASTM D4049 (water spray-off)	Loss = rapid washout in wet conditions.
Seal Compatibility Additives	Ester cobases, aromatic oils	0.1–0.5	✓	✓	✓	ASTM D4289, ISO 1817	Fail = seal shrink/ swell.

environments: fire. Steam turbines, gas turbines, nuclear plants, and hydroelectric facilities often operate hydraulic systems in close proximity to ignition sources such as hot surfaces, open flames, or electrical equipment. Conventional mineral oil-based fluids, while effective lubricants, can present unacceptable fire hazards under these conditions. To address this, a family of specialized fire-resistant fluids, classified as HFDU, HFDR, HFC, and HFAE, has been developed, each with unique chemistries, strengths, and limitations.

These fluids are engineered to balance hydraulic performance with fire safety, material compatibility, and long-term reliability. Ester-based HFDU fluids and phosphate ester HFDR fluids provide strong ignition resistance for turbine electrohydraulic controls, while water-based HFC and HFAE fluids leverage high water content to deliver outstanding flame suppression in steel mills, hydro plants, and other critical operations. The performance of each fluid type depends not only on its base chemistry but also on carefully selected additive systems that extend oxidative life, control corrosion, manage deposits, and ensure stability in demanding service conditions.

The following sections provide a detailed overview of each fire-resistant fluid type used in power generation, along with the additive strategies that enable them to meet the dual requirements of safety and hydraulic functionality.

HFDU – Synthetic Ester-Based

Water-free, non-aqueous fluids formulated from organic esters with high ignition resistance. Used in hydraulic systems in power plants where fire risk is present, such as turbine electrohydraulic controls and pressurized actuator systems near hot surfaces. Provide good lubricity and oxidation stability but require compatibility verification with seals and paints. Susceptible to hydrolysis if exposed to excessive moisture, so storage and system sealing are critical.

HFDR – Phosphate Ester-Based

Non-aqueous, high fire resistance, and self-extinguishing properties. Common in steam turbine EHC systems, gas turbine controls, and nuclear plant hydraulic systems. Resistant to ignition even in high-temperature spray conditions. Chemically aggressive to some elastomers and coatings, so material compatibility is a key selection factor. Require acid number monitoring and periodic fluid conditioning to prevent varnish and sludge formation.

HFC – Water Glycol-Based

Aqueous solutions containing water and glycol with additive packages for corrosion inhibition and wear protection. Offer strong fire resistance due to high water content. Used in hydraulic systems where leakage near ignition sources is possible. Lower lubricity compared to oil-based fluids; pump and component design must accommodate this. Limited upper temperature capability due to water evaporation.

HFAE – Oil-in-Water Emulsions

Contain a small percentage of mineral or synthetic oil emulsified in water. Excellent fire resistance due to high water content. Used in mining hydraulics, steel mills, and some hydroelectric gate actuation systems. Require continuous monitoring of oil concentration, pH, and microbiological growth. Limited lubricity and poor performance in freezing environments without antifreeze treatment.

ADDITIVES USED IN FIRE-RESISTANT FLUIDS

Fire-resistant fluids used in power generation particularly HFDU (phosphate esters or polyol esters), HFDR (phosphate ester), HFC (water-glycol), and HFAE (oil-in-water emulsions) are formulated to provide hydraulic performance while significantly reducing flammability risk in high-pressure, high-temperature, and ignition-prone environments such as steam turbine control systems, electrohydraulic governors, furnace operations, and hydroelectric plant gates. Because the base fluid type strongly influences performance and compatibility, the additive systems for these fluids are tailored to both the inherent chemistry and the specific demands of the application.

HFDU fluids, typically based on polyol esters or synthetic esters for modern systems, have high lubricity and thermal stability, but are susceptible to hydrolysis. Their additive packages often include phenolic and aminic antioxidants to extend oxidative life, metal deactivators (e.g., benzotriazole derivatives) to inhibit copper-catalyzed degradation, and robust rust inhibitors compatible with ester chemistry. Hydrolytic stability is reinforced with acid scavengers or specialized antihydrolysis additives. In turbine EHC (electrohydraulic control) systems, deposit control agents and varnish inhibitors are essential to prevent servo valve sticking.

HFDR fluids, traditionally phosphate esters, have outstanding fire resistance and excellent lubricity, but are inherently aggressive toward some elastomers, coatings, and paints. Additives in these systems focus on deposit control (especially neutralizing acidic degradation products), metal passivation for ferrous and non-ferrous surfaces, and antioxidants to slow thermal breakdown. Inhibited phosphate esters may also contain proprietary acid-neutralizing agents to extend fluid life in servo-controlled systems.

HFC fluids, or water-glycol types, typically contain 35–50% water, a glycol (ethylene or diethylene glycol), and a water-soluble additive package. Additives must control corrosion in a mixed-metal hydraulic circuit, so nitrite-, molybdate-, or amine-phosphate-based corrosion inhibitors are common, alongside biocides to prevent microbial growth in the aqueous phase. Antiwear additives are generally water-compatible phosphate esters or organic phosphates, with boundary lubrication enhanced by polymeric friction modifiers. Foam control is handled with water-compatible defoamants that avoid destabilizing the glycol-water emulsion.

HFAE fluids are oil-in-water emulsions with 95–99% water and 1–5% oil. Their additive packages must emulsify the oil phase while preventing corrosion in predominantly aqueous systems. They contain high levels of water-soluble rust inhibitors, biocides, and surfactants to stabilize the emulsion, along with low-toxicity antiwear additives compatible with both oil and water phases. In power generation, these are often used in hydraulic control systems where maximum fire resistance is needed, such as steel mill or turbine gate hydraulics.

Across all fire-resistant fluid types, additive formulation must balance **fire safety, corrosion control**, **oxidative stability**, **deposit management**, and **material compatibility**. In power generation environments, maintenance practices such as routine fluid analysis for acid number, water content, corrosion inhibitors, particulate contamination, and additive depletion are essential to ensure that the specialized additive systems remain effective over extended service intervals.

The following table (see Table 3.9) lists the tests run for quality control and assurance acceptance and in-service condition monitoring followed by functional additives used in fire-resistant fluids for power generation.

BIODEGRADABLE LUBRICANTS AND ENVIRONMENTALLY ACCEPTABLE LUBRICANTS (EALS)

Biodegradable lubricants are formulated to degrade naturally in the environment through biological activity, reducing persistence and ecological impact in the event of leakage or discharge. EALs meet defined criteria for biodegradability, low toxicity, and minimal bioaccumulation as specified in regulatory frameworks such as the U.S. EPA Vessel General Permit (VGP).

Base fluid types include synthetic esters, natural esters, PAGs, and certain highly refined vegetable oils. Synthetic esters offer high oxidative stability, good lubricity, and wide temperature capability, but require careful moisture management to avoid hydrolysis. Natural esters provide strong lubricity and biodegradability but have lower oxidation resistance, making them more suitable for low to moderate temperature applications. PAGs provide excellent lubricity and water solubility, which can aid in system cleanup, but may be incompatible with some seals and coatings.

EALs are used in applications with a high risk of environmental exposure, such as hydroelectric wicket gates, wind turbine pitch systems, marine propulsion systems, and dockside handling equipment. Selection depends on load-carrying capacity, viscosity stability, and compatibility with system metallurgy and sealing materials. Regular condition monitoring is essential to detect water ingress, oxidative degradation, or additive depletion that could shorten service life.

The following table (see Table 3.10) lists the tests run for quality control and assurance acceptance and in-service condition monitoring followed by functional additives used in Biodegradable lubricants for power generation.

Fluid Base Types in EALs/Biodegradable Lubricants (per ISO 15380)

- **HEES** – Synthetic esters
- **HETG** – Vegetable oils/triglycerides
- **HEPG** – Polyalkylene glycols
- **HEPR** – Polyalphaolefins and related synthetic hydrocarbons

TABLE 3.9
Functional Additives in Fire-Resistant Fluids

Additive Function	Typical Additive Chemistry	Typical Concentration Range (% wt)	Key Laboratory Tests (ASTM / ISO / IEC)	Notes
Antiwear/Extreme Pressure	Organophosphate esters, ashless dithiophosphates (Zn-free), synthetic esters, water-soluble phosphates (HFC/HFAE)	0.5–3.0	ASTM D4172 (4-ball wear), ASTM D2783 (EP), FZG (ISO 14635)	HFDU/HFDR use oil-soluble AW; HFC/HFAE use water-soluble chemistry.
Corrosion Inhibitors (Ferrous)	Alkaline sulfonates, succinic acid derivatives, amine phosphates	0.05–0.5	ASTM D665 (A/B), ISO 7120	Protects steel components; essential in water-containing fluids.
Corrosion Inhibitors (Non-Ferrous)	Benzotriazole, tolyltriazole	0.05–0.3	ASTM D130 (copper strip), ASTM D4048	Shields copper alloys, bronze bushings.
Oxidation Inhibitors	Aromatic amines, hindered phenols, phosphites	0.1–1.0	ASTM D2272 (RPVOT), ASTM D943 (TOST)	HFDU/HFDR rely heavily on antioxidants; HFC/HFAE have limited oxidative stress but still require some.
Rust Preventives	Petroleum sulfonates, fatty acid derivatives	0.05–0.2	ASTM D665, ISO 7120	Water ingress accelerates depletion; critical in all types.
Antifoam Agents	Polysiloxanes, polyacrylates	10–100 ppm	ASTM D892, ISO 6247	Low dose prevents aeration/cavitation.
Demulsifiers	Polyalkylene glycols, EO/PO copolymers	0.01–0.1	ASTM D1401 (demulsibility), ISO 6614	For HFDU/HFDR; HFC/HFAE are water-based so demulsibility not applicable.
pH Buffers	Alkanolamines, borates	0.05–0.3	ASTM D664 (acid number), ASTM E70 (pH)	Maintains corrosion protection in HFC/HFAE.

Biocides / Fungicides	Isothiazolinones, formaldehyde releasers	0.01–0.05	ASTM E2275 (microbial count)	Essential for HFC/HFAE to prevent microbial growth.
Viscosity Modifiers	High-MW polymers, esters	0.1–1.0	ASTM D445 (viscosity), ASTM D2270 (VI)	HFDU/HFDR only; not used in water-based fluids.
Fire-Resistance Enhancers	Halogen-free phosphorus compounds, synthetic esters	Variable (in base stock)	ASTM D92 (flash point), ISO 2592, FM 6930	Built into base chemistry; not separate for water-based types.
Thickeners (HFAE)	Polyacrylates, clays	0.1–1.0	ASTM D562 (consistency)	Stabilizes emulsion; prevents phase separation.
Antierosion Additives	Polyols, boundary lubricity improvers	0.05–0.3	Vane pump wear tests (OEM), ISO 20763	Reduces wear in high-flow water-based systems.

TABLE 3.10
Functional Additives in Biodegradable Lubricants and EALs – With Typical Concentrations and Lab Tests

Additive Function	Typical Additive Chemistry (Biodegradable / EAL-Compatible)	Typical Concentration Range (% wt)	Key Laboratory Tests (ASTM / ISO / OECD)	Notes
Antiwear / EP	Organophosphate esters, ashless dithiophosphates (Zn-free), borate esters, fatty acid esters	0.5–3.0	ASTM D4172 (4-ball wear), ASTM D2783 (EP properties), FZG A/8.3/90 (ISO 14635-1)	Must be non-toxic, meet biodegradability ≥60% (OECD 301B)
Corrosion Inhibitors (Ferrous)	Fatty amine phosphates, alkenyl succinic acids, aliphatic amine carboxylates	0.05–0.5	ASTM D665 (A/B), ISO 7120	Selected for low aquatic toxicity
Corrosion Inhibitors (Non-Ferrous)	Benzotriazole, tolyltriazole derivatives (low aquatic impact)	0.05–0.3	ASTM D130 (copper strip), ASTM D4048 (corrosion prevention)	Used at reduced levels to meet LC50/EC50 limits
Oxidation Inhibitors	Hindered phenols, aromatic amines (low-toxicity grades), tocopherol derivatives	0.1–1.0	ASTM D2272 (RPVOT), ASTM D943 (TOST), ASTM D6186 (PDSC)	Balances oxidative stability with biodegradability
Rust Preventives	Biodegradable petroleum sulfonates, fatty acid succinates	0.05–0.2	ASTM D665, ISO 7120	Water ingress accelerates depletion
Antifoam Agents	Polysiloxanes (low-ppm), polyacrylates	10–100 ppm	ASTM D892 (foam), ISO 6247	EALs often limit silicone use to meet biodegradability criteria
Demulsifiers	Polyalkylene glycols, EO/PO block copolymers	0.01–0.1	ASTM D1401 (demulsibility), ISO 6614	Must remain effective in synthetic ester base fluids
Friction Modifiers	Fatty acids, estolides, glycerol monoesters	0.05–0.5	ASTM D5183 (seizure load), SRV friction tests (DIN 51834)	Improves energy efficiency in gear/hydrostatic systems

Viscosity Modifiers	High MW biodegradable polymers (e.g., polyisoprene, PIB succinates)	0.1–1.0	ASTM D445 (viscosity), ASTM D2270 (VI)	Polymer shear stability tested per CEC L-45-99
Pour Point Depressants	Biodegradable alkylated naphthalenes, esters	0.05–0.3	ASTM D97 (pour point), ASTM D5949 (low-temp)	Must not impair biodegradability or hydrolytic stability
pH Buffers (for water-containing EALs)	Alkanolamines, amino alcohols	0.05–0.3	ASTM D664 (acid number), ASTM E70 (pH)	Maintains corrosion control in HETG/HSE fluids
Biocides (if water present)	Isothiazolinones, bronopol (low dose)	0.01–0.05	ASTM E2275 (microbial count)	Only in water-glycol or emulsion-based EALs; must be EPA-approved
Seal Swell Agents	Bio-based aromatic esters, dibasic acid esters	0.1–0.5	ASTM D471 (rubber immersion), ISO 1817	Used to maintain elastomer compatibility

APPLICATION MAPPING: WIND TURBINES, NUCLEAR ACTUATORS, GENSET VALVETRAINS, ETC.

Power generation equipment operates under diverse and demanding conditions, requiring lubricants and functional fluids tailored to each subsystem's unique mechanical, thermal, and environmental stresses. From the high-torque, variable-load gearboxes of wind turbines to the radiation-sensitive actuators of nuclear facilities, every application presents a distinct set of challenges that influence lubricant formulation and selection. No single fluid type can satisfy the wide spectrum of requirements, oxidation resistance, radiation stability, micropitting control, corrosion protection, fire safety, or biodegradability, that power plants must address across different technologies.

To meet these needs, engineers rely on carefully matched oils, greases, and specialty fluids that balance performance with safety and regulatory compliance. Synthetic EP gear oils extend the life of wind turbine drivetrains; phosphate ester fluids safeguard nuclear actuator systems against fire risk; modern engine oils protect genset valvetrains from wear, soot, and acid attack; EALs ensure hydroelectric wicket gates operate reliably in direct water contact; and ester-based transformer fluids deliver fire safety and biodegradability in electrical systems.

The following sections highlight key lubrication practices and fluid requirements across major power generation assets, emphasizing the role of chemistry, additive technology, and environmental compatibility in maintaining reliability and safety.

Wind Turbines

Main gearboxes use synthetic EP gear oils with high micropitting resistance, oxidation stability, and water tolerance. Generator bearings are lubricated with polyurea or calcium sulfonate complex greases for long service intervals and resistance to fretting corrosion. Pitch and yaw systems employ hydraulic fluids with strong antiwear protection, oxidation resistance, and low-temperature pumpability.

Nuclear Actuators

Control rod drive mechanisms, valve actuators, and containment isolation systems use greases and hydraulic fluids with radiation stability, oxidation resistance, and compatibility with stainless steels and elastomers used in nuclear service. Phosphate ester or synthetic ester fire-resistant fluids are common in electrohydraulic control systems where fire safety is critical.

Genset Valvetrains

Diesel and gas generator valvetrains rely on engine oils meeting API CK-4, FA-4, or OEM-specific gas engine oil standards. These oils must control wear in cam lobes, rocker arms, and valve guides while dispersing soot and neutralizing combustion acids. For biogas engines, low-ash formulations with high nitration and oxidation resistance are necessary to prevent pre-ignition and valve seat erosion.

Hydroelectric Wicket Gates and Kaplan Hubs

Require EALs or biodegradable greases that maintain lubricity in wet environments while resisting emulsification. Fluids must provide corrosion protection for bronze and ferrous components in constant water contact.

Power Plant Auxiliary Gearboxes and Bearings

Circulating and gear oils must match load profiles, operating temperatures, and expected contamination modes. In steam turbine-driven auxiliaries, turbine oils are used to maintain fluid uniformity and simplify inventory management.

Transformers and Switchgear

Insulating fluids are selected based on dielectric strength, fire safety, and environmental compliance. Ester-based fluids are favored in locations requiring higher fire points and biodegradability, while mineral oils remain common in high-load distribution transformers where fire safety is managed through containment.

4 Insulating Fluids and Dielectrics

MINERAL-BASED TRANSFORMER OILS (NAPHTHENIC, PARAFFINIC)

Mineral-based transformer oils remain the most widely used insulating fluids in power and distribution transformers, large reactors, and certain types of high-voltage switchgear. Their primary functions are to provide dielectric separation between energized conductors, to cool windings and cores by circulating heat to radiators or cooling fins, and to protect internal metallic surfaces from corrosion through the formation of a thin protective film.

Naphthenic Mineral Oils

Derived from crude sources with high naphthenic hydrocarbon content, these oils naturally exhibit low pour points without requiring pour point depressants. This characteristic makes them suitable for installations in cold climates where fluid flow at subzero temperatures is critical for maintaining cooling efficiency and preventing gas bubble entrapment in windings. Naphthenic oils have inherently higher solvency than paraffinic oils, allowing them to keep oxidation byproducts such as acids and sludge in suspension longer. This delays the formation of deposits on windings and radiator surfaces but can result in more contaminants being present in the fluid over time. Their oxidation stability is generally lower than that of highly refined paraffinic oils, requiring careful monitoring and antioxidant additive support to maintain service life.

Paraffinic Mineral Oils

Sourced from crude stocks rich in paraffinic hydrocarbons, these oils have higher natural oxidation stability due to lower levels of reactive sulfur and aromatic compounds. They typically offer longer service life in thermally stable conditions and generate fewer volatile acids during oxidation. However, paraffinic oils have higher natural pour points, often requiring pour point depressants in colder climates. Their lower solvency can allow sludge to precipitate and deposit more readily in transformer bottoms and coolers once oxidation progresses, potentially reducing heat transfer efficiency and impeding oil flow.

DOI: 10.1201/9781003775201-6

Common Properties and Performance Requirements

Regardless of type, transformer oils must maintain high dielectric strength, typically above 30 kV when tested per American Society for Testing and Materials (ASTM) D877 or D1816. Low moisture content is critical, as water dramatically reduces dielectric performance and accelerates cellulose insulation aging. Good heat transfer characteristics are essential to prevent localized hot spots. Dissolved gas behavior is important for fault detection; transformer oils must be compatible with dissolved gas analysis (DGA) to identify arcing, corona, or thermal faults through gas signature patterns.

Additive Systems and Refining

Modern transformer oils, whether naphthenic or paraffinic, are often hydrotreated or hydroprocessed to remove unstable components, improve oxidation stability, and minimize sulfur content. Antioxidant additives such as 2,6-di-tert-butyl para-cresol (DBPC) are commonly blended to slow oxidative degradation. These additives are consumed over time and require periodic oil testing to determine replenishment needs.

Mineral-based transformer oils, whether naphthenic or paraffinic, are formulated primarily from highly refined Group I base stocks and are optimized for electrical insulating performance, heat dissipation, and long-term stability within transformers and other oil-filled electrical equipment. While the base oil type largely determines physical properties, naphthenic oils having superior low-temperature fluidity and better solvency for oxidation byproducts, and paraffinic oils offering higher oxidation stability and longer service life, the additive system plays a critical role in preserving dielectric integrity and minimizing degradation over time.

The most common additive in both naphthenic and paraffinic transformer oils is a hindered phenolic antioxidant (such as 2,6-di-tert-butyl-p-cresol, DBPC), typically dosed at 0.1–0.3% by weight. This additive significantly slows oxidative polymerization, extending the oil's service life before acid number and sludge formation become problematic. In uninhibited grades, no antioxidant is added, and oxidative stability is entirely dependent on the inherent resistance of the base oil; inhibited grades provide much longer lifespans in high-load or high-temperature service.

Metal deactivators, often benzotriazole derivatives, are sometimes included to passivate copper windings and other non-ferrous metals, reducing catalytic oxidation. In some paraffinic oils, small amounts of pour point depressants may be added to improve low-temperature flow, though this is rarely needed in naphthenic oils due to their naturally low pour points. Certain formulations also include trace antistatic agents to help prevent charge accumulation during oil handling and circulation.

Unlike industrial lubricants, transformer oils do not contain detergents, dispersants, or antiwear agents, as these can impair dielectric properties or form conductive residues. Instead, the additive package is tightly restricted to maintain high dielectric strength, low dissipation factor, and minimal moisture affinity. In service, these oils are monitored for acidity, interfacial tension, dielectric breakdown voltage, DGA, and inhibitor depletion to determine maintenance intervals and inhibitor replenishment schedules.

TABLE 4.1
Common In-Service and Acceptance Tests for All Mineral-Based Transformer Oils

Test Parameter	ASTM/IEC/ISO Method	Purpose/Interpretation
Dielectric Breakdown Voltage	ASTM D1816, ASTM D877, IEC 60156	Measures insulation strength; reduced by moisture, particles, and oxidation products.
Moisture Content	ASTM D1533, IEC 60814	Indicates contamination; high levels accelerate oxidation and reduce dielectric strength.
Total Acid Number (TAN)	ASTM D974, IEC 62021-1	Monitors oxidation byproducts: rising TAN can signal inhibitor depletion.
Interfacial Tension (IFT)	ASTM D971, IEC 62961	Drops as polar oxidation products accumulate.
Color/Visual Appearance	ASTM D1500	Darkening indicates oxidation or contamination.
Furan Analysis	ASTM D5837	Measures paper insulation degradation by detecting furanic compounds.
DGA	ASTM D3612, IEC 60567	Detects fault gases from arcing, overheating, corona.
Oxidation Stability	ASTM D2440, IEC 61125	Evaluates additive performance over time.
Corrosive Sulfur	ASTM D1275B, IEC 62535	Identifies sulfur compounds harmful to copper.
Viscosity	ASTM D445	Changes with oxidation and contamination.

The following tables (see Tables 4.1 and 4.2) list the tests run for quality control and assurance acceptance and in-service condition monitoring followed by functional additives used in mineral-based transformer oils (naphthenic and paraffinic) for power generation (see Table 4.3).

These tables allow a reliability engineer to align transformer oil type with environmental conditions, operational profile, and maintenance resources.

Application Considerations

Naphthenic oils are favored in colder regions and for applications requiring high solvency, such as transformers with known contamination challenges. Paraffinic oils are often selected for long-life service in temperate climates or where sludge management practices are well established. In both cases, storage, handling, and filtration practices must prevent particulate contamination and moisture ingress. Oil reclamation, reconditioning, and full replacement strategies depend on DGA results, acidity levels, interfacial tension, and furan analysis of solid insulation degradation.

TABLE 4.2
Functional Additives in Mineral-Based Transformer Oils (Naphthenic and Paraffinic)

Additive Function	Typical Additive Chemistry	Typical Concentration Range (% wt)	Key Laboratory Tests (ASTM/IEC/ISO)	Notes
Oxidation Inhibitors	2,6-di-tert-butyl-p-cresol (DBPC), 2,6-di-tert-butylphenol (DBP), aromatic amines	0.2–0.4	ASTM D2440 (oxidation stability), ASTM D2112 (Rotating Pressure Vessel Oxidation Test – RPVOT), IEC 61125 (TOST)	Key for long-term stability; depletion rate is critical in service.
Metal Deactivators	Benzotriazole, tolyltriazole derivatives	0.005–0.05	ASTM D1275B (corrosive sulfur), IEC 62535 (DBDS detection)	Protects copper windings from catalytic oxidation.
Pour Point Depressants	Alkylated naphthalenes, polymethacrylates	0.02–0.2	ASTM D97 (pour point), ASTM D5950 (automatic PP)	Important for cold-climate transformers, especially with paraffinic base oils.
Moisture Inhibitors (rare in mineral oils)	Hydrophobic surface-active agents	<0.05	Karl Fischer moisture (ASTM D1533, IEC 60814)	Most mineral transformer oils rely on dehydration, not chemical inhibitors.
Antifoam Agents	Polysiloxanes, polyacrylates	5–50 ppm	ASTM D892 (foam), IEC 61125 foam tendency/stability	Low dose to avoid affecting dielectric strength.
Passivators/Corrosion Inhibitors	Benzotriazole derivatives, alkyl thiadiazoles	0.01–0.05	ASTM D130 (copper strip), ASTM D1275B (corrosive sulfur), IEC 62535	Often added to address corrosive sulfur issues in naphthenic oils.
Electrostatic Discharge Suppressants (special grades)	Conductive polymeric additives	<0.01	IEC 60247 (tan δ at 90 °C), ASTM D924	Rare; used in high-flow systems prone to static charge build-up.

TABLE 4.3
Naphthenic vs Paraffinic Transformer Oils – Performance and Lifecycle Comparison

Parameter/ Consideration	Naphthenic Mineral Oils	Paraffinic Mineral Oils	Reliability and Maintenance Impact
Base Crude Origin	High-naphthene crude fractions	High-paraffin crude fractions	Source chemistry drives pour point, oxidation stability, and solvency characteristics
Pour Point	Naturally low (often below −30 °C) without additives	Higher (often −10 °C to −15 °C) requiring pour point depressants in cold climates	Naphthenic oils flow reliably at low ambient temperatures, reducing cold-start thermal stress
Viscosity Index (VI)	Lower VI (~30–60), more viscosity changes with temperature	Higher VI (~90–100), less viscosity changes with temperature	Paraffinic oils maintain film thickness over broader temperature ranges
Oxidation Stability	Lower natural oxidation stability; requires antioxidant additive support	Higher natural oxidation stability; slower acid formation	Paraffinic oils generally provide longer service intervals before reclamation if kept clean
Sludge Formation	High solvency keeps oxidation products in suspension longer, delaying deposit formation	Lower solvency can cause sludge to precipitate earlier and settle in cool zones	Naphthenic oils maintain cleaner heat transfer surfaces initially; paraffinic oils require sludge monitoring during late-life
Heat Transfer Efficiency	Good, aided by high solvency and low pour point	Good, aided by stable viscosity across operating temperatures	Both types transfer heat effectively when clean and dry
Dielectric Strength	Comparable when clean and dry (>30 kV per ASTM D877)	Comparable when clean and dry (>30 kV per ASTM D877)	Regular dehydration and filtration critical for both
Gas Absorption/ DGA Compatibility	Good gas solubility; established DGA interpretation baselines	Good gas solubility; established DGA interpretation baselines	Both compatible with fault gas analysis; interpretation methods are transferable

Additive Requirements	Typically needs antioxidant package earlier in lifecycle	Often extended additive life due to inherent stability	Impacts oil testing intervals and additive replenishment schedule
Cold Climate Suitability	Excellent without chemical pour point depressants	Requires pour point depressants and more frequent low-temp viscosity checks	Influences fluid choice for transformers in subarctic or alpine regions
Maintenance Interval Triggers	Antioxidant depletion, acid number rise, IFT drop	Sludge deposition in coolers, acid number rise, IFT drop	Condition monitoring programs must be tuned to the dominant degradation pathway
Reclamation Feasibility	Highly reclaimable; responds well to Fuller's earth and adsorbent processes	Highly reclaimable; paraffinic oils often regain oxidation stability post-reconditioning	Selection affects reclamation economics and downtime planning
Typical Use Cases	Cold climates, high contamination risk environments, rapid-start transformers	Warm-to-temperate climates, long-life units with stable load cycles	Matching type to environment and operating profile improves service life and reduces intervention frequency

FIELD INSPECTION AND OIL SAMPLING PROTOCOL – MINERAL-BASED TRANSFORMER OILS

(Applicable to naphthenic and paraffinic oils with adjustments for chemistry and climate.)

1. **Preparation**
   - **Verify Documentation**
     Review transformer design, oil type (naphthenic or paraffinic), additive content, service history, and most recent oil analysis results.
   - **Inspection Equipment**
     Portable dielectric strength tester, moisture meter (in-oil), oil sampling kit with glass or metal sample bottles, syringe-type sampling device for low-volume draws, infrared thermometer, visual inspection tools (flashlight, inspection mirror).
   - **Environmental Check**
     Record ambient temperature, weather conditions, and load status at time of sampling.
2. **Visual and External Inspection**
   - Inspect transformer tank, radiators, and cooling fins for oil leaks, paint degradation, or signs of overheating.
   - Check breather condition; silica gel color must be within acceptable range. Replace if saturated.
   - Verify conservator tank oil level matches expected range for current temperature.
   - Inspect for physical damage, corrosion, or loose fittings.
   - For cold climates using paraffinic oil, check for clouding or wax formation during low temperatures.
   - For naphthenic oil systems, watch for signs of contamination in sight glasses due to higher solvency (oil may appear darker).
3. **Sampling Procedure**
   - **Sample Points**: Use dedicated oil sampling valves; avoid drain plugs unless designed for sampling.
   - **Oil Purge**: Drain at least 1–2 liters (or enough to flush stagnant oil) before collecting the test sample.
   - **Sample Containers**: Use pre-cleaned, dry, airtight containers; avoid plastics for DGA samples due to gas permeability.
   - **Labeling**: Record transformer ID, oil type, sample date, temperature, and load conditions.
   - **Special DGA Samples**: Fill syringes or glass bottles to eliminate head-space and seal immediately to prevent gas exchange.
4. **In-Field Testing**
   - **Dielectric Strength (ASTM D877 or D1816)**: Immediate indication of moisture or particulate contamination.
   - **Moisture Content**: Portable moisture meters (ppm basis) provide quick evaluation.

- **Temperature Differential**: Measure top oil temperature versus ambient; significant deviations from expected values can indicate cooling inefficiency.

5. **Laboratory Testing Panel**
   (Performed on every scheduled sample unless otherwise specified.)
   - **DGA**: Identify fault gases (H_2, CH_4, C_2H_2, C_2H_4, CO, CO_2) for arcing, overheating, or insulation degradation.
   - **Moisture Content (ASTM D1533 Karl Fischer)**: Keep below manufacturer-specified ppm limits (often ≤35 ppm for distribution transformers).
   - **Total Acid Number (TAN)**: Indicates oxidation level; naphthenic oils typically show earlier TAN rise than paraffinic.
   - **Interfacial Tension (IFT)**: Drop below 28 dynes/cm is a warning; correlates with oxidation byproducts and sludge risk.
   - **Antioxidant Content**: For inhibited oils, monitor DBPC or other additives; replenish if depleted.
   - **Color and Visual Sediment**: Early indicator of oxidation, sludge, or particulate contamination.
   - **Polychlorinated Biphenyl Content (if applicable)**: Confirm non-detectable if required for regulatory compliance.
6. **Frequency Guidelines**
   - **New Transformers**: Baseline sampling at commissioning, then at 6 months.
   - **Normal Service**: Annual sampling; semi-annual in high-load or critical service.
   - **Aged Units (>20 years)**: Semi-annual sampling regardless of load profile.
   - **Units with Adverse Results**: Increase to quarterly until stable.
7. **Interpretation Adjustments by Oil Type**
   - **Naphthenic Oils**: Expect higher solvency; darkening may not correlate directly to performance loss. Monitor TAN and antioxidant depletion closely.
   - **Paraffinic Oils**: Monitor for sludge deposition; IFT drops are often accompanied by visible sediment. Pour point depressant presence should be checked in cold climates.
8. **Corrective Actions**
   - **Minor Moisture or Particulate Contamination**: Vacuum dehydration and filtration.
   - **Oxidation/Acid Number Rise**: Antioxidant replenishment or partial oil replacement.
   - **Severe Degradation**: Full oil reclamation using Fuller's earth or similar adsorbent; full change-out if reclamation is not viable.
   - **DGA Abnormalities**: Initiate electrical diagnostics to locate fault source before re-energizing the unit.

SYNTHETIC ESTERS AND NATURAL ESTERS (FR3, MIDEL)

Synthetic esters and natural esters are insulating fluids engineered for high fire safety, biodegradability, and extended service life in transformers and other high-voltage equipment. Synthetic esters, such as Midel 7131, are manufactured through controlled chemical synthesis of mono- and polyol esters, giving them a consistent molecular structure with high oxidation stability and very low pour points, enabling service in cold climates without heating aids. They have high flash and fire points, typically over 300 °C, making them suitable for indoor or urban installations where fire risk mitigation is critical. Their dielectric strength remains stable at elevated moisture levels compared to mineral oils, allowing them to hold more water in solution without loss of insulating performance. They are resistant to oxidation when oxygen ingress is controlled, and inhibited variants can achieve very long service intervals. Material compatibility with cellulose, copper, and transformer seal materials is generally good, but elastomer selection should be verified to prevent swelling or hardening.

Natural esters, such as Cargill's FR3, are based on refined vegetable oils, typically soybean or rapeseed, processed to remove impurities and stabilized with antioxidants. They share the high fire point and biodegradability advantages of synthetic esters but have higher pour points, often requiring auxiliary heating in colder environments. A distinguishing feature of natural esters is their high water solubility relative to mineral oil, over ten times greater, allowing them to significantly slow the rate of cellulose insulation aging by reducing free water content in the paper. However, they are more prone to oxidation than synthetic esters, particularly in free-breathing transformer designs, so sealed tank configurations or nitrogen blankets are recommended to extend life. Over time, oxidation produces acids and polymers that can form sludge, particularly in vented units; this risk is managed through oxygen exclusion and periodic monitoring of acid number and interfacial tension. Both synthetic and natural esters require adapted dissolved gas analysis interpretation, as their fault gas generation patterns differ from mineral oil, especially in carbon monoxide and carbon dioxide levels from cellulose thermal degradation. Properly specified, installed, and maintained, ester fluids can deliver extended transformer life, higher overload capacity due to better thermal performance, and reduced environmental and fire risk across utility and industrial applications.

Synthetic esters and natural esters are insulating fluids used in transformers and other high-voltage apparatus where fire safety, environmental performance, and dielectric reliability are critical. Both fluid types share key common properties: high dielectric strength, elevated fire points typically above 300 °C, and strong biodegradability. They are non-toxic and present lower environmental remediation costs in case of leaks. Viscosity is higher than mineral oil at low temperatures, but synthetic esters can be engineered for low pour points, while natural esters generally have higher pour points requiring cold-weather mitigation. Moisture tolerance is a defining property, esters can hold several hundred ppm of water in solution without a sharp drop in breakdown voltage, reducing the risk of free water condensation in cellulose insulation. Oxidation stability varies: synthetic esters have inherently higher resistance to oxidation due to their molecular uniformity, while natural esters require stricter oxygen control to maintain performance.

Common Properties and Performance Requirements

Performance requirements for both ester types are governed by International Electrotechnical Commission (IEC) 61099 (synthetic esters) and IEC 62770 (natural esters) as well as ASTM D6871 and D6871M. Minimum dielectric strength is typically specified at ≥35 kV (ASTM D1816), with controlled limits on acidity (TAN), moisture content, and viscosity. Both must demonstrate stability under thermal aging, material compatibility with cellulose, copper, and transformer seal materials, and acceptable dissipation factor over the expected operating temperature range.

Additive Systems and Refining Methods

Additive systems and refining methods differ between the two. Synthetic esters are produced via chemical esterification of polyols and fatty acids, followed by purification to remove catalyst residues and unreacted materials. Antioxidants, metal deactivators, and pour point depressants may be added depending on application. Their controlled synthesis allows tight control of molecular weight distribution, enhancing thermal stability. Natural esters are refined from vegetable oils, commonly soybean or rapeseed, through degumming, bleaching, and deodorizing, then stabilized with food-grade antioxidants to delay oxidation. No pour point depressants are typically used, and their oxidative stability is directly influenced by the fatty acid profile of the source oil.

Synthetic esters and **natural esters** (such as FR3® from Cargill or Midel® fluids from M&I Materials) are used as alternative dielectric and cooling fluids in transformers, switchgear, and other oil-filled electrical equipment, offering **high fire resistance, biodegradability, and enhanced moisture tolerance** compared to mineral oils. While their base chemistry differs synthetic esters being purpose-built molecules with tailored chain length and branching, and natural esters being derived from refined vegetable oils such as soybean, rapeseed, or sunflower their additive systems share the goal of protecting the ester base from hydrolysis, oxidation, and corrosion while maintaining dielectric integrity.

In **synthetic ester transformer fluids**, the molecular structure is engineered for thermal stability, low pour point, and high oxidation resistance. Additive packages typically include **hindered phenolic antioxidants** (0.1–0.5% w/w) to extend service life under continuous high-temperature operation, **metal deactivators** (benzotriazole or derivatives) to prevent copper-catalyzed oxidation, and **antihydrolysis agents** to minimize acid formation from moisture exposure. Small amounts of pour point depressants may be used in certain synthetic esters to maintain cold-weather pumpability, especially in high-latitude installations.

Natural ester fluids such as FR3® and Midel® rely on the inherent oxidative stability and lubricity of the triglyceride base, but are more susceptible to oxidation at elevated temperatures. To address this, antioxidant levels are generally higher than in mineral oils, with a focus on **food-grade or environmentally compatible chemistries** to preserve the fluid's biodegradability. Moisture scavengers or passivating agents are often incorporated to manage the higher water solubility of natural esters, which while beneficial for reducing paper insulation embrittlement can accelerate oxidation if unmanaged. Natural esters also benefit from **corrosion inhibitors** compatible with cellulose insulation and transformer metals, and, in some cases, **pour point depressants** for cold-climate performance improvement.

Both ester types inherently have a **high fire point (typically > 300 °C)**, excellent lubricity, and strong film strength, reducing mechanical wear on moving components such as on-load tap changers. Their additive systems are minimalist to preserve dielectric performance, often focusing on just three functions: **antioxidation, metal passivation, and hydrolytic stability enhancement**. In service, ester fluids are monitored for **acid number, moisture content, interfacial tension, dielectric breakdown voltage, and inhibitor depletion**, with replenishment schedules designed to maintain both fluid life and insulation protection.

The following tables (see Tables 4.4 and 4.5) list the tests run for quality control and assurance acceptance and in-service condition monitoring followed by functional additives used in synthetic ester and natural ester lubricants for power generation (see Table 4.6).

Application Considerations

Application considerations include system design, cooling performance, maintenance strategy, and DGA interpretation. In sealed transformers or those with nitrogen blankets, both ester types offer long service life. In free-breathing designs, natural

TABLE 4.4
Common Acceptance and In-Service Tests for Both Synthetic and Natural Esters

Test Parameter	ASTM/IEC/ISO Method	Purpose/Interpretation
Dielectric Breakdown Voltage	ASTM D1816, IEC 60156	Insulation strength; reduced by moisture, oxidation, particulates.
Moisture Content	ASTM D1533, IEC 60814	High moisture accelerates acid formation; natural esters hold more water in solution without reducing breakdown voltage.
Total Acid Number (TAN)	ASTM D974, IEC 62021-1	Tracks oxidation and additive depletion; rising TAN signals intervention.
Interfacial Tension (IFT)	ASTM D971, IEC 62961	Declines as polar oxidation products accumulate.
Color/ Appearance	ASTM D1500	Darkening correlates with oxidation or contamination.
Oxidation Stability	ASTM D2112, IEC 61125	Measures remaining antioxidant reserve.
Corrosive Sulfur	ASTM D1275B, IEC 62535	Ensures no active sulfur species present.
Viscosity	ASTM D445	Monitors thickening from oxidation or polymerization.
Fire Point / Flash Point	ASTM D92, ISO 2592	Important for fire safety classification (K-class fluids).

TABLE 4.5
Functional Additives in Synthetic Esters and Natural Esters

Additive Function	Typical Additive Chemistry	Typical Concentration Range (% wt)	Key Laboratory Tests (ASTM/ IEC/ISO)	Notes
Oxidation Inhibitors	Hindered phenols (BHT, DBPC), aromatic amines, tocopherols (Vit. E in natural esters)	0.1–0.5	ASTM D2112 (RPVOT), IEC 61125 (TOST), ASTM D943	Natural esters inherently more oxidation-resistant due to fatty acids; antioxidants slow polymerization and acid buildup.
Metal Deactivators	Benzotriazole, tolyltriazole	0.005–0.05	ASTM D1275B (corrosive sulfur), IEC 62535	Protects copper windings from catalytic oxidation.
Pour Point Depressants	Alkylated naphthalenes, polymethacrylates (biodegradable grades)	0.02–0.2	ASTM D97, ASTM D5950	Critical for cold-weather applications; synthetic esters have better low-temp properties than natural esters.
Moisture Tolerance Modifiers	Polar additives, desiccant-compatible surfactants	<0.05	ASTM D1533, IEC 60814 (moisture content), water saturation curves	Used to improve water solubility management; natural esters have higher saturation point than synthetic esters.
Antifoam Agents	Polysiloxanes, polyacrylates	5–50 ppm	ASTM D892, IEC foam test	Low dose to avoid dielectric loss.
Passivators / Corrosion Inhibitors	Benzotriazole derivatives, alkyl thiadiazoles	0.01–0.05	ASTM D130 (copper strip), ASTM D1275B	Limits copper corrosion in high-load service.
Biocides *(Natural esters only)*	Isothiazolinones, bronopol (low dose)	0.01–0.05	ASTM E2275 (microbial count)	Used only where microbial contamination risk exists; more common in storage.
Viscosity Modifiers	High molecular weight esters, biodegradable polymers	0.1–1.0	ASTM D445 (viscosity), ASTM D2270 (VI)	Maintains viscosity index under high load.
Seal Swell Agents	Bio-based aromatic esters, dibasic acid esters	0.1–0.5	ASTM D471 (rubber immersion), ISO 1817	Ensures elastomer compatibility in retrofit conversions.

TABLE 4.6
Synthetic and Natural Ester Transformer Oils – Performance and Lifecycle Comparison

Parameter/ Consideration	Synthetic Esters (e.g., Midel 7131)	Natural Esters (e.g., FR3)	Reliability and Maintenance Impact
Base Source	Chemically engineered from synthetic acids and alcohols	Refined vegetable oils (soybean, rapeseed, sunflower)	Source chemistry affects oxidation stability, cold flow, and consistency
Fire Point	>300 °C	>300 °C	Both qualify as less-flammable fluids; high fire safety reduces secondary damage risk
Oxidation Stability	High; molecular structure resists oxidation	Moderate; triglycerides more reactive, rely on sealed systems or nitrogen blanket	Synthetic esters maintain properties longer in vented designs; natural esters require tighter oxygen control
Moisture Tolerance	High; maintains dielectric strength at elevated moisture	High; maintains dielectric strength at elevated moisture	Both slow cellulose insulation aging by absorbing and retaining moisture without dielectric collapse
Pour Point/Cold Climate Suitability	Can be engineered for low pour points (−20 °C to −30 °C)	Higher pour point (−10 °C to −15 °C typical); heating may be required in cold climates	Synthetic esters preferred for unheated outdoor service in subarctic or alpine climates
Dielectric Strength	Comparable to mineral oil when dry (>35 kV)	Comparable to mineral oil when dry (>35 kV)	Cleanliness and moisture control remain critical
Thermal Performance	Stable viscosity across operating range; high specific heat	Stable viscosity across operating range; high specific heat	Both types support effective winding and core cooling
Biodegradability	Readily biodegradable; low aquatic toxicity	Readily biodegradable; low aquatic toxicity	Meets EAL criteria; reduced environmental impact in case of leaks
Sludge Formation	Minimal; oxidative byproducts remain dissolved longer	Minimal if oxidation is controlled; can form sticky deposits if oxidized in vented units	Sealed or inerted systems reduce maintenance for both types

Additive Requirements	Usually contain oxidation inhibitors and metal passivators; long additive life	May rely more on natural antioxidants (tocopherols); additive depletion faster in high-load, oxygen-exposed service	Additive monitoring via TAN and inhibitor content testing
DGA Interpretation	Valid; gas generation patterns differ from mineral oil	Valid; gas generation patterns differ from mineral oil	Requires adjusted fault interpretation guidelines for ester chemistry
Retrofill Compatibility	Compatible with most cellulose and metals; some elastomer swelling possible	Compatible with most cellulose and metals; elastomer checks required	Pre-retrofill material compatibility review prevents sealing failures
Maintenance Interval Triggers	Acid number rise, inhibitor depletion, moisture accumulation	Acid number rise, moisture accumulation, oxidation signs in vented units	Monitoring intervals set by load profile and operating temperature
Typical Use Cases	Indoor substations, mobile substations, retrofills in high-load units, cold-climate service	Distribution transformers, urban/underground substations, retrofills for fire safety and environmental compliance	Application match driven by climate, fire risk, and insulation life extension objectives

esters require more frequent monitoring of TAN and interfacial tension to track oxidation, while synthetic esters are less affected but still benefit from oxygen exclusion. Cooling performance is slightly reduced compared to mineral oil due to higher viscosity, which must be factored into thermal design and load limits. Ester fluids allow higher moisture content in cellulose without loss of dielectric integrity, extending insulation life, but require ester-specific DGA interpretation since fault gas ratios differ from mineral oils, particularly in CO/CO_2 generation from paper aging and ethylene/methane levels from thermal stress. Compatibility with gaskets and paints must be verified to prevent swelling or softening. In retrofill projects, system flushing and removal of residual mineral oil are important to maintain ester fluid properties and meet performance specifications.

FIELD INSPECTION AND OIL SAMPLING PROTOCOL – SYNTHETIC AND NATURAL ESTER TRANSFORMER OILS

(Applicable to synthetic esters such as Midel 7131 and natural esters such as FR3.)

1. **Preparation**
   - **Verify Documentation**

   Confirm ester type (synthetic or natural), manufacturer, fire safety classification, date of last fill or retrofill, and most recent oil analysis report.
   - **Inspection Equipment**

   Portable dielectric strength tester, moisture meter calibrated for ester fluids, oil sampling kit with clean glass or metal containers, vacuum-tight DGA syringes, infrared thermometer, and tools for breather and seal inspection.
   - **Environmental Check**

   Record ambient temperature, transformer load, and weather conditions. In cold climates, note whether any oil heating systems are active before sampling.
2. **Visual and External Inspection**
   - Check tank, radiators, and conservator for leaks, especially around gaskets and seals known to swell in ester service.
   - Inspect breather for desiccant saturation; in natural ester systems, confirm oxygen control systems (sealed tank or nitrogen blanket) are intact.
   - Verify conservator oil level; ester fluids expand and contract more than mineral oil, so level changes are more pronounced with load and temperature.
   - Look for discoloration or haziness in sight glass; esters naturally darken faster than mineral oils but sudden changes can indicate contamination or oxidation.
3. **Sampling Procedure**
   - **Sample Points**: Use dedicated sampling valves; avoid high-disturbance points that could introduce particulates.
   - **Oil Purge**: Drain 1–2 liters to flush stagnant oil before taking sample.
   - **Sample Containers**: Use dry, clean glass bottles with airtight caps for standard tests. For DGA, use gas-tight syringes or special bottles with no headspace.

- **Labeling**: Include transformer ID, ester type, sample date/time, load, oil temperature, and ambient temperature.

4. **In-Field Testing (see Table 4.7)**
   - **Dielectric Strength**: ASTM D877 or D1816; high values (>35 kV) expected if moisture and particulates are controlled.
   - **Moisture Content**: Portable meters can provide quick ppm estimates, but confirm via laboratory Karl Fischer titration. Esters tolerate higher ppm moisture without loss of dielectric strength, but moisture still accelerates oxidation in vented systems.
   - **Temperature Check**: Compare top oil temperature to ambient and expected load-related rise; abnormal differentials may indicate cooling restriction.
5. **Laboratory Testing Panel**

(Performed on each scheduled sample.)

   - **DGA**: Gas generation patterns differ from mineral oil; use ester-specific interpretation tables. Monitor for acetylene, ethylene, methane, and carbon oxides.
   - **Moisture (ASTM D1533 Karl Fischer)**: Track trends; natural esters can hold >400 ppm before dielectric drop but moisture still promotes oxidation and insulation aging.
   - **Total Acid Number (TAN)**: Primary oxidation indicator. For natural esters, higher TAN can be tolerated before remedial action, but trending is critical. For synthetic esters, tighter TAN limits are advisable.
   - **Interfacial Tension (IFT)**: Less sensitive in esters than mineral oils but still useful for monitoring oxidation byproduct buildup.
   - **Color and Visual Sediment**: Natural esters may darken naturally; compare to baseline for abnormal change rates. Sediment may indicate oxidation sludge in vented units.
   - **Antioxidant/Inhibitor Content**: For synthetic esters with inhibitor packages, monitor depletion and replenish if necessary.
6. **Frequency Guidelines**
   - **New or Retrofilled Units**: Baseline sampling at commissioning and 6 months after.
   - **Normal Service**: Annual sampling; semi-annual for high-load or critical units.
   - **Vented Natural Ester Systems**: Semi-annual minimum due to increased oxidation exposure.
   - **Units with Adverse Results**: Increase to quarterly sampling until stability is verified.
7. **Interpretation Adjustments by Ester Type**
   - **Synthetic Esters**: Better oxidation resistance; TAN increases more slowly. Use tighter control limits to extend service life.
   - **Natural Esters**: Accept higher moisture content without dielectric loss; prioritize oxygen control and TAN trending to manage oxidation.

TABLE 4.7
Transformer Oil Field Protocol – Quick Reference Sheet ***(Mineral oil vs synthetic ester vs natural ester.)***

Parameter/Procedure Step	Mineral Oil (Naphthenic/ Paraffinic)	Synthetic Ester (e.g., Midel 7131)	Natural Ester (e.g., FR3)
Primary Functions	Dielectric insulation, cooling, corrosion protection	Same as mineral oil with higher fire point and biodegradability	Same as mineral oil with higher fire point, biodegradability, and greater moisture tolerance
Fire Point	140–170 °C typical	>300 °C	>300 °C
Cold Climate Flow	Naphthenic: good without additives; Paraffinic: requires pour point depressants	Can be engineered to –20 °C to –30 °C	Higher pour point (–10 °C to –15 °C); heating often required in cold climates
Moisture Tolerance (before dielectric loss)	Low (≤35 ppm typical)	High (>200 ppm)	Very high (>400 ppm)
Oxidation Stability	Paraffinic: high; Naphthenic: moderate	High	Moderate; requires oxygen control
Sampling Frequency – Normal Service	Annual; semi-annual for critical units	Annual; semi-annual for high-load units	Semi-annual minimum; quarterly in high-load or vented units
Sampling Frequency – Adverse Results	Quarterly until stable	Quarterly until stable	Quarterly until stable
Special Handling Notes	Avoid moisture ingress during sampling	Avoid cross-contamination; use ester-clean equipment	Same as synthetic ester; minimize oxygen exposure during handling
Key In-Field Tests	Dielectric strength, moisture (ppm), top-oil temperature	Dielectric strength, moisture, top-oil temperature	Same as synthetic ester
Lab Testing – Core Panel	DGA (mineral oil interpretation), TAN, IFT, moisture (KF), antioxidant content if inhibited	DGA (ester-specific interpretation), TAN (tight limits), IFT, moisture (KF), antioxidant if present	DGA (ester-specific interpretation), TAN (trend-based limits), IFT, moisture (KF)

DGA Interpretation	Standard mineral oil gas ratio methods	Adjusted patterns; some gases form at lower severity	Adjusted patterns; different CO/CO_2 ratios from cellulose aging
Maintenance Triggers	Low dielectric, high moisture, rising TAN/low IFT, sludge formation	Rising TAN, inhibitor depletion, moisture > design target	Rising TAN, moisture accumulation + oxygen exposure signs
Corrective Actions	Filtration, vacuum dehydration, antioxidant replenishment, reclamation or replacement	Vacuum dehydration, antioxidant replenishment, nitrogen blanketing if needed	Vacuum dehydration, nitrogen blanketing, sealing to reduce oxygen ingress
End-of-Life Indicators	TAN > 0.3–0.5 mg KOH/g, IFT < 28 dynes/cm, excessive sludge, abnormal DGA	TAN > manufacturer threshold, loss of dielectric margin, abnormal DGA	TAN > manufacturer threshold for natural esters, oxidation sludge, abnormal DGA

8. **Corrective Actions**
   - **Moisture Control**: Use vacuum dehydration; avoid excessive heat during treatment to prevent hydrolysis.
   - **Oxidation Control**: For vented natural ester units, improve sealing or add nitrogen blanket; replenish antioxidants in synthetic esters as per manufacturer guidance.
   - **Cooling Efficiency**: If viscosity increases due to oxidation, assess flow restrictions and plan partial or full fluid replacement.
   - **DGA Abnormalities**: Investigate winding hot spots or partial discharge sources; ester gas patterns may indicate overheating at lower gas generation levels compared to mineral oil.

Failure Modes and Degradation Pathways – Synthetic and Natural Ester Insulating Fluids with Criticality Weighting

Ester-based insulating fluids, both synthetic and natural, are increasingly used in transformers and high-voltage equipment due to their fire safety, biodegradability, and strong dielectric performance. However, like all insulating media, they are subject to chemical and physical degradation over time. Understanding the distinct failure modes of ester fluids, and their differences from mineral oil, is essential to maintaining reliability, preventing unplanned outages, and extending asset life.

Unlike mineral oils, esters exhibit unique pathways of oxidation, hydrolysis, and sludge formation, with natural esters generally more susceptible to rapid degradation when oxygen ingress or moisture contamination is present. Synthetic esters, while more oxidation-resistant, can still experience critical issues such as additive depletion or incompatibility with certain elastomers. Furthermore, diagnostic misinterpretation remains a risk, as DGA must be interpreted using ester-specific guidelines rather than mineral oil standards.

The table (see Table 4.8) that follows summarizes the primary degradation pathways for synthetic and natural ester insulating fluids, linking each failure mode to its typical causes, observable field indicators, and recommended field or laboratory tests. Corrective and preventive maintenance actions are suggested, and each pathway is assigned a criticality weighting on a 1–5 scale to help prioritize monitoring and intervention. This structured approach provides asset managers and reliability engineers with a practical framework to evaluate risks, implement corrective measures, and optimize life-cycle management of ester-filled power equipment.

Criticality Scale

- **1** – Minor impact on performance, low urgency
- **2** – Limited risk, can be addressed during scheduled maintenance
- **3** – Moderate operational risk, requires planned intervention
- **4** – High operational or asset life risk, requires prompt corrective action
- **5** – Severe risk to service continuity, asset failure likely if uncorrected

TABLE 4.8
Failure Modes and Degradation Pathways for Synthetic and Natural Ester Insulating Fluids

Failure Mode/ Degradation Pathway	Typical Cause	Observable Indicators	Field/Lab Tests	Maintenance / Corrective Action	Criticality (1–5)
Oxidation – Synthetic Ester	Oxygen ingress in partially sealed or vented units, high temperature operation	Gradual TAN increase, decrease in IFT, darkening of fluid, slight increase in viscosity	TAN (ASTM D974), IFT (ASTM D971), color (ASTM D1500), FTIR oxidation bands	Nitrogen blanketing, antioxidant replenishment if allowed by OEM, filtration to remove oxidation products	**3**
Oxidation – Natural Ester	Oxygen ingress, elevated operating temperature, insufficient antioxidant reserve	Rapid TAN rise, drop in IFT, sludge formation in free-breathing units, increased viscosity	TAN, IFT, visual inspection for sludge, antioxidant content analysis	Convert to sealed system, antioxidant top-up, periodic fluid reclamation, remove sludge by hot oil circulation	**5**
Thermal Decomposition	Overload operation, localized hot spots, poor cooling	Elevated CO and CO_2 in DGA, low levels of ethylene and methane, TAN rise in severe cases	DGA (ester-specific interpretation), hotspot temperature measurements	Address load management, improve cooling circulation, investigate insulation condition	**4**
Moisture Ingress	Seal degradation, desiccant failure, leak points, retrofill contamination	Moisture above design target but less impact on dielectric breakdown than mineral oil, possible paper degradation	Karl Fischer moisture (ASTM D1533), breakdown voltage (ASTM D1816)	Seal repair, dehydration via vacuum processing, ensure desiccant function	**4**
Contamination – Solid Particulate	Maintenance debris, wear particles, corrosion	Elevated particle count, reduced dielectric breakdown in extreme cases	ISO 4406 particle count, breakdown voltage	Filter to remove solids, inspect for wear sources, corrosion control	**3**

(continued)

TABLE 4.8 (Continued)
Failure Modes and Degradation Pathways for Synthetic and Natural Ester Insulating Fluids

Failure Mode/ Degradation Pathway	Typical Cause	Observable Indicators	Field/Lab Tests	Maintenance / Corrective Action	Criticality (1–5)
Contamination – Mineral Oil Residue in Retrofill	Incomplete flushing before ester fill	Lower fire point, altered viscosity, compromised environmental performance	Fire point test (ASTM D92), viscosity check, GC analysis for hydrocarbon profile	Drain and flush, refill with fresh ester, verify by lab confirmation	**3**
Sludge Formation – Natural Ester	Oxidation products polymerize in presence of oxygen	Deposits in cooler tubes, restricted oil flow, higher winding temperatures	Visual inspection, TAN, IFT	Hot oil flush, nitrogen blanket installation, periodic antioxidant replenishment	**5**
Chemical Interaction–Elastomer Incompatibility	Incorrect gasket or seal material	Swelling, softening, loss of sealing integrity	Visual inspection during maintenance, swelling index test	Replace with ester-compatible elastomers (nitrile variants, fluorocarbon, EPDM depending on ester type)	**4**
Additive Depletion	Long service intervals without replenishment	Declining oxidation stability, increased TAN	RPVOT (ASTM D2272), antioxidant content (HPLC)	Antioxidant replenishment if allowed, otherwise fluid replacement	**3**
Gas Generation Misinterpretation	Using mineral oil DGA fault limits for esters	False positives for thermal fault conditions	Apply ester-specific DGA guidelines (IEC 60599 adaptations)	Train analysts, update DGA interpretation software for ester fluids	**2**

SILICONE FLUIDS AND AROMATIC HYDROCARBONS

Silicone Fluids

Polydimethylsiloxane and related silicones are used where very high thermal stability and wide operating temperature range are required. Dielectric strength is comparable to transformer-grade mineral oil when dry. Viscosity changes little with temperature due to the low temperature coefficient of viscosity. Flash and fire points are high. Silicone fluids resist oxidation in air better than mineral oils but absorb dissolved gases differently, which affects dissolved gas analysis baselines. They are chemically inert toward copper and most steels. Compatibility checks are required for some elastomers and paints that can swell in silicone. Moisture tolerance is moderate: breakdown voltage drops in the presence of free water, so dehydration and tight sealing are still required. Performance requirements: dielectric strength at or above 30 to 35 kV per ASTM D1816, stable dissipation factor across the service temperature, minimal volatility in high-temperature compartments, acceptable gassing tendency under IEC and ASTM tests.

Additives and Processing

Base fluids are produced by controlled polymerization of siloxane monomers, then end-capped to control molecular weight and volatility. Additive packages are usually minimal to preserve dielectric loss characteristics: small amounts of passivators or antifoam agents may be used. Antioxidants are rarely required under normal electrical service, although high temperature zones can benefit from proprietary stabilizers. Refining focuses on removal of low molecular weight fractions and residual catalysts to reduce gassing and volatility.

Aromatic hydrocarbon insulating fluids: These include synthetic mono- and di-alkyl benzenes and selected hydrogenated aromatics formulated for dielectric use. They provide higher solvency than paraffinic mineral oils, which helps keep oxidation products in solution and maintains heat transfer surfaces cleaner early in life. Dielectric strength meets or exceeds mineral oil values when dry. Viscosity is moderate with higher viscosity index than naphthenic mineral oil in many grades. Thermal stability is good and volatility is lower than many light hydrocarbons. Moisture tolerance is similar to mineral oils: dissolved moisture reduces breakdown voltage, so dehydration remains a control point. Performance requirements mirror transformer oil standards: dielectric strength above 30 to 35 kV, controlled dissipation factor, acceptable corrosive sulfur status, and stability under thermal aging.

Fluids are produced by targeted catalytic alkylation or hydrogenation of aromatic streams followed by hydrotreating to remove reactive species. Additive packages are simple: antioxidants to slow peroxide formation, metal passivators where copper activity is a concern, and antifoam agents as required. Higher natural solvency means antioxidants can be consumed more quickly if contamination enters the system; trending of TAN and IFT is recommended. Unlike many mineral oils, these fluids often require tighter control of trace polar contaminants that can increase dielectric losses.

The following tables (see Tables 4.9 and 4.10) list the tests run for quality control and assurance acceptance and in-service condition monitoring followed by functional additives used in silicone fluids vs aromatic hydrocarbon insulating fluids for power generation (see Tables 4.11 and 4.12).

Application Considerations

Use cases include sealed transformers that benefit from high solvency and good low-temperature start-up, high-voltage direct current (HVDC) converter transformers where dissipation factor stability is important, and specialty capacitive equipment that specifies aromatic chemistry.

- **Benefits:** good oxidation resistance after hydrotreating, favorable cold flow in selected grades, cleaner heat transfer surfaces during early and mid-life due to solvency.
- **Risks:** potential for higher dielectric losses if moisture or polar contaminants accumulate, additive depletion rates that require closer monitoring in high-temperature service, and material compatibility checks for certain paints and sealants.

TABLE 4.9
Common Acceptance and In-Service Tests for Silicone Fluids and Aromatic Hydrocarbons

Test Parameter	ASTM / IEC / ISO Method	Purpose/Interpretation
Dielectric Breakdown Voltage	ASTM D1816, IEC 60156	Confirms insulating performance; silicone fluids retain dielectric strength even with some moisture.
Moisture Content	ASTM D1533, IEC 60814	Critical for aromatic hydrocarbons; less impact on silicone fluids but still monitored.
Total Acid Number (TAN)	ASTM D974, IEC 62021-1	Tracks oxidation; aromatics are more prone to TAN increase than silicones.
Interfacial Tension (IFT)	ASTM D971, IEC 62961	Drop indicates polar contaminants or oxidation products.
Color / Appearance	ASTM D1500	Change indicates contamination or oxidation.
Viscosity	ASTM D445	Change in silicone = polymer degradation or contamination; change in aromatics = oxidation or polymerization.
Oxidation Stability	ASTM D2112, IEC 61125	More critical for aromatic hydrocarbons; silicones tested primarily for comparison or in severe service.
Corrosive Sulfur	ASTM D1275B, IEC 62535	Rare concern in silicones; must be confirmed in aromatics.

TABLE 4.10
Functional Additives in Silicone Fluids and Aromatic Hydrocarbon Fluids

Additive Function	Typical Additive Chemistry	Typical Concentration Range (% wt)	Key Laboratory Tests (ASTM / IEC / ISO)	Notes
Oxidation Inhibitors	Aromatic amines (low volatility), hindered phenols	0.05–0.3	ASTM D2112 (RPVOT – modified), ASTM D943, IEC 61125	Silicone fluids are inherently oxidation-resistant; aromatic hydrocarbons benefit significantly from inhibitors.
Metal Deactivators	Benzotriazole derivatives	0.005–0.05	ASTM D1275B, IEC 62535	More relevant in aromatic hydrocarbons with copper contact; rare in silicone fluids.
Pour Point Depressants	Not typically needed in silicones; polymethacrylates in aromatics	0.02–0.1	ASTM D97, ASTM D5950	Silicone fluids have extremely low pour points; aromatic hydrocarbons may need improvement in cold climates.
Moisture Inhibitors	Hydrophobic agents (fluorinated silanes for silicone)	<0.05	ASTM D1533, IEC 60814 (Karl Fischer)	Silicone fluids tolerate water without dielectric loss; aromatics are more sensitive.
Antifoam Agents	Silicone-based defoamers (for aromatics), non-silicone organics (for silicone fluids)	5–50 ppm	ASTM D892, IEC foam tendency/stability	Silicone oils can foam if contaminated; aromatics prone to foaming under agitation.
Passivators/Corrosion Inhibitors	Benzotriazole, alkyl thiadiazoles	0.01–0.05	ASTM D130 (copper strip), ASTM D1275B	Used when metals are present in aromatic hydrocarbon systems.
Viscosity Modifiers	High-MW silicone polymers (for silicones), alkylated aromatics (for aromatics)	0.1–1.0	ASTM D445, ASTM D2270	Helps maintain thermal stability under shear; silicone viscosity is highly temperature stable.
Fire Resistance Enhancers	Not required for silicones (naturally high flash point); phosphorus-based in aromatics	Variable	ASTM D92, ISO 2592 (flash/fire point)	Silicones have self-extinguishing characteristics; aromatics may need additives for higher fire safety classes.

TABLE 4.11
Performance and Lifecycle Comparison Table for Silicone Fluids vs Aromatic Hydrocarbon Insulating Fluids

Parameter	Silicone Fluids	Aromatic Hydrocarbon Fluids
Dielectric Strength (New Fluid)	30–35 kV (ASTM D1816, 1 mm gap) when dry	30–35 kV (ASTM D1816, 1 mm gap) when dry
Fire Point	> 300 °C	150–180 °C
Viscosity Index	High, minimal change with temperature	Moderate, higher than naphthenic mineral oil
Thermal Stability	Excellent; minimal degradation below 200 °C	Good; stable under normal transformer service temperatures
Oxidation Resistance	High; low reactivity with oxygen	Good; requires antioxidant package to maintain stability
Moisture Tolerance	Moderate; moisture reduces dielectric strength	Similar to mineral oils; moisture reduces dielectric strength
Pour Point	–50 to –60 °C (excellent cold start)	–30 to –40 °C (grade dependent)
Sludge Formation Tendency	Very low; poor solvent for oxidation byproducts	Low to moderate; high solvency retains byproducts in solution longer
Additive System	Minimal; may contain passivators or antifoam agents	Antioxidants, metal passivators, antifoam; requires additive monitoring
Material Compatibility	May swell certain elastomers and paints; requires compatibility check	Generally compatible with transformer materials; verify with paints/sealants
Environmental Impact	Persistent in environment; special disposal requirements	Lower persistence; disposal similar to treated mineral oils
Typical Applications	Indoor transformers near ignition sources, high-temperature service, traction/furnace transformers	Sealed transformers, HVDC converter transformers, capacitive equipment
DGA Interpretation	Unique baseline gas patterns; must establish unit-specific trends	Similar to mineral oils but with different absolute gas generation rates
Service Life Expectancy	30–40 years with sealed system and proper moisture control	20–30 years with sealed system and antioxidant maintenance
Maintenance Considerations	Dry handling, desiccant breathers, gasket verification	Moisture control, antioxidant depletion monitoring, contamination control

TABLE 4.12
Failure Modes and Degradation Pathways Table for Silicone Fluids and Aromatic Hydrocarbon Fluids

Failure Mode/ Degradation Pathway	Description	Silicone Fluids Criticality (1–5)	Aromatic Hydrocarbon Fluids Criticality (1–5)
Oxidation	Reaction with oxygen leading to acid formation and dielectric loss	2 – High inherent oxidative stability, minimal byproduct formation under normal service	3 – Good stability with antioxidants, but depletion increases oxidation rate
Thermal Decomposition	Breakdown from sustained high temperatures producing volatile and solid residues	2 – Stable up to ~200 °C; decomposition above this	3 – Stable under transformer temps, but higher degradation above ~150–160 °C
Moisture Ingress	Water entry reducing breakdown voltage and promoting insulation aging	3 – Breakdown voltage drops in wet state, needs moisture control	3 – Similar sensitivity to mineral oils; moisture management critical
Contamination – Solid Particulate	Entry of dust, wear debris, or processing residue	2 – Low solvency means particles tend to settle; can cause local arcing	3 – Higher solvency keeps particles in suspension, increasing conductive risk
Sludge Formation	Insoluble oxidation/polymerization products deposit in windings and ducts	1 – Poor solvent for oxidation products prevents in-solution sludge buildup	2 – Higher solvency delays sludge formation but can carry more contaminants
Chemical Interaction – Elastomer	Swelling, shrinkage, or embrittlement of seals and gaskets	3 – Can swell certain rubbers; compatibility verification required	2 – Generally compatible with transformer-grade materials
Additive Depletion	Loss of antioxidant, passivator, or other performance chemicals	1 – Minimal additive use; depletion rarely a driver of failure	3 – Antioxidants are essential; depletion accelerates oxidation
Gas Generation Misinterpretation	DGA baseline misread leading to false fault diagnosis	4 – Different gas ratios vs mineral oils; requires redefined baselines	3 – More similar to mineral oils, but different absolute rates may mislead

The DGA interpretation follows mineral oil methods, but absolute gas generation rates can differ; establish unit-specific baselines after commissioning.

Field requirements include dry storage, clean transfer equipment dedicated to aromatic fluids, periodic verification of moisture, TAN, IFT, and dissipation factors to prevent insulation margin loss.

Field Inspection and Oil Sampling Protocol – Silicone Fluids and Aromatic Hydrocarbon Insulating Fluids

1. **Pre-Inspection Preparation**
   - **Review service history**: operating hours, load cycles, prior oil analysis results, last maintenance date, and fault events.
   - **Confirm oil type**: verify whether the unit contains silicone or aromatic hydrocarbon fluid through nameplate data, service logs, or prior lab reports.
   - **Gather equipment**: clean, dry sampling bottles (glass or HDPE rated for dielectric fluids), labels, chain-of-custody forms, portable moisture meter (if available), dielectric tester (if field testing), PPE per site requirements.
2. **Visual and Physical Inspection**
   - **Exterior survey**: check tank, conservator, and fittings for leaks, corrosion, paint blistering, or gasket extrusion.
   - **Breather condition**: for both fluid types, inspect silica gel or desiccant color change; replace if saturated.
   - **Pressure/vacuum gauges**: ensure readings are within design range; investigate persistent negative pressure.
   - **Oil level indicators**: compare to temperature-compensated range; low oil may indicate leaks or gas generation.
   - **Cooling systems**: verify radiator, fan, or pump operation; look for blockages or fouling.
3. **Sampling Preparation**
   - **Sampling point**: use the designated sample valve on the transformer or oil-filled equipment, typically mid-height drain or dedicated sampling port.
   - **Flush volume**: drain at least 1–2 liters of oil to purge stagnant fluid from the valve and piping before collecting the sample.
   - **Temperature note**: record oil temperature at sampling; critical for dielectric testing interpretation.
   - **Avoid aeration**: open valve slowly to prevent drawing in air bubbles.
4. **Sampling Procedure – Silicone Fluids**
   - Use **glass bottles** with PTFE-lined caps to prevent silicone adsorption into container walls.
   - Minimize headspace in the bottle to reduce oxygen ingress.
   - Avoid contamination from hoses or fittings not rated for silicone fluids.
   - For DGA samples, use glass syringes or evacuated glass vials per ASTM D3612.
   - Immediately label with fluid type, equipment ID, date/time, and sampler name.

5. **Sampling Procedure – Aromatic Hydrocarbon Fluids**
   - Use clean, dry glass or compatible HDPE bottles.
   - Maintain consistent sample volume for trending purposes.
   - Ensure all sampling equipment is free from residual mineral oil or other contaminants.
   - For DGA, follow ASTM D3612 protocols with appropriate gas-tight containers.
   - Label as per silicone fluid protocol.
6. **Field Tests (Optional but Recommended) (see Table 4.13)**
   - **Moisture content**: portable Karl Fischer or capacitance moisture meter.
   - **Dielectric breakdown voltage**: field oil tester per ASTM D877 or D1816.

TABLE 4.13
Field Inspection and Sampling – Silicone Fluids and Aromatic Hydrocarbon Insulating Fluids

Step	Silicone Fluids	Aromatic Hydrocarbon Fluids
1. **Verify Fluid Type**	Check nameplate, logs, or prior lab report; confirm silicone designation	Check nameplate, logs, or prior lab report; confirm aromatic hydrocarbon designation
2. **PPE and Tools**	PPE, clean glass bottles w/ PTFE caps, labels, moisture meter, dielectric tester	PPE, clean glass or HDPE bottles, labels, moisture meter, dielectric tester
3. **External Inspection**	Look for leaks, corrosion, gasket extrusion; check breather gel, oil level, pressure/vacuum	Same as silicone
4. **Cooling System**	Verify radiator/fan/pump operation; check for fouling	Same as silicone
5. **Sampling Prep**	Use mid-height or dedicated sampling port; purge 1–2 liters before collecting	Same as silicone
6. **Container Requirements**	Glass w/ PTFE cap; minimal headspace; avoid non-rated hoses/ fittings	Glass or HDPE; minimal headspace; avoid contamination from other fluids
7. **DGA Sampling**	Glass syringe or evacuated glass vial per ASTM D3612	Same as silicone
8. **Field Tests**	Moisture ppm, dielectric breakdown, visual clarity	Moisture ppm, dielectric breakdown, visual clarity
9. **Lab Tests – Core**	DGA, moisture, dielectric breakdown, viscosity, FTIR	DGA, moisture, IFT, acidity (NN), antioxidant level
10. **Trending and Review**	Watch for changes in DGA pattern, moisture spikes, viscosity shifts	Watch for antioxidant depletion, acidity rise, unusual DGA patterns
11. **Post-Sampling**	Reseal valve, wipe fittings, secure samples, complete chain-of-custody	Same as silicone

- **Visual clarity**: check against a white background for haze, particulates, or discoloration.

7. **Post-Sampling Actions**
   - Reseal sample valves, wipe down fittings, verify no leaks.
   - Return all waste oil to designated disposal containers.
   - Complete chain-of-custody documentation and dispatch samples to a qualified laboratory within 24–48 hours.
8. **Laboratory Test Recommendations – Silicone Fluids**
   - **DGA** (ASTM D3612) – establish equipment-specific gas profile.
   - **Moisture (Karl Fischer)** – ppm level trending.
   - **Dielectric breakdown** – voltage hold-off capability.
   - **Viscosity** – change may indicate contamination or degradation.
   - **Fourier transform infrared spectroscopy (FTIR)** – detect chemical changes and silicone backbone degradation.
9. **Laboratory Test Recommendations – Aromatic Hydrocarbon Fluids**
   - **DGA** – compare to mineral oil gas ratio charts, noting differences in absolute generation rates.
   - **Moisture content** – critical for dielectric retention.
   - **Interfacial tension (IFT)** – monitor for oxidation byproducts.
   - **Acidity (neutralization number)** – track additive depletion and oxidation.
   - **Oxidation inhibitors** – measure remaining antioxidant content.
10. **Data Review and Trending**
    - Compare results to baseline values established after commissioning or oil replacement.
    - Investigate rapid changes in gas generation, moisture, or acidity.
    - For silicone fluids, deviations in DGA patterns from baseline should trigger engineering review.
    - For aromatic hydrocarbons, antioxidant depletion or rising acidity should prompt planning for fluid reclamation or replacement.

GAS INSULATORS: SF_6, NOVEC 4710/5110 BLENDS, CO_2 BLENDS, DRY AIR

Gas-insulated systems are essential in modern power generation and transmission, where compact, high-voltage equipment must operate reliably under extreme electrical and thermal stress. The insulating medium is central to performance: it must provide high dielectric strength, effective arc quenching, chemical stability, and long service life while meeting increasingly stringent environmental and safety requirements. Traditionally, sulfur hexafluoride (SF_6) has been the dominant choice, valued for its unmatched dielectric properties and arc-quenching ability. However, its extremely high global warming potential (GWP) has driven the development and adoption of alternative solutions, including Novec™ 4710 and 5110 blends, CO_2-based mixtures, and even dry air in certain applications.

Unlike liquid lubricants or insulating oils, gases do not rely on dissolved additives for performance. Instead, their effectiveness depends on precise composition, moisture control, and the removal of arc byproducts and particulates through in-line purification systems. Molecular sieves, desiccants, scavengers, and high-efficiency filtration act as the functional equivalents of additives, preserving dielectric strength and preventing chemical degradation over time.

The following sections examine the major insulating gases used in power generation, SF_6, Novec 4710 and 5110 blends, CO_2-based mixtures, and dry air, highlighting their strengths, limitations, decomposition pathways, and the purity-maintenance practices required to ensure safe, reliable operation.

SF_6

Sulfur hexafluoride is an electronegative gas with very high electron capture cross section. Dielectric strength is roughly 2.5 times air at identical pressure and gap. Excellent arc quenching through electron attachment and rapid recombination to stable SF_6 in the presence of metal fluorides. Decomposes under high-energy arcs into SOF_2, SO_2F_2, HF, SO_2, and metal fluorides.

Novec 4710 Blends

C4F7N fluoronitrile used as a minority component blended with CO_2 or N_2. High dielectric strength per partial pressure, strong electron affinity, effective arc quenching when used at elevated pressure. Very low use concentration in mixtures controls toxicity and liquid formation risks. Decomposes to fluorinated fragments and HF under severe faults.

Novec 5110 Blends

C6 fluoroketone used as a minority component in CO_2 or air. High dielectric strength at moderate concentrations, good switching performance, favorable boiling characteristics compared with 4710 in some climates. Decomposition forms carbonyl fluoride, HF, and other short-chain fluorocarbons under severe energy input.

CO_2 Blends

Binary or ternary mixtures with CO_2 as carrier gas and small fractions of 4710 or 5110. CO_2 provides pressure stability and heat capacity. Alone, CO_2 has limited dielectric strength, so clearances or pressure must increase.

Dry Air

Filtered, oil-free, low dew point air. Lowest GWP and easiest logistics. Dielectric strength lower than SF_6, requires higher pressure, larger creepage and clearance, and careful surface cleanliness. Arc interruption capability is limited relative to fluorinated options.

Additives and Processing

- **Drying and cleanup**: molecular sieves for H_2O removal, alumina or fluoride-compatible media for acid scavenging in SF_6 service, high-efficiency particulate filters to capture metal fluorides.

- **Oxygen control**: keep O_2 low in SF_6 systems to limit SOx formation after faults.
- **Inhibitors**: proprietary corrosion inhibitors may be present in 4710 or 5110 blends at trace levels set by the original equipment manufacturer (OEM).
- **Gas make-up practice**: evacuation to sub-mbar, backfill with dry N_2 or CO_2, then charge with blend from certified premix cylinders. Avoid on-site hand mixing unless OEM-approved instrumentation is available.
- **Moisture specification**: typical dew point targets at service pressure are ≤ −40 °C for SF_6 and ≤ −20 to −40 °C for alternative blends and dry air, set by the equipment maker.

Gas Insulators

Gas insulators used in power generation such as SF_6 (sulfur hexafluoride), Novec™ 4710/5110 fluoronitrile blends, CO_2-based mixtures, and dry air are selected for their dielectric strength, arc-quenching capability, thermal stability, and compatibility with high-voltage equipment. While these are gases and thus do not use "additives" in the same sense as liquid lubricants, their performance in service is strongly influenced by **gas purity, stabilizing agents, moisture control, and handling protocols** that serve functions similar to an additive system in a fluid.

SF_6 (Sulfur Hexafluoride)

SF_6 has been the benchmark gas-insulating medium for decades due to its extremely high dielectric strength (~2.5× that of air) and superior arc-quenching properties. It is chemically inert under normal operating conditions but can form highly toxic and corrosive byproducts (e.g., SOF_2, SO_2F_2) when subjected to arcing. While no chemical additives are mixed into SF_6 for service, **moisture scrubbing, particulate filtration, and gas-phase purification systems** act as functional "additives" in the operational sense. Key control parameters include maintaining water content below ~10–20 ppmv and removing arc decomposition products through molecular sieve filters or activated alumina. Field handling often includes passivation of internal metal surfaces and the use of compatible lubricants for O-rings to reduce contamination risk.

Novec™ 4710/5110 Blends (Fluoronitrile/CO_2 or Fluoroketone/CO_2)

These next-generation insulating gases are designed to replace SF_6 while offering 90–99% lower GWP. Novec 4710 is a C_4-fluoronitrile often blended with CO_2 or N_2; Novec 5110 is a C_5-fluoroketone often blended similarly. While the active insulating gas provides dielectric and arc-quenching performance, **blend composition control is critical** proportioning, dryness (<100 ppmv H_2O), and particulate filtration are essential for maintaining dielectric strength. In some systems, stabilizers are added during manufacturing to inhibit long-term breakdown of the active molecule. The containment system often includes in-line molecular sieve dryers and particulate filters, which serve the functional role of an additive system by preserving purity and stability.

CO_2 Blends

CO_2 mixed with small amounts of fluoronitriles, fluoroketones, or other electronegative gases provides a cost-effective, lower-GWP insulating medium. CO_2 itself has lower dielectric strength than SF_6, so blends are engineered for optimal performance in specific gas-insulated switchgear (GIS) designs. Like Novec blends, there are no in-field "additives," but moisture and particulate control are critical especially because CO_2 can dissolve trace moisture and deposit ice under certain conditions. Equipment often incorporates desiccant cartridges and high-efficiency particle filters to keep insulation gaps clean and dry.

Dry Air

Dry air is used as an insulating medium in some medium-voltage switchgear and certain high-voltage applications where environmental impact must be minimal. The "additive" equivalent here is strictly **moisture control** dew points are maintained well below freezing (e.g., ≤ −40 °C) to prevent condensation and maintain dielectric integrity. No chemical stabilizers are added; instead, systems rely on continuous or periodic air drying using desiccants, refrigeration dryers, or vacuum conditioning.

For all gas insulators, the functional equivalent of an additive system is a purity maintenance protocol removing moisture, particulates, and arc byproducts while preserving correct gas composition and pressure. In practice, this involves molecular sieve dryers, activated carbon or alumina scavengers, particle filters, and gas quality monitoring (dew point, pressure, and gas chromatography for composition). While liquid lubricants use chemical additives dissolved in the base oil, gas systems rely on in-line conditioning and gas-phase purification to achieve the same end goal: stable, reliable dielectric performance over long service intervals in demanding power generation environments.

The following tables (see Tables 4.14, 4.15. 4.16, and 4.17) list the tests run for quality control and assurance acceptance and in-service condition monitoring followed by functional additives used in SF_6, Novec 4710/5110 blends, CO_2 blends, Dry Air for power generation.

Field Inspection and Gas Sampling Protocol for Gas Insulators

1. **Preparation**
   Confirm equipment type, rated filling gas, target service pressure, and temperature-compensated density setpoints. Review last dew point, GC composition, and PD data. Calibrate hand-held leak detector or IR camera.
2. **External checks**
   Walkdown for corrosion, chipped porcelain, paint failure near flanges, and signs of powdery white deposits that indicate fluoride salts. Verify density switches are in normal band for measured enclosure temperature. Check space heaters and thermostats on outdoor bays.

TABLE 4.14
Performance and Lifecycle Comparison for Gas Insulators

Attribute	SF_6	Novec 4710 Blend (with CO_2/N_2)	Novec 5110 Blend (with CO_2/air)	CO_2 Blend without Novec	Dry Air
Relative dielectric at 1 bar, 20 °C	~2.5 × air	~1.5–2.2 × air per total pressure	~1.4–2.0 × air per total pressure	~1.2 × air at high pressure	1.0
Typical service pressure	0.4–0.6 MPa	0.6–0.8 MPa	0.6–0.8 MPa	0.7–1.0 MPa	0.6–1.0 MPa
Arc interruption capability	High	High with nozzle redesign	High with nozzle redesign	Moderate	Low to moderate
Liquefaction margin in cold climates	Good	Tighter, manage mix ratio and heaters	Better than 4710 at equal pressure	Good	Good
Byproducts under faults	SOF_2, SO_2F_2, HF, metal fluorides	HF, fluorinated fragments	HF, carbonyl fluoride, fluorinated fragments	CO, CO_2, trace NOx	O_3, NOx, moisture effects
Monitoring complexity	Medium	High, mixture ratio and density windows	High, mixture ratio and density windows	Medium	Low
Environmental profile	Very high GWP	Low GWP blend, PFAS policy considerations	Low GWP blend, PFAS policy considerations	Low GWP	Zero GWP baseline
Enclosure size impact	Smallest	Slightly larger or higher pressure	Slightly larger or higher pressure	Larger	Largest
Expected service life with proper maintenance	30+ years	25–30 years	25–30 years	20–30 years	20–30 years

TABLE 4.15
Common Acceptance and In-Service Gas Testing Parameters

Test Parameter	Method (IEC / ASTM / CIGRÉ)	Purpose
Purity	IEC 60376 (new SF_6), IEC 60480 (in-service)	Confirms minimum % of insulating gas ($SF_6 \geq 99.9\%$, Novec blend per OEM spec).
Moisture/Dew Point	IEC 60480, ASTM D2029	Prevents dielectric strength loss and HF/H_2SO_4 formation.
Acid Gas Byproducts	IEC 60480, ASTM D2685, GC-MS	Identifies SO_2, HF, CF_4, COF_2 from arcing or overheating.
Gas Density/ Pressure	IEC 62271-4	Ensures proper dielectric spacing and quenching performance.
O_2 / CO_2 / N_2 Content	GC-TCD	Checks blend ratios for Novec and CO_2 blends; confirms no ingress air.
Particle Count	ISO 4406 (adapted for gases), IEC 60480	Detects debris from contact wear or insulation erosion.
Dielectric Withstand	IEC 61634 Annex, OEM methods	Validates breakdown voltage at operating pressure and temperature.
Leak Rate	IEC 60076-21 Annex, ASTM E493	Ensures compliance with emission limits (esp. SF_6 < 0.5%/year).

3. **Leak survey**
   Use IR SF_6 camera where permitted or a tuned leak detector for the specific gas. Inspect flanges, O-rings, manways, density monitor capillaries, and instrumentation tees. Record ppm readings and locations.
4. **Dew point and pressure-temperature check**
   Connect a clean, dry sampling line with particulate filter. Purge the line. Measure dew point at operating pressure using a calibrated hygrometer. Log absolute pressure and temperature. Convert to density or mass fraction using the OEM chart.
5. **Gas sampling for laboratory analysis**
   - Use stainless or glass-lined cylinders with known internal volume.
   - Purge the cylinder with at least three volume exchanges.
   - For 4710 or 5110 blends, capture a second sample for composition ratio verification.
   - Seal, label with equipment ID, gas type, pressure, temperature, and sampler ID.
6. **Field functional checks**
   Exercise mechanism where safe. Monitor operating times and coil currents. Abnormal timing may indicate density loss or liquefaction effects.
7. **Laboratory tests**
   - **GC or GC-MS**: verify SF_6 purity or Novec blend composition, quantify O_2, N_2, CO_2, and trace fluorinated byproducts.

TABLE 4.16
Functional Additives in Gas Insulators

Functional Role/ Additive Type	Typical Additive or Conditioning Agent	Typical Concentration Range	Key Laboratory Tests (IEC / ASTM / CIGRÉ)	Notes
Moisture Suppressants	None added in SF_6; drying via molecular sieves or desiccant cartridges; for Novec/CO_2 blends use integrated dryer filters	N/A (physical removal, not chemical dosing)	IEC 60376 (SF_6), ASTM D2029 (dew point), IEC 60480 (in-service moisture)	Moisture must be < –50°C dew point to prevent breakdown and HF formation.
Decomposition Scavengers	Metal oxide getters (Al_2O_3, CuO, ZnO) in filter packs	Fixed mass in filter cartridge	IEC 60480 gas decomposition analysis (SO_2, HF, CF_4), FTIR, GC-MS	Used in SF_6, Novec blends to absorb acidic by-products from arcing.
Stabilizers (Novec blends)	Proprietary arcing stabilizers to reduce by-product reactivity	< 1% (proprietary)	OEM acceptance test, GC-MS, ASTM D5504 (sulfur compounds)	Helps suppress formation of toxic/acidic species under high-energy faults.
Leak Detection Agents	Tracer gases (helium, CF_4 in ppm levels for new equipment tests)	100–500 ppm (temporary for factory test)	ASTM E493 (helium leak), IEC 60076-21 Annex (SF_6 tightness test)	Used during acceptance pressure/leak tests, not in service.
Buffer Gases in Blends	CO_2, N_2, or O_2 (balancing Novec dielectric strength vs. temperature performance)	10–70% depending on blend	IEC 60376, IEC 62271-4, GC-TCD	Gas ratio affects dielectric strength, boiling point, and operating temperature.
Particle Traps/ Filters	Not an additive, but in-line filtration removes particulates and by-products	N/A	IEC 60480 (particle count, visual), ISO 4406 adapted for gas	Required to maintain dielectric strength in all gas systems.
Drying/ Conditioning Agents (Dry Air)	Silica gel, molecular sieves in circulation loop	N/A	IEC 60480 moisture analysis, ASTM D2029	Keeps dew point below –40°C; crucial in high humidity environments.

TABLE 4.17
Failure Modes and Degradation Pathways for Gas Insulators

Failure mode	Typical cause	Indicators	Field or lab tests	Corrective action	Criticality 1–5
Moisture ingress	Leaks, desiccant exhaustion, poor evacuation	Dew point rises, reduced dielectric margin	Dew point measurement, pressure-density correlation	Leak locate and repair, evacuate, dry, recharge	5
Density loss	Seal leakage, temperature sensor error	Low density alarm, frequent refills	Density monitor check, helium or SF_6 sniffer, IR camera	Tighten or replace seals, verify sensors, recharge	4
Gas decomposition after faults	Repetitive switching, internal PD, flashover	HF detection, corrosion traces, byproduct peaks	FTIR or GC, PD survey, visual inspection	Remove byproducts with filters, repair root cause, replace gas	4
Mixture off-ratio in 4710/5110 systems	Wrong fill, valve cross-connection, temperature stratification	Abnormal interrupter performance, changed liquefaction margin	Gas composition analysis by GC, density vs temperature cross-check	Recover gas, charge with certified premix, update nameplate	4
Particulate contamination	Erosion of nozzles, manufacturing debris	PD inception at lower voltage, local heating	Particle counting, borescope, PD test	Filter or flush, replace worn parts	3
Corrosion from acidic byproducts	HF and SOx attack of metals and elastomers	Etching, pitting, sticky mechanisms	Surface wipe tests, coupons, HF indicator tubes	Neutralize, replace affected parts, improve scavenging media	3
Sensor or monitor drift	Aging density switch, miscalibrated PT100	False alarms or missed low-density events	Calibration check against reference, loop test	Replace or recalibrate instruments	3
Liquefaction in cold weather	High pressure with rich blend, poor heating	Slow operation, trip-to-lockout in cold start	Temperature and pressure trend vs phase line, visual frosting	Increase heater capacity, adjust mix ratio, de-rate	3
Air contamination	Improper purge, open fitting	Lower dielectric, higher PD	Oxygen measurement, GC, breakdown test on sample	Evacuate, dry, recharge with certified gas	3

- **Acid gas screening**: HF indicator, SOx measurement for SF_6 units.
- **Moisture by chilled mirror or electrolytic method**: confirm dew point trend.
- **Particulate**: if PD is suspected, run particle count on extracted gas path or inspect filter media.

8. **Acceptance criteria and actions**
   - Dew point above limit: dehydrate by evacuation and dry backfill.
   - Off-ratio blend: recover, charge with certified premix, update density settings if the OEM procedure requires.
   - High byproducts: filter with fluoride-compatible media, replace nozzles if erosion is present, schedule internal inspection.
   - Recurrent low-density alarms: leak repair, instrument replacement, pressure test, and hold.
9. **Documentation**
   Log all measurements with pressure and temperature corrections. Attach GC printouts. Update the computerized maintenance management system with leakage rate estimate, parts replaced, and next inspection interval.

Application Considerations

- GIS and GIL: SF_6 remains the reference for compact designs and very high voltage classes. 4710 or 5110 blends can reach comparable dielectric margins at higher pressure and with tuned nozzle designs. CO_2 or dry air systems require larger enclosures and stricter cleanliness.
- Dead-tank breakers and live-tank breakers: alternative blends are viable with nozzle and puffer redesign, higher density thresholds, and upgraded monitoring.
- Ambient temperature: verify liquefaction margin. 4710 blends have higher liquefaction temperature than 5110 blends at the same pressure. Cold regions often need heaters or reduced mixing ratios to avoid condensation.
- Environmental and compliance: SF_6 has very high GWP. Alternatives lower total CO_2e but may be subject to regional PFAS policy. Validate labeling, recovery, and end-of-life handling with current regulations.
- Diagnostics: byproduct signatures differ. SF_6 faults produce SOF_2, SO_2F_2, HF. 4710 and 5110 blends produce HF and carbonyl species. Gas chromatography with FTIR support is recommended for root-cause analysis.

CABLE DIELECTRIC FLUIDS AND HVDC COOLING

High-voltage transmission systems depend on more than conductors and insulation alone, specialized liquid dielectrics and cooling media are vital to ensuring reliable, long-term performance of underground cables, submarine links, and HVDC converter stations. These fluids must simultaneously deliver electrical insulation, thermal management, and chemical stability under operating conditions that often extend for decades without replacement. Their role is especially critical in power generation and transmission infrastructure where outages are costly and system repair is complex.

Cable dielectric fluids are designed to maintain high dielectric strength, low dissipation factor, and chemical compatibility with cellulose or polymer insulation systems. Traditional high-pressure fluid-filled cable oils and mass-impregnated compounds remain in use, but newer synthetic alkylbenzene and ester fluids offer improved oxidative stability, biodegradability, and cold-temperature performance. Additive chemistry is deliberately minimal to preserve dielectric properties, with antioxidants, metal passivators, and antihydrolysis agents applied selectively to extend service life. Moisture control and purity management are paramount, since even trace contamination can compromise dielectric reliability.

HVDC cooling fluids, by contrast, are optimized for heat transfer in high-power electronics and cable terminations. These include deionized water, water-glycol mixtures, synthetic esters, silicone oils, and emerging fluoroketone-based coolants for combined insulation and cooling. Here, the additive focus shifts toward corrosion inhibition, antioxidant stability, and material compatibility with copper, aluminum, and elastomers used in converter and cable systems. As with dielectric oils, inline conditioning, through drying, filtration, and inhibitor monitoring, is central to maintaining fluid integrity and ensuring predictable service performance.

The following sections examine the principal categories of cable dielectric and HVDC cooling fluids, the additive chemistries employed, and the processing and conditioning practices that preserve electrical and thermal performance in mission-critical power networks.

High-Pressure Fluid-Filled Cable Oils (HPFF)

Low-viscosity naphthenic mineral insulating oils refined for low gassing, low sulfur, and high dielectric strength. Inhibited with phenolic antioxidants. Designed to remain mobile at moderate temperatures and to circulate under small pressure gradients.

Mass-Impregnated and Non-Draining Compounds (MI and MIND)

Highly viscous impregnants for paper-insulated DC cables and some submarine links. Based on highly refined mineral fractions, synthetic alkylbenzenes, or polybutenes. Very low volatility, high resistivity, and high DC breakdown strength.

Synthetic Cable Fluids

Alkylbenzene and synthetic ester formulations used where low pour point, improved oxidative stability, or specific DC dielectric behavior is required. Higher solvency than mineral oil, stable dissipation factor if moisture is controlled.

HVDC Valve Cooling Media

Deionized water for thyristor and IGBT valve heat exchangers. Water-glycol where freeze protection is required. Dielectric liquids for direct-to-chip or immersion loops in specialty converters. Two-phase refrigerants and CO_2 systems in compact high heat flux hardware when specified by the OEM.

HVDC Cable Accessory Cooling

Water or water-glycol jacket cooling at landfalls and terminations in high-ampacity export circuits. Some stations use dielectric liquids in closed plates near joints to avoid leakage current paths.

Additives and Processing
Cable Dielectric Fluids

- Antioxidants: DBPC or similar phenolic inhibitors to slow acid formation.
- Metal passivators where copper activity is a risk.
- Pour point depressants for mineral oils in cold climates.
- Processing: vacuum dehydration to single-digit ppm water, multi-stage degassing, fine filtration to submicron levels. Nitrogen blanketing in storage. Certified low corrosive sulfur per ASTM D1275 and DBDS screening.

HVDC Cooling Media

- Deionized water: phosphate-free corrosion inhibitors matched to mixed metals, biocide program, oxygen control via degassing or nitrogen sweep.
- Water-glycol: nitrite-free or hybrid organic acid technology packages for aluminum and copper alloys, pH buffers, azole passivators for yellow metals.
- Dielectric liquids: minimal additive load to preserve dielectric loss factor; selected antifoam and corrosion inhibitors when approved by the OEM.
- Processing: resin bed polishing for water conductivity control, side-stream particulate filtration, air purge and vacuum fill to remove entrained gas, leak-free quick connects.

Cable dielectric fluids and HVDC cooling fluids in power generation and transmission are specialized liquid systems designed to provide both electrical insulation and thermal management for high-voltage cables, terminations, and converter equipment. Their formulation and additive chemistry are dictated by the need to maintain high dielectric strength, low loss factor, chemical compatibility with cable materials, and stable viscosity across the operating temperature range, while also controlling oxidation, moisture, and deposit formation over decades of service.

Cable Dielectric Fluids

Historically, mineral oil-based dielectric fluids (refined naphthenic oils) have been used in mass-impregnated paper-insulated cables and fluid-filled pipe-type transmission cables. These oils are chosen for high dielectric strength (typically > 40 kV, ASTM D877), low pour point, and good compatibility with cellulose paper insulation. Modern systems also use synthetic alkylbenzenes or synthetic esters for improved oxidation stability, biodegradability, and low-temperature performance.

Additive systems for cable dielectric oils are minimal to avoid altering electrical properties. A hindered phenolic antioxidant (0.05–0.3% w/w) is typically the only deliberate additive, extending oxidative life without increasing conductivity. In long-line cable systems, metal passivators such as benzotriazole derivatives are sometimes added to inhibit copper or lead sheath catalyzed degradation. For ester-based dielectric fluids, antihydrolysis agents help prevent acid formation from trace moisture ingress. Moisture control is critical: while ester fluids tolerate higher dissolved water content without dielectric failure, mineral oils require ppm-level moisture suppression to maintain dielectric strength.

HVDC Cooling Fluids

In high-voltage direct current (HVDC) converter stations and certain submarine cable systems, cooling fluids are used in liquid-cooled power electronics, transformers, and sometimes in cable thermal management jackets. These can be synthetic hydrocarbons (PAO, alkylbenzenes), synthetic esters, or silicone fluids selected for low viscosity, high specific heat capacity, and compatibility with electrical components. In emerging HVDC applications, fluoroketone-based dielectric coolants are being introduced for combined insulation and cooling functions with ultra-low GWP.

Cooling fluid additive systems are similar in philosophy to turbine oils, though far simpler to protect dielectric integrity. A small percentage of antioxidants slows oxidative breakdown under thermal load. Corrosion inhibitors, chosen for compatibility with aluminum, copper, and polymer seals, protect against galvanic and chemical attack in mixed-metal cooling loops. In ester-based coolants, antihydrolysis agents are added, while in silicone-based coolants, moisture scavengers are sometimes included to prevent electrical tracking. For sealed systems, defoamants are rarely required; for vented or high-flow systems, silicone-free antifoam agents may be used sparingly to avoid air entrapment.

Common Conditioning and Maintenance Practices

For both cable dielectric and HVDC cooling fluids (see Tables 4.18, 4.19, 4.20, and 4.21), the functional equivalent of a "life-extending additive system" is often an **in-line conditioning loop**:

- **Molecular sieve dryers** to control moisture.
- **Fullers earth or activated alumina filters** to remove acids and polar degradation products.
- **Fine filtration (≤ 1 μm)** to keep particulate levels below thresholds that could cause electrical discharge initiation.
- **Periodic inhibitor replenishment** based on antioxidant depletion testing (ASTM D6971 RULER, ASTM D2440 oxidation stability).

TABLE 4.18
Common Acceptance and In-Service Tests for Cable Dielectric Fluids and HVDC Cooling Fluids

Test Parameter	ASTM/IEC/ISO Method	Purpose/Interpretation
Dielectric Breakdown Voltage	ASTM D1816, IEC 60156	Confirms insulation strength for cable dielectric fluids.
Moisture Content	ASTM D1533 (Karl Fischer), IEC 60814	Prevents dielectric loss and corrosion.
Total Acid Number (TAN)	ASTM D974, IEC 62021-1	Rising TAN signals oxidation and additive depletion.
Interfacial Tension (IFT)	ASTM D971	Decline indicates polar contaminants/oxidation.
Oxidation Stability	ASTM D2112 (RPVOT), IEC 61125	Measures remaining antioxidant reserve.
Viscosity	ASTM D445	Ensures correct cooling flow and dielectric coverage.
Corrosive Sulfur	ASTM D1275B, IEC 62535	Particularly for cable oils in contact with copper.
Foam Test	ASTM D892	Detects foaming tendency in pumped systems.
Particulate Cleanliness	ISO 4406 adapted	Ensures no solid debris that could block flow or affect breakdown voltage.
Thermal Conductivity	ASTM D2717, transient hot wire methods	For HVDC cooling, verifies heat transfer performance.

Field Inspection and Sampling Protocol for Cable Dielectric Fluids and HVDC Cooling Fluids

A. **Cable dielectric fluids**
   1. **Preparation.** Confirm cable type, design pressure, fluid spec, last TAN, IFT, moisture, and pressure logs. Verify calibrated gauges at head tanks or pumping stations.
   2. **Walkdown.** Inspect sheath, link boxes, valve pits, and pipe runs for stains, soil wet spots, or catchment sump levels. Check regulator setpoints and surge volume levels.
   3. **Pressure and temperature.** Record pressure at standard locations and ambient soil or vault temperature. Compare to seasonal trend curves.
   4. **Sampling.**
      - Use dedicated stainless valves. Purge at least 1–2 liters.
      - Collect in clean glass bottles with PTFE caps. Minimize headspace.
      - For MI or MIND, take a warmed sample at the accessory pot where OEM allows.
      - Label cable ID, location, temperature, pressure, time.

TABLE 4.19
Functional Additives in Cable Dielectric Fluids and HVDC Cooling Fluids

Additive Function	Typical Additive Chemistry	Typical Concentration Range (% wt)	Key Laboratory Tests (ASTM/IEC/ISO)	Notes
Oxidation Inhibitors	DBPC, hindered phenols, aromatic amines	0.1–0.3	ASTM D2112 (RPVOT), ASTM D943 (TOST), IEC 61125	Prevent sludge and acid buildup in long-life sealed cable systems and HVDC loops.
Metal Deactivators	Benzotriazole, tolyltriazole	0.005–0.05	ASTM D1275B, IEC 62535	Protects copper conductors, heat exchanger tubes from catalytic oxidation.
Pour Point Depressants	Polymethacrylates, alkylated naphthalenes	0.02–0.1	ASTM D97, ASTM D5950	Ensures fluid flow in cold climates; more relevant for HVDC cooling loops in outdoor installations.
Antifoam Agents	Polysiloxanes, polyacrylates	5–50 ppm	ASTM D892, ISO 6247	Prevents aeration in circulation loops; foam reduces cooling efficiency and can lower dielectric strength in cables.
Corrosion Inhibitors	Alkenyl succinic acids, fatty amine phosphates	0.05–0.3	ASTM D665 (A/B), ASTM D130 (copper strip)	Protects steel cable sheaths, converter cooling circuit metals.
Viscosity Modifiers	High-MW polymers, alkylated aromatics	0.1–1.0	ASTM D445 (viscosity), ASTM D2270 (VI)	Maintains thermal transfer performance over wide temp ranges.
Biocides *(water-based HVDC cooling only)*	Isothiazolinones, bronopol	0.01–0.05	ASTM E2275 (microbial count)	Prevents microbial fouling in water-glycol HVDC cooling loops.
Dye / Leak Tracers	Fluorescent or colored dyes	ppm-level	Visual inspection, UV leak detection	Used for maintenance leak tracing in HVDC systems, not in dielectric cables.

TABLE 4.20
Performance and Lifecycle Comparison for Cable Dielectric Fluids

Attribute	HPFF Mineral Oil	MI or MIND Impregnant	Synthetic Cable Fluid
Viscosity at 40 °C	3–10 mm²/s	1000+ mm²/s	3–15 mm²/s
Pour point	−30 to −45 °C with depressant	Typically high; not intended to flow	−40 to −60 °C possible
AC dielectric strength	High when dry	Very high	High to very high
DC resistivity	High	Very high	High
Oxidation stability	Moderate with inhibitor	High due to low volatility	High
Solvency	Moderate	Low	High
Leakage risk	Moderate if sheath compromised	Low drip MIND, moderate for MI	Moderate
Typical use	Pipe-type and self-contained AC cables	HVDC land and submarine	AC and DC where low temp and stability are required
Expected service life	30+ years with pressure control	30+ years when dry and gas-free	30+ years with proper sealing

TABLE 4.21
HVDC Cooling Media

Attribute	Deionized Water	Water-Glycol	Dielectric Liquid	Two-Phase CO_2 or Refrigerant
Heat capacity	Very high	High	Moderate	High effective via phase change
Electrical properties	Conductive if not polished	Conductive if not polished	Insulating	Insulating vapor/liquid
Viscosity at 25 °C	Very low	Low to moderate	Moderate	Low effective
Corrosion risk	High without inhibitors	Moderate with proper inhibitors	Low if approved fluid	Low in stainless systems
Biofouling risk	Moderate to high	Moderate	Low	Low
Freeze protection	Poor	Good	Good	Good
Maintenance load	High polishing and monitoring	Moderate	Low to moderate	Moderate to high commissioning complexity
Typical use	Valves with plate exchangers	Cold climates and outdoor loops	Immersion or near live parts	Compact high flux systems

5. **Field checks.** Visual clarity, odor, and color. If available, spot dielectric test and moisture screening.
6. **Laboratory panel.** Karl Fischer moisture, breakdown voltage per ASTM D1816, TAN, IFT, particle count, corrosive sulfur and DBDS, antioxidant content, GC for contaminant screen when leakage is suspected.
7. **Actions.**
   - Moisture high: schedule vacuum dehydration and nitrogen blanket verification.
   - TAN or IFT trending adverse: reclaim or partial changeout, review hot-spot loading.
   - Pressure instability: regulator service, leak locate and repair, recharge volumes.
   - Corrosive sulfur: full fluid replacement and hardware inspection.

B. HVDC cooling loops

1. **Preparation.** Confirm medium, loop volume, target conductivity or resistivity, inhibitor type and setpoints, last microbial and corrosion data.
2. **Inspection.** Check pumps, strainers, quick connects, hose condition, expansion tanks, and vent points. Verify flow and delta-T at heat exchangers.
3. **Sampling.**
   - Water and water-glycol: draw from flowing line post-filter; use clean plastic bottles for chemistry and glass for TOC if required.
   - Dielectric liquids: use glass with PTFE caps; purge line to remove air.
4. **Field tests**. Conductivity or resistivity, pH, temperature, and visual turbidity.
5. **Laboratory panel.**
   - Water and water-glycol: pH, alkalinity, inhibitor residual, metals (Fe, Cu, Al, Zn), hardness, chloride, sulfate, microbiology or ATP, TAN for glycol, viscosity at service temperature.
   - Dielectric liquids: breakdown voltage, resistivity, moisture by Karl Fischer, particle count, FTIR for oxidation markers.
6. **Actions.**
   - High conductivity: resin bed service, partial drain and DI makeup, oxygen control.
   - Low inhibitor or off-pH: redose to OEM setpoints, verify materials compatibility.
   - Biofouling: biocide shock, mechanical cleaning, add side-stream filtration.
   - Dielectric liquid off-spec: drain, flush, refill with approved fluid, investigate cross-leak.

This section supports specification, commissioning, and maintenance planning for fluid-filled cable systems and HVDC cooling circuits.

Application Considerations

HPFF pipe-type and self-contained fluid-filled cables. Maintain pressure above the cavitation and partial discharge threshold with head tanks or pumping plants. Thermal expansion drives seasonal pressure swings; surge volumes and regulators must be sized for soil and load temperature spans. Oil must remain gas-free to prevent void formation in paper insulation. Leak detection and spill containment are primary risks in urban corridors.

MI and MIND DC cables. High DC field stresses favor impregnates with low ionic mobility and high resistivity. Space charge accumulation is a design constraint, fluid purity and paper dryness control field grading reliability. MIND compounds reduce drip risk during hot operations on steep grades.

Synthetic cable fluids. Useful where lower pour point and higher oxidation resistance are required, or where DC loss characteristics are specified. Verify compatibility with paper binders, lead sheaths, and joint sealing compounds.

HVDC valves and reactors. Deionized water loops offer high specific heat and low viscosity for plate heat exchangers. Maintain very low conductivity to prevent stray current. Water-glycol improves freeze protection but raises viscosity and can foul if oxidized. Dielectric liquids or engineered refrigerants are specified where live parts proximity or high voltage clearances rule out water.

Terminations and joints. Localized cooling may be needed for high current accessories. Select fluids that do not create conductive films, that are compatible with EPDM and silicone rubbers, and that meet site fire protection requirements.

INSULATING GREASES AND SOLID-STATE DIELECTRIC CONSIDERATIONS

Insulating greases are non-curing compounds that fill interfacial voids, repel moisture, and raise surface withstand at terminations and interfaces. Base fluids include polydimethylsiloxane silicone, perfluoropolyether, synthetic hydrocarbons such as polyalphaolefin (PAO), and fluorosilicone for fuel and solvent exposure. Thickeners include silica, lithium soaps, polyurea, and polytetrafluoroethylene (PTFE). Functional fillers tailor properties. Alumina and boron nitride increase thermal conductivity. PTFE reduces friction at sliding contacts. Hydrophobic silica tunes bleed and structure. Typical dielectric strength is on the order of 10 to 20 kV per mm in the bulk, but system benefits from gap filling and water barrier action rather than bulk hold-off.

Solid-state dielectrics carry the field in the bulk and at the surface. Common classes include thermoset epoxies with mica or glass reinforcement, epoxy mica tapes for rotating machines, cycloaliphatic epoxy for outdoor bushings, polyethylene terephthalate (PET) and polyimide films for slot liners, cross-linked polyethylene (XLPE) and ethylene propylene rubber (EPR) for cables, PTFE for high frequency, polyether ether ketone (PEEK) where temperature and chemical resistance are critical, and

TABLE 4.22
Failure Modes and Degradation Pathways for Cable Dielectric Fluids and HVDC Cooling

Failure mode	Typical cause	Indicators	Tests	Corrective action	Criticality 1–5
Moisture ingress in cable oil	Seal or sheath leak, poor drying	Dielectric strength drop, TAN rise, gas bubbles on hot load	Karl Fischer, breakdown voltage, dissolved gas	Locate leak, dry-out, vacuum refill, reseal	5
Pressure loss HPFF	Head tank failure, leak, temperature drift	Low pressure alarm, PD inception, oil pump cycling	Pressure log vs ambient, patrol leak survey	Repair leak, restore regulator function, replenish oil	5
Oxidation and sludge in cable oil	Aged inhibitor, high hot-spot	TAN increase, IFT decrease, filter plugging	TAN, IFT, color, particle count	Reclaim or replace oil, add inhibitor when approved	3
Corrosive sulfur in cable oil	Off-spec supply or contamination	Copper sulfide, dark paper	ASTM D1275, DBDS	Replace oil, clean, replace affected parts	4
Space charge in DC paper	Moisture, ionic contamination	Elevated DC losses, heating under load	DC tan delta, polarization-depolarization	Dry-out, fluid replacement, lower load until stable	4
Water loop corrosion	Low pH, oxygen ingress, wrong inhibitor	Flow restriction, leak, metal loss	pH, conductivity, corrosion coupons, Fe/Cu ppm	Rebalance chemistry, replace sections, oxygen control	4
Water loop biofouling	Warm stagnant zones, weak biocide	Slime, delta-P rise, HX efficiency loss	ATP count, microbiology, differential pressure	Biocide shock, mechanical clean, add side-stream filter	3
Glycol oxidation	Overheat, air ingress	Acid number rise, varnish, flow loss	TAN, FTIR, viscosity	Partial changeout, improve sealing, add antioxidant package	3
Dielectric coolant contamination	Cross-leak, wrong fluid top-up	Dielectric loss rise, sensor alarms	Resistivity, breakdown voltage, GC	Drain, flush, refill with approved fluid	4
Two-phase maldistribution	Poor charging, trapped gas	Hot spots, unstable temperatures	Mass flow and superheat checks	Recommission loop, fix distribution hardware	3

ceramics such as alumina for very high field strength and thermal conductivity. Key properties are dielectric strength, partial discharge inception voltage, dissipation factor, tracking and erosion resistance, moisture absorption, coefficient of thermal expansion, and thermal class.

Additives and Processing

Insulating greases rely on antioxidants for hydrocarbon bases, metal passivators for copper and brass, corrosion inhibitors for steel interfaces, and fungicides where warm moisture is expected. Formulation balances worked penetration, bleed, and oil separation. Processing uses vacuum mixing to limit entrained air, followed by controlled milling to set thickener structure. Packaging must be moisture tight. Shelf life depends on base oil oxidation stability and thickener integrity.

Solid dielectrics depend on resin chemistry and cure discipline. Epoxies use stoichiometric hardeners and post cure to raise glass transition temperature. Silane coupling agents improve filler bonding. Vacuum pressure impregnation of windings drives resin into mica and glass tapes, then a full cure closes voids. Stress grading systems use semiconductive or resistive coatings based on zinc oxide or silicon carbide to smooth field gradients near ends and screens. Room-temperature-vulcanizing (RTV) silicone top coats limit surface tracking on outdoor polymeric housings.

Insulating greases and **solid dielectrics** serve specialized but critical roles in power generation, transmission, and distribution systems. Both are designed to maintain dielectric integrity under high electrical stress while providing additional physical or environmental protection. The formulation or composition of these materials is tailored to electrical properties (dielectric strength, volume resistivity, dissipation factor), mechanical robustness, and long-term stability under thermal, chemical, and environmental stress.

Insulating Greases

Insulating greases are non-curing, water-repellent, and electrically non-conductive compounds used at cable terminations, connectors, bushings, and high-voltage outdoor interfaces to prevent tracking, moisture ingress, and corrosion (see Tables 4.23 and 4.24). The most common base is **polydimethylsiloxane (PDMS) silicone oil** thickened with an inert filler such as fumed silica. These greases exhibit high dielectric strength (>20 kV/mm), excellent thermal stability, and wide operating temperature ranges (−50 °C to +200 °C).

The "additive" system in insulating greases is minimal, as conductive or polar additives could impair insulation. Key functional ingredients include the following:

- **Thickener/filler**: Fumed silica, treated clay, or PTFE to give mechanical body and prevent oil bleed.
- **Corrosion inhibitors**: Compatible organosilanes or neutral inhibitors to protect metallic contact surfaces without compromising dielectric strength.
- **UV stabilizers** (for outdoor service): Hindered amine light stabilizers (HALS) or inorganic UV absorbers to prevent surface degradation.
- **Water repellents**: Organosilicon or fluorosilane treatments to maintain hydrophobicity.

TABLE 4.23
Insulating Greases

Parameter	Silicone grease	PFPE grease	Synthetic hydrocarbon grease
Service temperature range	Very wide	Wide at high end	Moderate to wide
Dielectric strength, bulk	High	High	Moderate to high
Water resistance	Excellent	Excellent	Good
Oxidation stability	High	Very high	Moderate to high with antioxidants
Plastics and elastomer compatibility	May swell some elastomers and polycarbonate	Broad compatibility	Good with common elastomers, verify for ABS and PC
Thermal conductivity with fillers	Moderate to high	Moderate to high	Moderate to high
Typical use	Outdoor terminations, boots, O rings, thermal interfaces	Chemical exposure, oxygen service, nuclear	Indoor terminations, low cost interfaces

TABLE 4.24
Insulating Greases – Typical Use

Base Type	Typical Use
Silicone grease	Outdoor terminations, boots, O-rings, thermal interfaces
PFPE grease	Chemical exposure, oxygen service, nuclear
Synthetic hydrocarbon grease	Indoor terminations, low-cost interfaces

Specialized insulating greases for **EHV (extra-high-voltage) terminations** may include surface-active additives to reduce corona inception at stress points while still maintaining bulk insulating properties. Because these greases are non-curing, maintenance involves periodic inspection for displacement, contamination, or surface tracking.

Solid Dielectrics

Solid dielectric materials in power generation and transmission equipment form the primary insulation in components such as bushings, cable insulation, transformer pressboard, HV stator windings, and switchgear barriers. Common solid dielectrics include:

- **Cross-linked polyethylene (XLPE)** and **ethylene-propylene rubber (EPR)** for high-voltage cable insulation: chosen for high dielectric strength (~25–35 kV/mm), low dielectric loss, and resistance to partial discharge.
- **Epoxy resins** (with silica or alumina fillers) for cast insulation in switchgear and instrument transformers: tailored for thermal class ratings and mechanical strength.
- **Mica paper with epoxy or silicone binders** in stator coils: offering outstanding dielectric endurance and partial discharge resistance in rotating machines.
- **Ceramics (alumina, steatite)** for bushings and support insulators: providing excellent mechanical stability and arc resistance.
- **Polyesters, polyimides (e.g., Kapton®)** for slot liners and flexible insulation in rotating machines and transformers.

Solid dielectric performance is dictated by base polymer purity, filler dispersion, and moisture resistance. In polymeric dielectrics, **antioxidants** (for organic materials), **UV stabilizers**, and **antitracking fillers** (aluminum hydroxide, ATH; silica; titanium dioxide) act as built-in "additives" to extend life. In filled epoxy systems, **coupling agents** (organosilanes) improve bonding between resin and mineral filler, enhancing mechanical strength and partial discharge resistance.

Maintenance and Degradation Considerations

- **Insulating greases** degrade primarily via contamination (dust, moisture, conductive particles), oil separation, and surface erosion in UV or corona-rich environments.
- **Solid dielectrics** fail via partial discharge erosion, thermal aging, tracking, cracking, or moisture ingress, with detection through insulation resistance (IR), dissipation factor (tan δ), and partial discharge testing (see Table 4.25).
- For both, periodic inspection, dielectric testing, and in some cases reapplication or recoating are required to maintain service reliability.

PERFORMANCE AND LIFECYCLE COMPARISON FOR GREASES

Application Considerations

Greases are used in high voltage cable terminations, elbow and T body interfaces, switchgear boots, outdoor lighting connectors, current transformer secondary terminals, and sealing O rings that must remain lubricated and water tight (see Tables 4.26, 4.27, and 4.28). Electrical function is to displace air and water at interfaces and to prevent tracking across surfaces. Thermal greases couple power electronics modules to heatsinks. Selection hinges on dielectric need, temperature window, solvent exposure, and plastics compatibility. Silicone greases suit wide temperature and water ingress. PFPE suits oxygen service and chemical exposure. Hydrocarbon greases fit cost-sensitive indoor terminations with moderate temperature.

TABLE 4.25
Solid Dielectrics – Performance and Lifecycle Comparison

Parameter	XLPE or EPR Cable Insulation	Epoxy Mica VPI System	Cycloaliphatic Epoxy Bushing	Alumina Ceramic
Dielectric strength	Very high	Very high	Very high	Highest
PD inception behavior	Good with clean void-free extrusion	Good with full impregnation	Sensitive to voids and contamination	Excellent if glaze intact
Tracking and erosion resistance	Good with jackets and sheds	Good indoors	High outdoors with RTV coat	Not applicable, hard and inert
Moisture uptake	Low to moderate	Low	Low	Very low
Thermal class	90–105 °C typical for cable	130–155 °C common	105–120 °C typical	200 °C+ equivalent
Mechanical toughness	High flexibility	Rigid, tough	Rigid	Very rigid, brittle
Typical use	HV and EHV cables and joints	Generator and motor insulation	Outdoor bushings and insulators	Feed-throughs, vacuum bushings

TABLE 4.26
Quality Control Tests for Insulating Grease and Solid-State Dielectrics

For Insulating Greases:	For Solid-State Dielectrics:
Dielectric Breakdown Voltage – ASTM D149	**Dielectric Strength** – ASTM D149, IEC 60243
Dielectric Constant / Dissipation Factor – ASTM D150	**Volume Resistivity** – ASTM D257
Oil Separation / Bleed – ASTM D6184	**Tracking / Erosion Resistance** – IEC 60587, ASTM D2303
Drop Point – ASTM D566 / D2265	**Partial Discharge Inception Voltage** – IEC 60270
Worked Penetration – ASTM D217	**Thermal Aging Resistance** – IEC 60216
Moisture Content – ASTM D1533 (Karl Fischer)	**Arc Resistance** – ASTM D495
Thermal Stability – ASTM D3336	**Mechanical Tensile / Elongation** – ASTM D638

TABLE 4.27
Functional Additives in Insulating Greases and Solid-State Dielectrics

Additive Function	Typical Additive Chemistry	Typical Concentration Range	Key Laboratory Tests (ASTM / IEC / ISO)	Notes
Thickener (Grease Structure)	Silica, PTFE, clay, lithium complex (non-melting), polyurea	5–15% (w/w)	ASTM D217 (worked penetration), ASTM D566 (dropping point)	Determines mechanical stability and bleed rate; for insulation, thickeners must be non-melting.
Base Fluid	Silicone oil (PDMS), synthetic hydrocarbon (PAO), ester fluids	Balance	ASTM D445 (viscosity), ASTM D92 (flash point)	Silicone base for wide temp and dielectric stability; esters for biodegradability in outdoor use.
Antioxidants	Hindered phenols, aromatic amines	0.1–1.0%	ASTM D942 (oxidation stability), ASTM D2272 (RPVOT)	Protects base oil from thermal oxidation in service.
Corrosion Inhibitors	Benzotriazole, alkenyl succinic acid derivatives	0.05–0.5%	ASTM D4048 (corrosion), ASTM D130 (copper strip)	Protects metallic contacts, terminals, and housings.
Antiwear Agents	PTFE particles, molybdenum disulfide (non-conductive grades), borate esters	0.5–5.0%	ASTM D2266 (4-ball wear), ASTM D2596 (EP properties)	Improves film strength; only non-conductive AWs used in insulating greases.
Dielectric Strength Enhancers	High-purity fillers (alumina, silica, boron nitride)	2–10%	ASTM D149 (dielectric breakdown), ASTM D150 (dielectric constant)	Raises breakdown voltage, reduces tracking.
Moisture Scavengers	Molecular sieve powder, hydrophobic silica	0.5–3.0%	IEC 60814 (moisture), Karl Fischer (ASTM D1533)	Minimizes moisture-induced dielectric loss.
Thermal Conductivity Enhancers *(for solid-state heat dissipation)*	Boron nitride, alumina, aluminum oxide	10–40%	ASTM D5470 (thermal conductivity), ASTM D150	Used in greases for power electronics or solid-state insulation cooling.

UV Stabilizers	Benzotriazoles, HALS (hindered amine light stabilizers)	0.05–0.3%	ASTM G154 (UV exposure)	Extends outdoor service life of polymeric insulation and grease in exposed terminations.
Flame Retardants	Halogen-free phosphorus compounds, melamine cyanurate	1–5%	ASTM D2863 (LOI), UL 94	Required for greases and polymers in fire-critical locations.
Antitracking Additives (Solid-State)	Alumina trihydrate (ATH), silicone fillers	20–50% in solid insulation	IEC 60587 (tracking resistance), ASTM D2303	Reduces surface carbonization under high voltage stress.
Plasticizers (Solid-State)	Phthalate-free esters, phosphate esters	5–20%	ASTM D792 (density), ASTM D2240 (hardness), DSC analysis	Maintains flexibility in cold climates; must not migrate excessively.
Crosslinking Agents (Solid-State)	Peroxides, silanes (for XLPE/EPR)	ppm to % level depending on polymer	ASTM D2765 (gel content), FTIR	Enhances thermal and electrical stability in polymeric insulation.

TABLE 4.28
Failure Modes and Degradation Pathways for Insulating Greases and Solid-State Dielectrics

Failure mode	Population affected	Cause	Indicators	Tests	Corrective action	Criticality 1 to 5
Grease oil separation and bleed	All grease types	Thermal cycling, overfilling, incompatible plastics	Oil pooling, drip, reduced fill at interface	Worked penetration, ASTM D6184 bleed, visual	Remove excess, select lower bleed grade, change base oil	3
Grease moisture washout	Low adhesion grades	Rain, spray, pressure washing	Loss of bead, surface tracking marks	Visual, insulation resistance drop	Clean, dry, reapply higher adhesion grade, add boots	4
Grease dielectric contamination	All	Dust, carbonized debris, metallic fines	Darkened paste, local PD	Wipe test, PD survey	Clean and replace grease, add shields or filters	4
Grease incompatibility with elastomer	Silicone or certain hydrocarbons	Swelling of EPDM, NBR, or PC softening	Seal extrusion, cracking over time	Shore hardness, mass change test	Change grease chemistry or seal material	4
Grease thermal dry out	Hydrocarbon grades	High skin temperature, long dwell	Chalky residue, loss of wetting	FTIR oxidation markers, penetration change	Replace with higher stability base, add shade or heatsink	3
Solid PD and treeing	XLPE, epoxy	Voids, sharp edges, contamination	PD inception at lower voltage, rising tan delta	IEC 60270 PD, VLF or AC withstand, dissipation factor	Re terminate or re pot, improve stress grading	5
Solid tracking and erosion	Outdoor polymer housings	Salt fog, UV, pollution, wetting	Surface carbon tracks, audible noise	Salt fog test, visual, infrared for hot spots	Clean, RTV recoat, replace sheds	4
Solid delamination and voiding	Epoxy mica, cast epoxies	Poor cure, thermal cycling	Ultrasound echo, partial discharge	Ultrasonic C scan, PD, thermography	Re impregnate if possible, replace unit	5
Solid thermal cracking	Ceramics, epoxies	Thermal shock, CTE mismatch	Visible cracks, corona	Visual, UV camera	Replace, redesign interface for CTE match	4
Surface contamination flashover	All solids with surfaces	Soot, dust, salt, grease residues	Audible corona, streaking	Leakage current, visual at night, UV	Clean, improve shielding and creepage	4

Solid dielectrics form the primary insulation in cables, machines, bushings, and power electronics. XLPE and EPR carry DC and AC fields in cables and joints. Epoxy mica systems handle slot and end winding stress in generators and motors. Cycloaliphatic epoxy and silicone rubber housings resist UV and salt fog on outdoor apparatus. Ceramics carry very high fields in feed-throughs and vacuum bushings. Application choice depends on field stress, operating temperature, pollution class, mechanical load, and maintenance access. Interfaces between solids and metals need matched expansion and cured-in stress control. Surface design must control creepage, water film formation, and pollution flashover.

Field Inspection and Sampling Protocol for Insulating Greases and Solid-State Dielectrics

Grease at Electrical Interfaces

1. Preparation. Identify grease type, location, and service environment. Review last outage notes and any corona or insulation resistance anomalies. Assemble clean plastic or glass jars with PTFE lined lids, lint free wipes, solvent approved by the OEM, and PPE.
2. Visual inspection. Check terminations, boots, elbows, and gland seals for gaps, oil separation, washout, and contamination. Inspect for tracking marks and carbon paths around the interface.
3. Sampling. Lift a pea sized sample from the interface edge with a clean PTFE or stainless tool. Avoid metal debris. Place in labeled jar. Do not over disturb the primary seal.
4. Field checks. Note color, texture, and presence of free oil. Measure insulation resistance to ground if the asset is de energized and cleared.
5. Laboratory panel. Worked penetration, bleed and oil separation, FTIR for oxidation and base identification, volume resistivity, dielectric strength on thin films, plastics compatibility coupon test where issues were observed.
6. Actions. If bleed or washout is confirmed, clean the interface with approved solvent, dry thoroughly, re apply a low bleed high adhesion grade suited to the plastics present. If contamination is present, restore creepage shields and barriers to limit dust and salt deposition. If incompatibility is detected, change either grease chemistry or seal material.

Solid Dielectrics in Service

1. Preparation. Gather nameplate data, service temperature history, PD records, and prior tan delta trends. Confirm test equipment calibration for PD and insulation measurements.
2. External inspection. Inspect sheds and housings for chalking, erosion, carbon tracks, and bio growth. Inspect epoxy castings for cracks and discoloration. Verify creepage distances are not compromised by deposits.
3. Electrical tests. Perform partial discharge test per IEC 60270 where feasible. Measure insulation resistance and dissipation factor at specified voltage and

temperature. For cables, apply VLF or AC withstand per procedure and record PD inception and extinction if allowed.

4. Non-destructive imaging. Use infrared to locate surface heating at degraded tracks. Use UV corona camera in low light for surface discharge pinpointing. For rotating machines, perform offline PD during outage.
5. Sampling and materials tests. Do not remove bulk material from live assets. For failed parts, send coupons for microscopy, dielectric strength per ASTM D149, CTI per IEC 60112, and moisture content for hygroscopic laminates.
6. Actions. Clean contaminated surfaces with approved detergent and deionized water, dry, then apply RTV silicone top coat on outdoor polymer housings where tracking was observed. Re terminate cable ends with correct stress control components if PD inception is low. Replace cracked ceramics or delaminated epoxies. Adjust shielding and sheds to improve creepage and water shedding in high pollution sites.

This guidance sets material selection, inspection intervals, and corrective steps for interface greases and primary solid insulation across power equipment.

5 Cooling Fluids and Thermal Transfer Media

COOLING FLUIDS AND THERMAL TRANSFER MEDIA: WATER-BASED SYSTEMS

Water-based coolants remain the highest heat-capacity, most economical thermal transfer media in industry. This examination covers three categories: deionized water, water-glycol solutions, and nanofluids. For each, chemical overview, additives and processing, application considerations, and refining of the additive package are outlined. In conclusion, a performance and lifecycle comparison table (see Table 5.1), a failure modes and degradation pathways table (see Table 5.2), and a field inspection and fluid sampling protocol are presented.

Deionized (DI) Water

DI water is water stripped of ionic species to the sub-ppm range using mixed-bed ion exchange and polishing. It delivers the best specific heat and thermal conductivity of any practical coolant, rapid boundary-layer heat transfer, very low viscosity, and low cost. Risks arise from its electrical conductivity rise after pick-up of dissolved ions, aggressive leaching of metals when unbuffered, and high dissolved oxygen solubility. In copper, aluminum, and soldered systems, uninhibited DI can drive pitting and galvanic couples. In carbon steel, oxygen ingress promotes flash rust and magnetite sludging.

Additives and Processing

DI water by itself is almost never acceptable. Standard practice adds a low-foaming blended inhibitor package: nitrite or molybdate for ferrous protection, azoles for copper alloys, phosphate or silicate films where compatible, and a pH buffer targeting 8.5 to 9.5 for mixed-metal circuits. A non-ionic dispersant controls scale precursors and fines. A non-oxidizing biocide program (isothiazolinones, DBNPA, or SH stash) is used in closed loops. Processing steps: charcoal and submicron filtration post-DI, oxygen scrubbing or headspace blanketing where feasible, and final cleanliness verification by conductivity, total organic carbon (TOC), and particle count.

DOI: 10.1201/9781003775201-7

Application Considerations

DI water excels in high heat-flux electronics, laser cooling, polymer extrusion rolls, and closed hydronics where freezing is not a risk. Maintain conductivity below the original equipment manufacturer (OEM) threshold, commonly 0.1 to 10 µS/cm depending on insulation spacing. Control dissolved oxygen and CO_2 ingress at expansion tanks, mechanical seals, and quick-disconnects. Avoid phosphate or silicate if the system cannot tolerate film deposition.

Additives and Refining

Refine the package by metals present and operating temperature. Copper alloys: add tolyltriazole or benzotriazole at 100 to 500 ppm active. Aluminum: avoid silicate starvation, consider 150 to 300 ppm silicate with periodic replenishment and silica watch. Stainless circuits: emphasize oxygen control and dispersant, minimize high-pH exposure that can promote stress corrosion in chlorinated environments. Validate compatibility with elastomers and solders; replace nitrite with molybdate where microbiological nitrite reduction is a known issue. Typical control tests: American Society for Testing and Materials (ASTM) D1121 reserve alkalinity, D1384 simulated service corrosion, conductivity, and metals by inductively coupled plasma (ICP).

Water-Glycol (EG or PG blends)

Ethylene glycol (EG) and propylene glycol (PG) depress the freezing point and raise the boiling point as concentration increases. Compared with water, heat capacity and thermal conductivity are lower and viscosity is higher. EG oxidizes to glycolic, formic, and oxalic acids in the presence of oxygen and heat, consuming inhibitors and lowering pH. PG oxidizes more slowly but still acidifies. Both can form organic acids and polymeric gums that foul heat exchangers.

Additives and Processing

Modern, low-silicate inhibitor packages for water-glycol include phosphate, carboxylates, azoles, and sometimes borate for buffering. Heavy-duty formulations may include nitrite or molybdate for cast iron and wet liners. Use demineralized make-up water to prevent hardness scale and silicate drop-out. Blend with slow agitation, add inhibitors last, and filter through 10 to 25 µm during fill to capture assembly debris. Balance antimicrobial strategy: closed loops often rely on biostatic properties at high glycol content plus periodic biocide shots.

Application Considerations

Water-glycol is preferred where ambient freezing is possible, where burst protection is required, or where systems see intermittent operation. EG offers better thermal performance than PG at a given percent but PG is favored for food, pharma, or environmental sensitivity. Typical set points: 30 to 50% glycol by volume. Verify pump sizing due to higher viscosity. Confirm seal compatibility and oxygen ingress control to minimize oxidation. Monitor with refractometer or density to verify concentration, pH, inhibitor residual, and anion balance.

Additives and Refining

Refine by duty and metallurgy. High aluminum content: limit silicate and maintain organic acid technology that is Al-friendly. High solder or yellow metals: maintain azole residuals. Diesel jacket water or wet sleeve engines: consider nitrite-molybdate blends and cavitation suppressants, verify ASTM D6210 or equivalent heavy-duty spec conformance. Track total acid number growth as an oxidation sentinel and refresh inhibitors before total acid number (TAN) inflects upward. Useful tests include ASTM D3321 freeze protection, D1384 corrosion, D4340 hot surface aluminum corrosion, D1121 reserve alkalinity, and ICP metals.

Nanofluids

Nanofluids are base fluids, typically water or water-glycol, loaded with nanoparticles such as Al_2O_3, CuO, graphene, or SiC at 0.01 to 2% volume. Claimed benefits include enhanced thermal conductivity and altered boundary-layer behavior. In practice, conductivity gains of 5 to 30% are common at room temperature but are offset by viscosity increases, stability challenges, and potential erosion if particles de-stabilize. Surface functionalization and surfactant systems are required for colloidal stability.

Additives and Processing

Stability is the central design variable. Particles are milled to narrow size distributions, surface-treated to match zeta potential targets, and dispersed using high-shear or ultrasonics. A dispersant package and pH control minimize agglomeration. Corrosion inhibitors appropriate to the metals in service are still required. Final polishing filtration is limited to avoid stripping particles; use magnetic traps and low-shear strainers instead. Shelf-life and temperature swing testing are mandatory before adoption.

Application Considerations

Nanofluids can be useful in constrained heat exchangers where area cannot be increased and where filtration is carefully engineered. They are not drop-in replacements. Expect tighter housekeeping, more frequent monitoring, and a defined reclaim or bleed-and-feed plan to manage particle size drift. Avoid in systems with tight clearances, small orifices, or elastomers that may adsorb surfactants.

Additives and Refining

Refine by selecting particle chemistry to match the metallurgy. Copper alloys dislike hard oxide impacts if de-stabilization occurs. Aluminum benefits from silica or alumina with silane surface treatment and pH around 8 to 9. Choose dispersants that do not foam and that tolerate biocides. Validate with accelerated stability, sedimentation rate, and heat-flux rig tests. Because standards are still maturing, qualify to internal performance gates in addition to general coolant tests.

TABLE 5.1
Performance and Lifecycle Comparison for Cooling Fluids and Thermal Transfer Media

Attribute	DI Water (inhibited)	Water-Glycol 30–50% (EG or PG)	Nanofluid in Water or Water-Glycol
Specific heat at 25 C	~4.18 kJ/kg·K	~3.2 to 3.7 kJ/kg·K	Base minus 1 to 10% due to viscosity rise
Thermal conductivity at 25 C	~0.6 W/m·K	~0.3 to 0.4 W/m·K	Base plus 5 to 30% typical
Viscosity at 25 C	~0.9 mPa·s	~2 to 7 mPa·s	Base times 1.2 to 2.0 typical
Freeze protection	None	Excellent, tuneable	Follows base fluid
Boiling margin (pressurized)	Moderate	Good	Follows base fluid
Corrosion control need	High, must inhibit	High, must inhibit	High, plus surfactant compatibility
Biofouling risk	Moderate to high	Lower at high glycol, still non-zero	Moderate, surfactants can feed films
Erosion risk	Low	Low	Elevated if de-stabilized particles
Filtration	Standard fine filtration	Standard fine filtration	Limited; use strainers and traps
Monitoring intensity	Medium	Medium	High
Typical service life	1 to 3 years between partial recharges if tight	2 to 5 years with inhibitor maintenance	Pilot dependent; plan frequent refresh or bleed
Cost	Lowest	Low to medium	Medium to high plus development effort
Environmental profile	Benign; biocides add complexity	EG toxic, PG food-grade; disposal rules apply	Depends on nanoparticle chemistry and disposal plan

Field Inspection and Fluid Sampling Protocol for Cooling Fluids and Thermal Transfer Media

Safety and Cleanliness

Lockout and tagout the circuit. Verify pressure and temperature are within safe limits. Use sterile or new bottles for microbiology and acid-washed bottles for metals. Rinse sample lines and bottles with the process fluid except for microbiology samples which should not be rinsed.

Sampling Points and Method

Install dedicated, laminar-flow sampling valves on the main return header. Avoid dead-legs and low points. Purge at least three line volumes before capture. Fill bottles

TABLE 5.2
Failure Modes and Degradation Pathways for Cooling Fluids and Thermal Transfer Media

Mode	Mechanism	Triggers	Diagnostics	Consequences	Mitigation
General corrosion	Metal dissolution from low pH or depleted inhibitors	Oxygen ingress, CO_2 absorption, additive depletion	pH, reserve alkalinity, ICP Fe/Cu/Al, ASTM D1384 coupons	Leaks, sludge, reduced heat transfer	Maintain inhibitor residuals, seal air leaks, nitrogen blanket
Pitting and crevice attack	Local depassivation on Cu, Al, steels	Chlorides, stagnant zones, under-deposit	Chloride test, pit coupons, boroscope	Rapid wall loss and failures	Control halides, flow velocity, dispersant use, clean fouling
Galvanic corrosion	Dissimilar metals in conductive water	Mixed metallurgy, high conductivity	Conductivity rise, mixed-potential mapping	Focused attack on less noble metal	Azoles for Cu, filmers, electrical isolation, keep conductivity low
Scaling and deposition	$CaCO_3$, silicate gels, iron oxide deposits	Hard make-up, oversilicated blends, poor filtration	Hardness, silica, turbidity, deposit analysis	Hot-side fouling, ΔP rise, hotspots	Demineralized water, dispersant, proper silicate control, filtration
Oxidation of glycol	Acid formation, viscosity rise, sludge	Heat, oxygen, metals acting as catalysts	TAN, color, FTIR, anions by IC	Acidic corrosion, gelation, pump strain	Oxygen control, antioxidant package, timely inhibitor top-up
Microbiological growth	Biofilms, under-deposit corrosion	Warm loops, stagnant segments, nutrients	ATP, dip slides, biofilm probes	Fouling, odor, MIC pitting	Biocide program, velocity control, dead-leg removal
Cavitation and erosion	Vapor bubble collapse, particle impingement	Low NPSH, high velocity elbows, hard particles	Pump vibration, ultrasonic, wear metals	Liner pitting, impeller damage	NPSH margin, remove gas, smooth hydraulics, particle control
Nanofluid instability	Agglomeration and sedimentation	pH drift, ionic contamination, thermal cycling	DLS size, turbidity, settling tests	Orifice plugging, erosion, loss of benefit	Surface treatment, zeta control, gentle filtration, periodic redispersion

midstream without aeration and without headspace for volatile measurements. Record system operating conditions, makeup fluid used, recent chemical additions, and time since last maintenance.

Frequency

Commissioning: weekly until baselines are stable. Steady state: quarterly for closed DI or water-glycol loops, monthly if high heat or frequent make-up, monthly for any nanofluid pilot. After any major intervention, resample within 72 hours.

Field Tests

- Appearance, odor, and color index
- Temperature and pressure at draw
- Conductivity or resistivity. DI targets are typically 0.1 to 10 μS/cm per OEM
- pH at 25 C, target per chemistry. DI inhibited 8.5 to 9.5 typical. Water-glycol often 8.0 to 10.0 depending on inhibitors
- Refractometer or density for glycol percent and freeze point
- Biocide residual where applicable

Laboratory Tests

- Corrosion control: reserve alkalinity, specific inhibitor residuals, azole content
- Metals by ICP for Fe, Cu, Al, Zn, Pb, Sn as corrosion indicators
- Anions by ion chromatography for chloride, sulfate, nitrate, formate, glycolate
- Total acid number for glycol systems as oxidation indicator
- Particle count and gravimetric solids. Use membrane patch microscopy to identify rust, silica, gasket, or polymer fines
- Microbiology: adenosine triphosphate (ATP) or culture methods. For problem loops, biofilm probe or coupon program
- For DI quality: TOC, silica, and sodium as cleanliness sentinels
- For nanofluids: dynamic light scattering for particle size, zeta potential, and accelerated thermal cycling stability

Acceptance Limits and Actions

- pH out of range by more than 0.3 units: adjust buffers or partial drain and recharge
- Conductivity above OEM limit or rising trend: locate make-up leaks, polish through ion exchange side-stream, or partial drain and refill with treated water
- Inhibitor residual below minimum: dose per vendor curves, confirm within 24 hours
- TAN rising trend in glycol: identify oxygen ingress, consider partial drain, replenish inhibitors, add antioxidant package if supported
- Metals above baseline: search for oxygen entry, stagnant legs, or galvanic couples. Inspect strainers and magnets for evidence

- Microbiological growth: shock biocide dose, purge low-flow zones, verify velocity and temperature cycling do not favor biofilm adhesion
- Nanofluid instability: redisperse under controlled shear, correct pH and ionic strength, or refresh a portion of the inventory

Documentation

Maintain a control chart for pH, conductivity, inhibitor residuals, TAN, metals, glycol percent, particle count, and key microbiology metrics. Tie every change to a dated corrective action. Keep a configuration log of metallurgy, elastomers, biocides, and inhibitor chemistry to avoid incompatible future additions.

Useful references when selecting tests and limits include ASTM D1384 for simulated corrosion in engine coolants, ASTM D4340 for hot surface aluminum corrosion, ASTM D1121 for reserve alkalinity, ASTM D3321 or refractometry methods for freeze point and glycol percentage, and ICP spectrometry for wear and corrosion metals. Where vendor specifications exist for proprietary blends, prioritize those limits and methods.

ADVANCED COOLING AND DIELECTRIC FLUIDS: NANOFLUIDS, FLUOROKETONES, AND SYNTHETIC HYDROCARBONS

As power electronics, high-voltage systems, and digital infrastructure push the limits of heat flux and dielectric stress, traditional fluids such as mineral oils, water-glycol mixtures, and ester-based coolants are increasingly supplemented, or even replaced, by advanced chemistries engineered for specific performance demands. Among these, nanofluids, fluoroketones, and synthetic hydrocarbons represent three important categories, each offering distinct advantages in thermal management, dielectric strength, and environmental impact.

Nanofluids introduce suspended nanoparticles into conventional base liquids, enhancing thermal conductivity and heat transfer efficiency. While promising, they demand careful attention to stability, dispersion, and compatibility to avoid risks such as agglomeration or erosion. Fluoroketones, such as 3M Novec™ 649, provide low-global-warming-potential alternatives for immersion cooling and fire-safe dielectric service, with gas-phase variants (4710 and 5110) developed as next-generation replacements for SF_6 in high-voltage switchgear. Synthetic hydrocarbons, including polyalphaolefins (PAOs) and alkylbenzenes, continue to serve in specialized thermal management and refrigeration roles where low pour points, oxidative stability, and broad compatibility are required.

Unlike legacy fluids, these advanced systems often operate at the edge of material and thermal limits. Their long-term reliability depends on not only intrinsic chemistry but also rigorous conditioning practices: moisture and gas control, antioxidant preservation, filtration, and compatibility testing with metals, elastomers, and system surfaces. Together, these fluids represent the frontier of cooling and dielectric technology, enabling higher efficiency, improved safety, and reduced environmental footprint in the next generation of power and electronic systems.

Nanofluids

Nanofluids are base liquids: typically water or water-glycol loaded with nanoparticles such as Al_2O_3, CuO, SiC, or carbon allotropes like graphene. Reported gains in thermal conductivity often sit in the 5 to 30% range at modest volume fractions, with higher values possible at greater loadings. Gains trade off against viscosity rise and stability limits that raise pumping power and filtration burden. The dominant engineering risks are agglomeration, sedimentation, surfactant loss, and particle-induced erosion if the dispersion destabilizes.

Additives and Processing

Stability is the control knob. Particles require surface functionalization to set zeta potential; dispersants and pH control keep Brownian stabilization dominant. High-shear or ultrasonic dispersion is followed by thermal cycling qualification and dynamic light scattering to confirm size and polydispersity. Corrosion inhibitors still follow the metallurgy of the loop. Filtration must be rethought: fine filter media can strip surfactant or the particles themselves; use staged strainers and magnetic traps with verified beta ratios that preserve the engineered distribution.

Application Considerations

Best value appears where heat-transfer area is constrained, two-phase boiling is undesirable, and the system can tolerate tighter housekeeping. Avoid orifice-rich hydraulics and microclearances. Expect higher monitoring cadence: particle size, zeta potential, viscosity, and deposit microscopy. When using graphene-based dispersions, target minimal effective loading to limit viscosity penalties while capturing conductivity uplift.

Refinements

Match particle chemistry to metals: alumina or silica with silane treatment for Al-rich circuits; avoid hard oxides against soft copper if loss of stability is plausible. Select non-foaming dispersants compatible with chosen biocides. Write end-of-life rules in advance: bleed-and-feed or reclaim via centrifugation, then redope surfactant and verify stability before return to service.

Fluoroketones and Related 3M Novec Lines

3M Novec 649 is a liquid fluoroketone, chemical name perfluoro(2-methyl-3-pentanone), with boiling point near 49 C, low viscosity, good dielectric strength, and very low global warming potential (GWP) relative to HFCs. It is non-flammable and widely used for single and two-phase immersion cooling of power electronics and servers. Novec 4710 and Novec 5110 are fluorinated insulating gases used in high-voltage equipment as SF_6 alternatives. They offer high dielectric strength and reduced environmental impact in gas-insulated switchgear. These are not liquid immersion coolants; they are gas media for insulation and arc quenching.

Additives and Processing

Novec 649 is typically run neat without corrosion inhibitors; it is chemically inert and water-immiscible but can carry dissolved gases and trace moisture. Processing focuses on moisture control, degassing, and particulate cleanliness. Dry the system hardware, bake out as feasible, and circulate through depth media to submicron levels before commissioning. Gas variants 4710 and 5110 are handled as sealed gas systems: quality assurance focuses on moisture, gas purity, and tightness testing to prevent dilution by air.

Application Considerations

For Novec 649 immersion, use size condensers for the 49 °C normal boiling point if using two phase; understand that heat capacity is far lower than water, so performance comes from latent heat and nucleate boiling rather than sensible heat. Verify elastomer compatibility and avoid materials that trap liquid in pores. For 4710 and 5110, treat them as dielectric gases in gas-insulated line (GIL) and gas-insulated bus (GIB). Thermal transport is incidental; the primary requirement is dielectric coordination, pressure control, and leak integrity.

Refinements

For liquid fluorinated ketone (FK) loops, emphasize dryness: inline molecular sieve dryers sized for the expected moisture ingress and periodic Karl Fischer checks. Control non-condensable load in two-phase rigs and implement sight-glass or acoustic void fraction monitoring. For gas media, adopt OEM purity specs and pressure-temperature compensation logging to detect slow losses. Where Novec 649 is unavailable, FK-649 replacements from third parties exist; qualify with full compatibility and heat-flux testing.

Synthetic Hydrocarbons: PAO and Alkylbenzenes

PAO heat transfer fluids are linear or lightly branched oligomers of alpha-olefins. They deliver wide liquid ranges, very low pour points, low volatility, and high oxidation resistance compared with mineral oils. Thermal stability supports extended service life in closed systems when oxygen is controlled (see Tables 5.3 and 5.4).

Alkylbenzenes are synthetic mono-aromatic base stocks with excellent solubility for traditional refrigeration additives and good miscibility with CFC and HCFC refrigerants; they are still used in retrofit or legacy chillers and can serve as dielectric hydrocarbon coolants in some electronics where non-polarity and low moisture solubility are valued.

Additives and Processing

PAO systems usually run with a lean antioxidant package: phenolic and amine AO, metal deactivator if copper is present, and antifoam. They are not water-based, so corrosion is typically under-deposit or from oxygen-driven acid formation. Commission with hot-oil flush, 3 to 10 μm filtration, and nitrogen blanketing.

TABLE 5.3
Performance and Lifecycle Comparison for Nanofluids, Fluoroketones, and Synthetic Hydrocarbons

Attribute	Nanofluid in Water or Water-Glycol	Fluoroketone 649 (liquid)	Fluoroketone 4710 or 5110 (gas, GIS)	PAO Heat Transfer Oil	Alkylbenzene (AKB)
Primary role	Enhanced liquid heat transfer	Immersion cooling, single or two phase	Dielectric insulation and arc quench gas	Hot-oil and dielectric cooling	Refrigeration and sealed hydrocarbon cooling
Thermal conductivity vs base	+5 to +30% typical at low vol%	Low vs water; relies on latent heat in two phase	Not a liquid coolant	Moderate vs mineral oils; stable at high T	Similar to mineral oils
Viscosity impact	Increases with loading; raises pump power	Low viscosity	Gas properties govern	Low at cold, stable at hot	Moderate
Electrical behavior	Conductive per base fluid; not dielectric	High dielectric strength	High dielectric strength gas	Excellent dielectric	Good dielectric for hydrocarbon duty
Fire risk	Follows base fluid and surfactant	Non-flammable	Non-flammable	Low flash at high grades; manage hot spots	Flammable hydrocarbon; sealed service
Stability risks	Agglomeration, surfactant loss	Moisture pickup, NCGs in two phase	Gas purity, leaks	Oxidation to acids, varnish	Moisture, additive depletion
Typical service life	Pilot dependent; manage by bleed-and-feed	Multi-year with dryness control	Multi-year if tight and pure	5 to 10 years under inert blanket	3 to 8 years in clean sealed systems
Environmental profile	Depends on particle chemistry and surfactants	Very low GWP, no ODP	Reduced GWP vs SF_6	Hydrocarbon; good biodegradability variants	Hydrocarbon; refrigerant specific
Notes	High monitoring burden	Boiling point ~49 C; two-phase design critical	Used in GIS, not immersion	Excellent low-temp pumpability	Strong legacy tie to CFC/HCFC systems

TABLE 5.4
Failure Modes and Degradation Pathways for Nanofluids, Fluoroketones, and Synthetic Hydrocarbons

Mode	Mechanism	Triggers	Diagnostics	Consequences	Mitigation
Nanofluid agglomeration	Loss of surfactant or zeta potential; particles cluster	pH drift, ionic contamination, thermal cycles	DLS size, zeta, turbidity, settling tests	Plugging, erosion, loss of benefit	Redope dispersant, correct pH and ionic strength, controlled redispersion
Particle-induced erosion	Hard particles strike surfaces	Destabilization, poor filtration	Patch microscopy, wear metals	Orifice and impeller wear	Keep dispersion stable, use strainers and magnetic traps
FK-649 moisture uptake	Dissolved water elevates corrosion risk at interfaces	Poor dryness control	Karl Fischer, dew point of headspace	Dielectric stress, film breakdown	Inline dryers, bakeout, tight hardware
Two-phase NCG loading	Non-condensables reduce condensation efficiency	Leaks, poor degassing	Void fraction, condenser ΔT-LMTD loss	Hot spots, capacity loss	Degas, leak-test, maintain condenser approach
GIS gas dilution (4710/5110)	Air or moisture lowers dielectric strength	Seal leaks, service errors	Gas chromatography, dew point, pressure-temperature trending	PD and flashover	Tightness testing, purity specs, humidity limits
PAO oxidation	Radical chain oxidation to acids and oligomers	Oxygen ingress, high film temps	TAN, RPVOT, FTIR, MPC patch, RULER	Varnish, fouling, pump strain	Nitrogen blanket, AO maintenance, hot-oil velocity control
AKB moisture and acid rise	Moisture ingress and additive depletion	Poor sealing, refrigerant breakdown	Moisture ppm, TAN, color, metals	Bearing wear, copper plating	Drying, additive refresh, seal upkeep

Alkylbenzene refrigeration oils may include antiwear and corrosion inhibitors tuned for compressor metallurgy. Keep moisture very low to avoid hydrolysis of any additive chemistries and to protect bearings.

Application Considerations

Use PAO where electrical insulation is required, where ambient freeze protection is needed without glycols, and where long hot-side residence times exist. Check pump sizing for viscosity at the coldest start temperature. Alkylbenzenes fit older refrigerants and some specialty sealed cooling loops. For modern hydrofluoroolefin (HFO) refrigerants or CO_2, compatibility must be confirmed.

Refinements

For PAO, control oxygen and temperature: nitrogen blanket, low-leak seals, and oversized expansion space reduce oxidation rate. Track varnish precursors with membrane patch colorimetry (MPC) patch rating and antioxidant health with voltammetry. For alkylbenzenes, maintain low moisture and monitor color, acid number, and metals; where electrical duty exists, proof dielectric strength at service temperature after moisture control.

Field Inspection and Fluid Sampling Protocol for Nanofluids, Fluoroketones, and Synthetic Hydrocarbons

Safety and Cleanliness

Lockout, depressurize, or de-energize as appropriate. For gases, follow GIS standards for gas handling and dew point safety. Use clean, dry sampling hardware. Moisture control is paramount for FK-649 and GIS gases.

Sampling Points and Method

- **Nanofluids:** sample from a main return header with steady flow. Purge lines, collect in glass or compatible polymer bottles. Capture separate aliquots for DLS and chemistry.
- **FK-649 liquid:** draw from circulating loop into dry, sealed glassware. Avoid headspace when testing for dissolved gases and water.
- **4710/5110 gases:** use calibrated gas sampling bulbs or cylinders. Measure dew point and gas composition on site if possible.
- **PAO and AKB:** hot-drain midstream samples into dry bottles. Avoid aeration. Record temperature and load.

Frequency

Commissioning: weekly until trends stabilize. Steady state: quarterly for PAO and AKB; monthly for FK-649 two-phase rigs and any nanofluid trials; semi-annual for GIS gases unless OEM requires tighter intervals.

Field Tests

- Nanofluids: appearance, turbidity, pH, conductivity, quick DLS if available.
- FK-649: visual clarity, moisture via portable KF if available, non-condensable check by condenser approach temperature.
- GIS gases: dew point, pressure-temperature, gas purity.
- PAO and AKB: visual, odor, dielectric strength where relevant, quick TAN spot if kits available.

Laboratory Tests

- **Nanofluids:** DLS size and zeta potential, viscosity vs shear and temperature, ICP metals, deposit patch microscopy, base-fluid corrosion tests as applicable.
- **FK-649:** Karl Fischer water, GC for dissolved gases if two-phase, materials compatibility microscopy, dielectric breakdown where specified.
- **4710/5110:** GC for composition, moisture, acidity, and byproducts per OEM.
- **PAO:** viscosity index, TAN by D664, MPC varnish rating, FTIR oxidation, RPVOT, particle count ISO 4406.
- **AKB:** moisture ppm, TAN, metals, compatibility with refrigerant where applicable.

Acceptance Limits and Actions

- Nanofluids: if mean particle size shifts by more than 30% or zeta potential magnitude drops below the validated threshold, redisperse or refresh.
- FK-649: moisture above vendor limit triggers inline drying or partial drain and dry-refill. Non-condensable rise demands degas and leak hunt.
- GIS gases: dew point out of spec or composition drift mandates leak correction and gas reclamation per OEM.
- PAO and AKB: TAN inflection upward, MPC increase, or metals rise calls for oxygen control checks, filter upgrades, AO top-treat, or partial refresh.

Notes

- FK-649 boil point near 49 °C is central to two-phase design. Plan condenser approach temperature and non-condensable control accordingly.
- 4710 and 5110 are insulating gases for switchgear, not immersion coolants. Treat them under GIS maintenance rules; thermal transport is a secondary property.
- PAO providers now position PAO coolants as per- and poly-fluoroalkyl substances (PFAS)-free alternatives for dielectric immersion in some applications; vet claims with dielectric and compatibility testing in your hardware.

Phase-Change Fluids for Data Centers and Electronics

Two-phase cooling leverages latent heat at the boiling surface and condensation away from the heat source. Compared with single-phase liquids, boiling enables much higher heat flux at small temperature deltas when nucleation is well managed and non-condensables are minimized. In data centers these fluids must be electrically insulating, low toxicity, and low GWP, while remaining compatible with server materials and seals. Industrial practice splits into immersion tanks and direct-to-chip evaporative cold plates.

Fluorinated Ketones and Legacy PFAS Lines

Perfluoroketones such as the compound historically sold as Novec 649 are non-flammable dielectrics with normal boiling points near 49 °C, low viscosity, and strong dielectric strength. They have been popular for two-phase immersion and power electronics because latent heat compensates for modest liquid thermal conductivity. 3M announced a global exit from PFAS manufacturing by December 31, 2025, so Novec and Fluorinert availability sunsets, forcing qualification of alternates or stock-down strategies (see Tables 5.5 and 5.6).

TABLE 5.5
Performance and Lifecycle Comparison for Nanofluids, Fluoroketones, and Synthetic Hydrocarbons

Attribute	FK-type dielectric, e.g., legacy 649	Low-GWP HFO/ HFE e-cooling fluids	Notes
Boiling point at 1 atm	~49 °C typical for 649-class	Vendor-set, often 30–60 °C	Select to match chip junction targets.
Dielectric behavior	Excellent, non-flammable	Excellent, typically non-flammable	Confirm breakdown voltage at service temps.
Latent heat utilization	High, enables compact condensers	High, similar principle	Condenser sizing and NCG control dominate.
Materials compatibility	Generally inert to metals and polymers	Vendor specific	Always run coupon packs and seal soaks.
Environmental profile	Low GWP for FK 649 but PFAS class scrutiny	Low GWP marketing focus	Verify product-specific GWP and EHS sheets.
Availability risk	Discontinued by end of 2025	Actively marketed	Plan migration off PFAS-based fluids.
Typical service interval	Multi-year if dry and tight	Multi-year if dry and tight	Fluid losses mainly from leaks and handling.

TABLE 5.6
Failure Modes and Degradation Pathways for Nanofluids, Fluoroketones, and Synthetic Hydrocarbons

Mode	Mechanism	Triggers	Diagnostics	Consequences	Mitigation
Moisture pickup	Dissolved water lowers dielectric strength and alters boiling behavior	Humid air ingress, wet hardware	Karl Fischer water, breakdown voltage	PD events, arcing risk, unstable boiling	Inline dryers, sealed headspace, dry assembly.
Non-condensable gas loading	NCGs blanket condenser area, reduce heat rejection	Leaks, poor degas, decomposition of trapped air	Condenser approach T rise, pressure drift, gas sampling	Hot spots, capacity loss	Vacuum or membrane degas, leak hunt.
Nucleation site poisoning	Surface films or additives suppress boiling	Manufacturing residues, additive overdose	Heat flux test, contact angle, SEM of surfaces	Higher wall temps, premature CHF	Cleanliness specs, no unqualified additives.
Particulate fouling	Debris pins bubbles or abrades channels	Assembly debris, pump wear	Particle count, patch microscopy	Flow maldistribution, erosion	Depth filtration during commissioning, strainer.
Chemical loss or supply obsolescence	Fluid no longer available or reclaimed	PFAS phase-out, vendor changes	Market alerts, vendor letters	Forced fluid migration	Dual-source and qualify alternates early.
Gas-circuit dilution in D2C	Air dilution lowers dielectric and changes saturation curve	Imperfect seals, service events	*P-T* trending, GC of loop gas	Flashover risk, unstable control	Tightness testing, service SOPs.

HFO/HFE Families Marketed for e-Cooling

Vendors have positioned low-GWP hydrofluoroolefins and related ethers for two-phase e-cooling, including Honeywell's Solstice E-Cooling and Chemours' Opteon branded two-phase immersion fluids. These are dielectric, non-flammable or difficult to ignite, and designed to boil near electronics-friendly temperatures while offering materially lower GWP than legacy PFAS. Real-world deployments focus on direct-to-chip evaporative plates and two-phase immersion pilots. Availability, handling, and environmental profiles are vendor specific and must be verified.

Additives and Processing

Most two-phase dielectrics are run neat. The control variables are dryness, particulate cleanliness, surface energy of wetted materials, and removal of non-condensable gases. Additive packages are minimal because surfactants or corrosion inhibitors can poison nucleation sites, foam during vigorous boiling, or degrade dielectric strength. Commission with dehydrated hardware, bakeouts where safe, submicron depth filtration, nitrogen, or vacuum degas, and verified condenser performance at the target approach temperature. Where the OEM provides a stabilizer or metal deactivator, dose only within that program and trend breakdown voltage after any change.

Application Considerations

- **Immersion tanks**: Fluid choice is set by boiling point, dielectric strength, materials compatibility, and condensers sized to reject latent heat with minimal non-condensables. Moisture must be kept well below vendor limits because dissolved water degrades dielectric strength and disrupts nucleation.
- **Direct-to-chip evaporative plates**: Channel geometry and surface finish govern critical heat flux and onset of dry-out; two-phase HFO fluids are marketed specifically for this duty. Keep gas management tight and provide a reliable condenser or CDU to avoid pressure drift.
- **Supply chain**: The PFAS exit compresses legacy options, so programs should qualify non-PFAS low-GWP fluids early, with side-by-side rigs to validate heat flux, materials compatibility, and serviceability. Regulatory posture on PFAS also affects reclamation and waste.

Field Inspection and Fluid Sampling Protocol for Nanofluids, Fluoroketones, and Synthetic Hydrocarbons

Safety and Cleanliness

De-energize per facility LOTO. Use baked-dry glassware and metal hardware. Keep exposure to air minimal during draws. Moisture control is the program's backbone.

Sampling Points and Method

Install dedicated sample ports on the liquid return and, for two-phase tanks, a head-space takeoff for non-condensable checks. Purge several line volumes. Collect separate aliquots for moisture and for dielectric testing. For direct-to-chip systems, include a port at the condenser outlet to assess NCGs and dissolved gas.

Frequency

- Commissioning: weekly until trends flatten.
- Steady state: monthly for the first quarter, then quarterly if stable.
- After any breach or service event: sample within 24 to 72 hours.

Field Tests
Appearance and odor, operating pressure and saturation temperature, condenser approach temperature, quick moisture screening if available, and leak-rate inference from *P-T* drift at steady load.

Laboratory Tests

- **Moisture**: Karl Fischer to vendor limit.
- **Dielectric strength**: breakdown voltage at operating temperature.
- **Gas analysis**: headspace GC for air fraction or composition as provided by the OEM.
- **Non-volatile residue and particulates**: gravimetric plus membrane patch microscopy to find fibers, solder mask, or gasket crumbs.
- **Materials checks**: periodic seal and polymer coupons, especially during first-year operation.

Acceptance Limits and Actions

- Moisture above the OEM limit: put the loop on dryers, check desiccant capacity, consider partial drain and dry refill.
- Rising condenser approach temperature at constant load: degas and investigate for NCGs or fouled coils.
- Breakdown voltage below minimum: quarantine, dry, and retest before return to service.
- Particle uptick after maintenance: recirculate through depth media until counts return to baseline.

Program Notes and Migration Planning

- If you are still on PFAS-based fluids, set a dated migration plan. Track last-buy dates and qualify at least one non-PFAS dielectric in a pilot loop with identical hardware, then lock a supply and reclamation plan.
- When evaluating alternatives, run identical heat-flux rigs and coupon packs, and score fluids on latent heat performance under non-condensable injection, wettability stability after thermal cycling, and ease of reclamation. Third-party selection guides and white papers can help structure the test slate.

- Vendors advertising two-phase e-cooling fluids include Honeywell Solstice and Chemours Opteon. Validate claims for your geometry and operating envelope.

COMMISSIONING CHECKLIST: TWO-PHASE DIELECTRIC COOLING

(1) **Pre-commission**
  - Final design review: heat flux map per zone, target saturation temperature, condenser duty and approach target, pump headroom at design mass flow.
  - Instrumentation check: calibrated *P-T* sensors at evaporator inlet/outlet, condenser inlet/outlet, headspace or receiver; sight glass or void fraction meter; moisture test kit access; sampling ports.
  - Materials verification: metals, seals, PCB finishes, potting compounds, cable jackets validated against chosen fluid.

(2) **Hardware preparation**
  - Degrease and flush all wetted parts; final rinse with reagent-grade alcohol or vendor-approved solvent; dry with heated, filtered air or nitrogen.
  - Bakeout where safe: 60 to 80 °C for 8 to 24 h to drive off moisture from crevices and laminates.
  - Filters and strainers: depth media staged to submicron; differential pressure transducers installed; spare elements on site.
  - Leak test with dry nitrogen or vacuum decay to OEM tightness. Record baseline leak rate.

(3) **Fluid preparation**
  - Certificate of analysis on receipt lot. Verify moisture and non-volatile residue meet spec.
  - Pre-dry in a side vessel if needed using molecular sieve dryer; circulate through depth media.
  - Keep all vessels and lines capped; avoid ambient air exposure.

(4) **Controlled fill and degas**
  - Evacuate loop to <5 mbar absolute where hardware allows or purge with dry nitrogen.
  - Bottom-up fill to minimize vapor traps. Recirculate through filters.
  - Active degas: vacuum or membrane degasser online until non-condensables stabilize at baseline.
  - Dryness confirmation by Karl Fischer on a drawn sample before heat load is applied.

(5) **Thermal ramp and tuning**
  - Step heat flux in ~20% increments to design. At each step log:
    - o Saturation temperature and pressure
    - o Condenser inlet and outlet temperatures, approach to ambient or secondary loop
    - o Differential pressure across filters
    - o Power in vs condenser duty
    - o Headspace gas bleed rate if used

- Trim condenser fan or CDU controls to hit approach target. Confirm stable nucleation without overshoot.

(6) Baseline qualification

- Draw baseline samples: moisture, non-volatile residue, particle membrane patch, optional breakdown voltage at operating temperature.
- Establish control chart baselines over a 72-hour steady run at design heat.
- Lock tamper-proof settings on CDUs, set alarm thresholds.

(7) Acceptance

- All control limits in the plan below met for 24 continuous hours at design heat.
- Leak rate within spec; make-up additions negligible.
- Documentation package signed: *P-T* logs, moisture data, approach data, photos of sight glasses, filter ΔP traces.

Control Plan

A. Limits by fluid family

- **FK-type dielectric liquids**: moisture ≤ 50 ppm operational, ≤ 20 ppm stretch; non-condensables at condenser ≤ 0.5% equivalent by volume; condenser approach at design load ≤ 5 °C, alarm at 7 °C; pressure drift at steady load ≤ 2 kPa over 30 min.
- **Low-GWP HFO/HFE e-cooling liquids**: moisture ≤ 100 ppm operational, ≤ 50 ppm stretch; non-condensables ≤ 0.5% equivalent; condenser approach ≤ 6 °C, alarm at 8 °C; pressure drift ≤ 3 kPa over 30 min.

If your design heat flux is unusually high, tighten approach limits by 1 to 2 °C and require headspace NCG ≤ 0.3% (see Tables 5.7, 5.8, and 5.9).

B. Immersion Tanks: control matrix

C. Direct-to-Chip Evaporative Plates: control matrix

D. Monitoring cadence

E. Trigger-based actions

- **Approach rises by ≥3 °C at constant load**: assume NCG or condenser fouling. Put degasser at max, inspect condenser surfaces, verify fans or CDU flow, sample headspace for air fraction. Do not raise setpoints to mask the symptom.
- **Moisture exceeds limit**: switch to a fresh dryer bed, confirm seals and headspace gaskets, check make-up handling. If above limit for 24 h despite drying, perform a 10 to 30% dry-refill.
- **Plate ΔT creeps upward by ≥3 °C**: check filter ΔP, microchannel fouling, pump NPSH, and confirm no surfactants or residues were introduced during service.
- **Recurrent alarms**: hold load at 70%, run a controlled degas, replace filters, reestablish baselines, and only then return to full load.

F. Documentation

- Control charts: moisture, condenser approach, saturation pressure, plate ΔT, filter ΔP, make-up volume.

TABLE 5.7
Immersion Tanks: Control Matrix

Parameter	Sensor or Test	Baseline Target	Control Limit	Action if Out of Control
Moisture (ppm)	Karl Fischer	FK: 15–30; HFO/ HFE: 30–60	FK: ≤50; HFO/ HFE: ≤100	Put loop on dryers, verify dryer capacity, partial drain and dry-refill, retest in 24 h
Condenser approach, °C	*T* in/out on condenser and ambient or secondary loop	3–4	≤5 (FK) or ≤6 (HFO/HFE)	Degas loop, check condenser fouling and airflow, verify CDU controls
Non-condensables	Headspace GC or inferred via approach rise at constant load	"Not detected" or stable approach	≤0.5% eq. vol.	Vacuum or membrane degas, leak hunt, seal service
Saturation *P-T* stability	*P-T* at tank and evaporator	Flat at steady load	≤2 kPa drift (FK) or ≤3 kPa (HFO/ HFE)	Check NCG, verify PID tuning, inspect pump cavitation
Particles/NVR	Membrane patch + gravimetric	Light, no fibrous or solder-mask residue	No rising trend after maintenance	Recirculate through depth filters until baseline restored
Fluid losses	Level over time	<0.1% per week	≤0.3% per week	Leak survey, tighten fittings, inspect quick connects
Visual boiling quality	Sight glass or borescope	Stable nucleate, no film boiling	No sustained dry-out on hotspots	Reduce heat flux temporarily, improve wetting, clean surfaces

- Change log: every chemical, mechanical, or control change stamped with date, time, person, reason.
- Annual review: revalidate limits against actual heat flux and any hardware changes. Update alarm setpoints and dryer sizing accordingly.

GAS COOLING MEDIA FOR GENERATORS AND ELECTRONICS: HYDROGEN, HELIUM, CO_2, AND AIR

Hydrogen (Generator Cooling)

Hydrogen provides the highest thermal conductivity of any practical gas along with very low density, which slashes windage losses in large turbogenerators. Typical

TABLE 5.8
Direct-to-Chip Evaporative Plates: Control Matrix

Parameter	Sensor or Test	Baseline Target	Control Limit	Action if Out of Control
Plate ΔT (wall minus sat), °C	Embedded diode or TCs + *P-T*	≤10 at design flux	≤12; alarm at 15	Purge NCGs, inspect microchannels for debris, verify pump flow
Channel DP, kPa	DP taps across cold plate	As-designed	±10% of baseline	Check filter loading, channel fouling, pump curve shift
Condenser approach, °C	As above	3–5	≤6–8	Degas, clean condenser, verify secondary loop capacity
Moisture (ppm)	Karl Fischer	FK: 15–30; HFO/ HFE: 30–60	FK ≤50; HFO/HFE ≤100	Dryers online, investigate ingress points
Nucleation consistency	IR camera or embedded sensors	Uniform	No hot streaking	Clean plate surfaces, avoid additives, relap if required

TABLE 5.9
Monitoring Cadence

Activity	Commissioning	Steady State	Event-Driven
Moisture KF	Daily	Monthly; quarterly if rock-solid	Immediately after any breach or service
P-T and approach logs	Continuous trending	Continuous	After any alarm adjust and annotate
Degas rate or NCG check	Daily	Weekly	After approach rise >2 °C at constant load
Particles/NVR	Weekly	Quarterly	After maintenance or filter change
Visual inspection of boil quality	Daily	Weekly	After load or configuration changes

operation uses 2 to 6 bar absolute to raise gas density and heat capacity. Hydrogen is non-toxic but flammable across a wide range in air; safe operation depends on tight enclosures, controlled purging, continuous purity monitoring, and reliable seal-oil systems that prevent air ingress. Hydrogen carries little moisture at service dew points, reducing internal corrosion and improving stator cooling effectiveness (see Tables 5.10 and 5.11).

TABLE 5.10
Performance and Lifecycle Comparison – Hydrogen, Helium, CO_2, and Air

Attribute (approx. 25°C, 1 atm unless noted)	Hydrogen	Helium	CO_2	Air
Thermal conductivity, W/m·K	~0.18	~0.15	~0.016	~0.026
Specific heat, kJ/kg·K	~14.3	~5.2	~0.84	~1.0
Density, kg/m³	~0.09	~0.18	~1.98	~1.2
Viscosity, µPa·s	~9	~20	~15	~18
Windage losses at equal speed	Lowest	Low	Highest	High
Flammability	High in air	None	None	Supports combustion
Typical generator use	Primary cooling gas	Commissioning, special cases	Purge gas between air and H_2	Open/TEWAC cooling medium
Gas additives	None	None	None	None
O&M focus	Purity, dew point, seal-oil	Dryness, leak minimization	Dryness, purge discipline	Filtration, humidity control
Relative operating cost	Medium	High	Low	Lowest

Additives and Processing

No chemical additives are used in the hydrogen itself. Processing focuses on gas purity, dryness, and cleanliness: bulk supply verification, dryer operation, coalescing filtration on recirculation, and oxygen monitoring. Seal-oil systems receive the "chemistry" attention: antifoam to control entrained H_2, antioxidant package to resist aeration-driven oxidation during mode changes, and demulsifier to release condensed water. Dryers are sized to maintain a subzero dew point margin at operating pressure. CO_2 is used for purge during shutdown or startup to avoid passing directly through the explosive region; helium may be used for leak-testing and rapid diagnostics.

Application Considerations

Hydrogen excels in very large generators where windage reductions and high heat-flux removal justify the gas handling plant. The enclosure must be gas-tight, with shaft seals, seal-oil pressure control, hydrogen analyzers, flame arrestors, and purge valving that enforces the Air→CO_2→H_2 or reverse sequence. Cooler bundles and gas paths require smooth hydraulics; any fouling or damaged fins degrade unit rating. The control plan centers on purity, dew point, differential pressure across coolers, seal-oil differential to hydrogen, and verified purge interlocks.

TABLE 5.11
Failure Modes and Degradation Pathways for Hydrogen, Helium, CO_2, and Air

Mode	Mechanism	Triggers	Diagnostics	Consequences	Mitigation
Air ingress into H_2 space	Explosive mixture formation; loss of cooling purity	Seal failure, low seal-oil differential, poor purge	H_2 purity meter, O_2 analyzer, rate-of-change alarms	Trip risk, explosion hazard, rating loss	Maintain seal-oil > H_2 pressure, validate purge valves, leak tests
Moisture contamination (all gases)	Corrosion, reduced heat transfer	Dryer failure, gasket leaks, humid intake air	Dew point trending, corrosion coupons	Cooler fouling, winding degradation	Service dryers, fix leaks, dehumidify intake
CO_2 acidification in oil	Carbonic acid formation, foaming	CO_2 purge with wet oil, long CO_2 holds	Oil TAN, water ppm, foam tests, visual	Seal-oil level instability, corrosion	Dry CO_2, strong demulsifier, controlled degas time
Hydrogen loss by leakage	Rating drop, air ingress risk	Shaft-seal wear, flange leaks	Make-up rate, pressure decay tests	Frequent purges, safety events	Seal overhaul, gasket program, helium leak-check
Particulate fouling of coolers	ΔP rise, heat-rate loss	Inadequate filtration, dirty intake	DP across coolers, visual, IR thermography	Hot spots, forced derate	Filter upgrades, cleaning schedule, upstream housekeeping
Oil aeration and foam	Gas dissolution and breakout	Mode changes, CO_2 to air transitions, high return turbulence	Air-release tests, foam tendency, sight-glass	Pump cavitation, level trips	Proper antifoam, calm returns, degas coils
Bearing-oil oxidation (air systems)	Viscosity rise, varnish	High O_2, high temp, long drains	TAN, RPVOT, MPC patch	Deposits, overheating	Better oxidation package, cooler maintenance, shorter drains
Hydrogen analyzer drift	Hidden purity risk	Sensor aging, contamination	Cross-check with lab bottle tests	False security, unsafe operation	Calibration schedule, dual sensors, bump tests

Additives and Refining

Refine the **oil** package rather than the gas. In seal-oil reservoirs, prefer Group II or III base oils with low volatility and strong air-release; keep silicone antifoam lean and stable. In bearing oils, maintain oxidation resistance and foam control to handle any hydrogen entrainment episodes. For older systems with copper alloys in coolers, use a metal deactivator in the bearing oil if copper pickup appears. Align oil selection with OEM elastomers and shaft seal materials.

Helium

Helium combines very high thermal conductivity with chemical inertness and non-flammability. Density is higher than hydrogen but far below air; windage is modest. Helium does not support combustion or corrosion reactions and is widely used for leak-detection because of its very low background in air. In rotating machinery it can serve as a temporary or specialty coolant, although cost and supply logistics limit routine use in power stations.

Additives and Processing

No additives are used. Processing mirrors hydrogen without the flammability concerns: secure supply, dryer service to maintain low dew point, and high-efficiency particulate filtration to protect narrow gas passages. Validate analyzer ranges for helium if purity trending is required, or rely on pressure-rate and thermal performance indicators.

Application Considerations

Helium is chosen when absolute inertness is required or during commissioning and maintenance when hydrogen service is not yet established. It is also used to chase small leaks. Expect slightly lower cooling performance versus hydrogen at equal pressure because thermal conductivity and specific heat are a bit lower. Operating costs are higher due to gas cost and makeup losses.

Additives and Refining

None in the gas phase. For associated lubricants, the same air-release and foam control concerns apply, although helium solubility in oil is lower than hydrogen. Use dry-gas mechanical seals and high-integrity gaskets to minimize expensive losses.

Carbon Dioxide (CO_2)

CO_2 is non-flammable, chemically stable, and much denser than air. In hydrogen generator service its principal role is the **intermediate purge gas** that safely displaces air before hydrogen admission and safely displaces hydrogen before opening to atmosphere. As a coolant, gaseous CO_2 has low thermal conductivity and adds windage due to its density; it is not used as the operating gas in large hydrogen-class generators. In other industries CO_2 is useful for spot cooling and, in closed cycles, as a working fluid at higher pressures.

Additives and Processing

No additives are used. Processing stresses dryness and purity since CO_2 plus water forms carbonic acid that can attack copper and steel surfaces, especially in stagnant zones and seal-oil return paths. Use desiccant dryers and keep dew point comfortably below ambient. Verify pressure regulators and purge valves seal tightly to avoid air dilution.

Application Considerations

In generator duty the critical factor is purge sequence integrity. CO_2 is introduced to push out air prior to hydrogen admission, and later to push hydrogen to a safe receiver prior to opening to atmosphere. CO_2 can dissolve in seal oil and later evolve, causing foaming and level instability if control valves or reservoirs are not designed for controlled degassing. Plan sufficient time and venting for CO_2-offgassing from oil before returning to service on air or hydrogen.

Additives and Refining

Focus on **seal-oil** demulsifiers and antifoam choice to manage CO_2 breakout during transitions. Keep neutralization capacity in the oil sufficient to buffer any carbonic acid formation traced to moisture ingress during long CO_2 holds.

Air

Air is the default coolant for open or TEWAC machines. It is inexpensive, readily available, and familiar, but offers low thermal conductivity with higher density than hydrogen or helium, which increases windage losses. Air contains oxygen and moisture that drive corrosion, oxidation of oils, and deposition if filtration is inadequate. For many medium generators and motors, carefully managed air cooling is entirely adequate.

Additives and Processing

The gas itself is unmodified. Processing means **conditioning**: particulate filtration, mist removal, optional dehumidification, and positive-pressure management in TEWAC systems to keep contaminants out. Bearing and seal oils carry the additive load: detergency for deposit control, oxidation inhibitors, rust inhibitors, and robust antifoam for aeration control.

Application Considerations

Air wins on simplicity and cost, loses on efficiency and maximum rating. It is best for smaller machines, moderate ambient temperatures, and sites where hydrogen systems would be disproportionate. Attention to inlet filtration, cooler hygiene, and moisture control preserves heat-rate and extends winding life. TEWAC designs shift heat to a secondary water loop, reducing room heat and permitting tighter particulate control.

Additives and Refining

Tailor bearing-oil antioxidants to ambient and load, increase rust inhibitor emphasis in coastal or humid climates, and adjust filter change intervals based on differential pressure trends and deposit audits. Where copper corrosion is observed in coolers, consider a metal deactivator in the oil and review air humidity control.

Values are indicative for selection and control planning rather than spec limits.

Field Inspection and Sampling Protocol for Hydrogen, Helium, CO_2, and Air

Safety and Readiness

Lockout and tagout electrical drives and auxiliaries as required. Verify gas-hazard controls: area monitors, analyzer operation, purge interlocks. For hydrogen work, confirm the Air↔CO_2↔H_2 sequence and instrument availability. Treat helium work as confined-space inerting risk. Use baked-dry bottles and metal fittings for moisture-critical samples.

Gas-Side Checks (All Media)

- **Purity**: online analyzer for hydrogen; bottle grab sample for laboratory verification as needed.
- **Oxygen**: continuous for hydrogen spaces; spot-check during purges.
- **Dew point**: at enclosure and dryer discharge. Target a comfortable margin below coldest surface.
- **Pressure and leakage**: pressure-decay or helium mass-spec survey during outages; weekly makeup rate review.
- **Cooler health**: differential pressure, approach temperature, IR survey of casings and headers.

Seal-Oil and Bearing-Oil Sampling

- **Points**: main seal-oil reservoir, return header before degas, bearing oil reservoir, and a post-filter bearing supply point.
- **Method**: sample midstream from live, clean valves after flushing several line volumes. Avoid aeration. Use dry bottles; for dissolved gas or moisture, fill completely and cap immediately.
- **Frequency**: monthly in hydrogen units for seal oil; quarterly for bearing oil in steady service; within 24 to 72 hours after any purge event or gas upset.

Field Tests

- Visual clarity, color, and odor.
- Temperature, system pressures, hydrogen purity, oxygen, and dew point at time of draw.
- For oils: quick water screen, foam tendency check if available, and conductivity for contamination flags.

Laboratory Tests: Seal-Oil and Bearing-Oil

- **Viscosity** at 40 and 100 C; **V100/V40 trend** for thickening.
- **Water ppm** by Karl Fischer; keep very low in hydrogen service.

- **TAN** for oxidation and CO_2 acid formation during purge cycles.
- **Air release and foam tendency** to verify antifoam performance after CO_2 transitions.
- **Particle count** and **patch microscopy** to find rust, elastomer crumbs, or seal wear.
- **Metals by ICP** for wear and corrosion fingerprints.
- **RULER/voltammetry** or **RPVOT** for antioxidant health on longer drains.

Control Limits and Actions

- **Hydrogen purity**: hold at or above the OEM minimum, commonly 98 to 99.5%. If purity falls or O_2 rises, initiate controlled purge with CO_2, investigate seal-oil differential, and leak-survey.
- **Dew point**: if dew point rises toward ambient, service dryers and inspect ingress points; do not ignore a trending increase even if still "acceptable."
- **Seal-oil**: water above baseline or TAN climbing after CO_2 use calls for dewatering, partial drain-and-fill, and a review of purge dwell time. Persistent foaming requires antifoam adjustment and calmer return hydraulics.
- **Cooler performance**: rising DP or approach temperature calls for cleaning, filter changes, and upstream housekeeping improvements.

Documentation
Maintain control charts for purity, O_2, dew point, pressure makeup rate, cooler DP, and all oil key performance indicators (KPIs). Tie each variance to a dated corrective action. Keep a purge log that records start and end concentrations for each stage: Air→CO_2→H_2 and H_2→CO_2→Air.

Program Refinements That Pay Off

- Set seal-oil differential pressure targets that exceed hydrogen pressure by a fixed margin under all loads, and alarm on excursions.
- Calibrate hydrogen analyzers and dew point sensors on a strict interval, with a second method available for spot verification.
- During CO_2 purges, schedule an oil de-gas period before switching to air or hydrogen, and sample TAN and water within 24 to 72 hours.
- Add a routine helium leak-check during planned outages. Even small leaks erode performance and raise safety risk.
- In air-cooled plants, invest in better inlet filtration and humidity control. It is the cheapest life-extension lever for windings and oils.

NUCLEAR REACTOR COOLANTS: LIGHT WATER, HEAVY WATER, SODIUM, HELIUM, AND MOLTEN SALTS

The choice of coolant is one of the most defining design decisions in nuclear reactor engineering. Beyond simply removing heat from the core, the coolant governs neutron moderation, materials compatibility, chemistry control, safety margins, and overall plant efficiency. Over decades of nuclear development, five major coolant families have emerged, light water, heavy water, liquid sodium, helium gas, and molten salts, each offering distinct thermophysical and nuclear properties that shape reactor design and operation (see Tables 5.12 and 5.13).

Light and heavy water systems dominate today's commercial fleet, benefiting from established operating experience, mature chemistry controls, and robust supply chains. Liquid sodium has long been favored in fast breeder and advanced reactors for its unmatched thermal conductivity and low operating pressure, though it demands stringent containment and oxygen control. Helium-cooled reactors enable very high outlet temperatures for Brayton-cycle efficiency, while avoiding chemical reactivity, at the expense of challenging gas dynamics. Molten salts, whether fluoride- or chloride-based, represent a frontier option with high heat capacity, passive safety features, and chemical complexity tied to redox control.

Unlike conventional industrial coolants, nuclear reactor coolants require continuous chemistry management, impurity control, and system-specific conditioning to maintain both thermal and nuclear performance. Additives are minimal or absent; instead, operational chemistry strategies, boric acid and hydrogen dosing in light water, cold traps in sodium, getter beds in helium, and redox buffering in salts, serve as functional "additives" to stabilize coolant properties and protect materials.

LIGHT WATER (H_2O)

Light-water reactors rely on excellent heat capacity, high latent heat, and established materials experience. In pressurized water reactors (PWRs) the primary circuit operates as subcooled liquid at high pressure to avoid bulk boiling, while boiling water reactors (BWRs) boil in-core. Radiolysis generates H_2 and O_2; chemistry control suppresses oxidizing species to protect alloys. Chloride and sulfate at trace levels can drive stress corrosion cracking under radiation and high temperature.

Additives and Processing

PWRs use boric acid as a soluble neutron absorber with lithium hydroxide to control pH at temperature. Dissolved hydrogen is maintained to recombine oxidizing radicals from radiolysis. Cation conductivity, degassed conductivity, chloride, sulfate, and dissolved oxygen are tightly controlled. BWRs use hydrogen water chemistry or noble metal chemistry to shift electrochemical potential negative and mitigate intergranular stress corrosion cracking (IGSCC), with strict oxygen and halide limits. Demineralizers and letdown cleanup control crud and corrosion products.

Application Considerations

Light water provides the most mature supply chain and analytics. Economics are favorable, but materials must tolerate irradiation-assisted corrosion and, in BWRs, two-phase flow with crud transport. Feedwater and condensate polishing prevent impurity ingress. Transients such as shutdown chemistry and startup oxygen spikes need active management.

TABLE 5.12
Performance and Lifecycle Comparison Nuclear Reactor Coolants: Light Water, Heavy Water, Sodium, Helium, and Molten Salts

Attribute (Typical Order-of-Magnitude)	Light Water	Heavy Water	Sodium	Helium	Molten salts
Coolant phase	Liquid, subcooled or boiling	Liquid, subcooled	Liquid metal	Gas	Ionic liquid
Operating pressure	High (PWR ~15 MPa)	High	Low	High	Low
Outlet temperature range	~290 to 330 °C (LWR)	Similar to LWR	~450 to 550 °C	~750 to 950 °C (HTGR)	~550 to 720 °C (concept dependent)
Thermal conductivity	Moderate	Moderate	Very high	Low vs liquids	Moderate
Heat capacity per mass	High	High	Moderate	Low	Moderate to high
Neutronic effect	Strong moderator	Very strong moderator	Minimal moderation	Minimal	Minimal to moderate depending on salt
Chemistry program complexity	High but mature	High plus tritium	High, sodium-specific	Medium, gas purity control	High, redox control
Typical service life of coolant	Indefinite with cleanup	Indefinite with cleanup	Indefinite with purification	Indefinite with purification	Indefinite with purification and make-up
Primary hazards	SCC, crud, radiolysis gases	Tritium handling	Na–air/Na–water reactions	Air ingress, dust	Oxidation, moisture, freeze events

TABLE 5.13
Failure Modes and Degradation Pathways Nuclear Reactor Coolants: Light Water, Heavy Water, Sodium, Helium, and Molten Salts

Mode	Mechanism	Triggers	Diagnostics	Consequences	Mitigation
PWR corrosion and SCC	Halides, low pH at temperature, radiation	Chloride/sulfate ingress, poor hydrogen control	Conductivity, chloride ppb, ECP, ICP metals	Leak, dose rise, component replacement	Tight chemistry limits, hydrogen control, zinc injection, materials upgrades
BWR IGSCC	Oxidizing chemistry raises ECP	Oxygen spikes, poor HWC/NMCA	ECP probes, DO, sulfate	Crack initiation in sensitized austenitics	Hydrogen or noble metal chemistry, impurity control
Heavy water tritium permeation	T diffuses through metals	Neutron capture on D	Tritium monitors, smear surveys	Dose, environmental release burden	Permeation barriers, detritiation systems, seal upgrades
Sodium–water reaction	Rapid exothermic reaction forms NaOH, H_2	Leak in steam generator	Hydrogen in secondary, acoustic monitors	Tube rupture, fire risk	Double-wall HX, fast detection, dump and isolation systems
Sodium oxygen pickup	Na_2O formation, corrosion, plugging	Air ingress, wet cover gas	Plugging meter, oxygen analyzers	Flow restriction, corrosion	Cold traps, tight cover gas control
Helium air ingress	Oxidation of graphite and alloys	Leak, depressurization	O_2/H_2O analyzers, EPR on graphite	Graphite oxidation, dust, dose	Leak prevention, getters, emergency purge
Graphite dust transport	Erosion, plate-out	Flow transients, abrasion	Particle monitors, filter DP	HX fouling, dose redistribution	Filtration, surface treatments, flow control
Salt oxidation	Alloy dissolution, plate-out	Moisture or O_2 ingress, redox drift	Redox probes, corrosion coupons	Thinning, HX fouling	Strict dryness, redox buffering, inert gas cover
Salt freeze and thaw	Blockage, thermal stress	Loss of heat, cold spots	Temperature mapping, ΔP	Plugging, cracking	Heat tracing, freeze plug design, controlled thaw
Chloride-salt corrosion	Alloy attack via chlorides	High redox potential, impurities	Coupon mass loss, SEM/EDS	Rapid wall loss	Ultradry handling, alternative alloys/coatings

Additives and Refining

Refine lithium dosing to minimize Alloy 600 and 690 corrosion while maintaining pH at temperature. Adjust hydrogen injection to the minimum required to hold electrochemical corrosion potential (ECP) targets. For BWRs, select noble metal on-line or during outage based on dose and ECP targets. Track zinc injection programs to lower dose rates by reducing Co incorporation.

HEAVY WATER (D_2O)

Heavy water offers exceptional neutron economy for natural-uranium or low-enrichment designs. Thermal-hydraulic properties are close to light water with slightly lower thermal conductivity and higher viscosity. Tritium forms via neutron capture on deuterium and requires collection and handling infrastructure.

Additives and Processing

Chemistry mirrors light-water practice but with isotopic purity control. Oxygen removal and tight halide control remain essential. Ion exchange resins and vacuum degassing circuits are selected to minimize D_2O loss and preserve isotopic enrichment. Leak detection and recovery systems are integral.

Application Considerations

Heavy water enables specific core physics advantages but increases capital and O&M due to tritium management, isotopic makeup, and strict leakage control. Materials compatibility aligns with LWR alloys, but tritium permeation through metals and polymers requires barrier strategies.

Additives and Refining

Optimize dissolved hydrogen and pH control to balance corrosion mitigation against tritium production and transport. Maintain deuteration levels in any added chemicals to avoid isotopic dilution.

LIQUID SODIUM (NA)

Sodium provides very high thermal conductivity, low neutron moderation, and operation at near-atmospheric pressure for high core outlet temperatures. It reacts vigorously with water and air, forming NaOH and Na_2O with heat and hydrogen evolution. Impurities such as oxygen and hydrogen determine corrosion and plugging behavior.

Additives and Processing

No additives are dissolved in sodium. Purification relies on cold traps and getters to remove oxygen as Na_2O and hydrogen as NaH. Plugging meters infer oxygen by measuring solid formation temperature. Inert gas cover systems, vacuum handling, and high-integrity welded boundaries limit contamination. All sampling and maintenance occur in dry, oxygen-free environments.

Application Considerations

Sodium enables compact heat exchangers, high efficiency cycles, and fast spectra. The plant must segregate sodium from water or steam using double-wall or leak detection to prevent sodium–water reactions. Instrumentation and maintenance require sodium-wetted expertise and specialized decontamination.

Additives and Refining

Refine cold-trap operating temperature to balance impurity removal and sodium loss. Maintain argon cover gas dryness to single-digit ppm H_2O and O_2. Specify compatible steels and nickel alloys with proven sodium corrosion data; avoid dissimilar-metal crevices.

HELIUM (GAS-COOLED, HTGR, AND HTR)

Helium is inert, single-phase across the plant, and supports very high outlet temperatures without corrosion products from the coolant itself. Thermal conductivity is high for a gas but far lower than liquids; heat transfer relies on high mass flow, pressure, and efficient core geometry. Graphite moderator and reflectors dominate materials interactions.

Additives and Processing

No chemical additives are used in helium. Gas chemistry control is impurity control: ppb to ppm limits for H_2O, O_2, CO, CO_2, CH_4, N_2. Getters and catalytic beds maintain a mildly reducing environment to avoid graphite oxidation and alloy oxidation. High-efficiency filtration manages graphite dust.

Application Considerations

Helium supports Brayton or indirect cycles with high efficiency. Plant design must control leakage because helium is costly and permeates seals. Graphite dust transport and plate-out on heat exchangers must be monitored. Transient air ingress events are design-basis scenarios.

Additives and Refining

Tune getter beds to maintain CO/CO_2 ratios that keep graphite redox-neutral. Specify seal technologies and coatings that limit permeation. Maintain moisture well below graphite oxidation thresholds.

MOLTEN SALTS

Fluoride salts such as FLiBe and FLiNaK and chloride salts for fast-spectrum concepts offer low vapor pressure, wide liquid range, and strong heat capacity at elevated temperatures. They are ionic, conductive, and chemically active with oxides and moisture. Redox state governs corrosion: oxidizing salts dissolve alloy constituents while properly reduced salts are compatible with nickel-based alloys and certain graphite components.

Additives and Processing

No traditional additives are used. Chemistry is controlled by redox buffering, typically using metallic beryllium in FLiBe or controlled HF/H_2 sparging to set potential. Purification removes moisture, oxygen, and metallic impurities by vacuum drying, HF treatment, and cold trapping. All handling uses inert cover gas and strictly dried systems.

Application Considerations

Molten salts enable high outlet temperatures and passive safety via large thermal capacity and freeze plugs. They require compatible structural alloys, leak-tight welded systems, and heaters to prevent freezing. Fission product management in MSRs uses on-line processing; for solid-fuel salt-cooled reactors, salt purification and plate-out control are central.

Additives and Refining

Maintain redox within a narrow band verified by electrochemical probes or metal coupon response. Select alloys such as Hastelloy-N derivatives for fluoride salts and evaluate advanced alloys or coatings for chloride systems. Minimize oxygen and moisture ingress by double seals and inerted maintenance.

Field Inspection and Sampling Protocol for Nuclear Reactor Coolants: Light Water, Heavy Water, Sodium, Helium, and Molten Salts

Safety and Cleanliness

Define radiological boundaries and personal protective equipment (PPE). All sampling hardware must be scrubbed of oxygen and moisture for sodium, molten salts, and helium circuits. For water circuits, follow primary/secondary boundary procedures. Use dedicated lines with purge capability and capture representative, flowing samples. For any ancillary lubricating oils connected to coolant pumps or circulators, sample separately using clean, dry bottles.

Coolant Sampling: Light and Heavy Water

- **Points**: primary coolant letdown, reactor vessel head space or recirc header, feedwater/condensate polishing outlets, waste streams for balance closure.
- **Field checks**: temperature, pressure, dissolved oxygen, cation and degassed conductivity, pH at temperature proxy, radiochemistry quick screen.
- **Lab slate**: chloride and sulfate at sub-ppb, lithium and boron (PWR), zinc if injected, dissolved hydrogen (PWR) or ECP trend (BWR), total organic carbon, corrosion product metals by ICP, radionuclide speciation for dose control.
- **Frequency**: continuous DO and conductivity, hourly during startups, daily to weekly during steady-state per plant chemistry specification.

Coolant Sampling: Sodium

- **Points**: hot leg loop with isokinetic probe, cold-trap discharge, cover gas.
- **Field checks**: plugging temperature, cover gas dew point.
- **Lab slate**: oxygen ppm, hydrogen ppm, carbon and nitrogen as required, trace metals, particle morphology after controlled solidification. All samples handled in sealed, inerted ampoules.
- **Frequency**: per shift for plugging meter, daily to weekly for oxygen and hydrogen, after any event or maintenance.

Coolant Sampling: Helium

- **Points**: main loop header, return from core, getter outlet.
- **Field checks**: moisture and oxygen by on-line analyzers, total hydrocarbon, CO/CO_2 ratio, particulate monitoring for graphite dust.
- **Lab slate**: confirmatory GC for impurities, dust composition and size distribution, plate-out coupons.
- **Frequency**: continuous analyzers with daily review, lab confirmation weekly or after transients.

Coolant Sampling: Molten Salts

- **Points**: bypass loop with electrochemical cell, drain tank, and heat exchanger outlet.
- **Field checks**: temperature, ΔP across filters, in-situ redox potential.
- **Lab slate**: redox by electrochemistry, moisture by Karl Fischer on extracted aliquots prepared in inert glovebox, dissolved metals by ICP, particulate content, alloy coupon mass change and surface analysis.
- **Frequency**: redox continuous or hourly during transients, full slate weekly to monthly, immediately after any maintenance exposing salt to air.

Ancillary Oil Sampling for Coolant Pumps, Circulators, and Seal Systems

- **Points**: bearing oil reservoir, return line pre-degas, post-filter feed.
- **Field checks**: visual, temperature, vibration context.
- **Lab slate**: viscosity, water by KF, TAN, particle count and patch microscopy, metals by ICP, foam and air release. For sodium plants, screen for Na or Na_2O ingress; for helium plants, watch air entrainment indicators due to low-density gas.
- **Frequency**: monthly routine, within 24 to 72 hours after coolant-side chemistry upsets or purge events.

Control Limits and Actions

- **Light/heavy water**: if chloride or sulfate exceed action levels, initiate power-dependent limits, isolate sources, and increase letdown cleanup. If dissolved oxygen rises, correct deaeration and adjust hydrogen injection or HWC.

- **Sodium**: if plugging temperature rises or oxygen exceeds limit, increase cold-trap flow, locate ingress, verify cover-gas purification. Hydrogen rise suggests steam leak; perform leak localization and isolate the affected circuit.
- **Helium**: if O_2 or H_2O trend upward, check leak-tightness, swap or regenerate getters, and adjust purge. Dust increase triggers filtration check and potential flow-path inspection.
- **Molten salts**: if redox drifts oxidizing, add reductant or run HF/H_2 sparge within approved windows. If moisture detected, quarantine loop, redry salt, and investigate ingress.
- **Oils**: water or TAN rise after purge or maintenance requires dewatering, partial change, and review of seals and venting practices.

Documentation

Maintain control charts for each chemistry sentinel: conductivity, DO, DH_2 or ECP, plugging temperature, impurity ppm, redox potential, dust count, plus oil KPIs. Tie alarms to immediate corrective actions and keep outage debriefs with coupon and component inspections linked to chemistry history.

Program Refinements That Pay Off

- Couple chemistry limits to power and temperature windows rather than single fixed numbers to avoid overcorrection at low load.
- Keep a validated chain for impurity tracing from warehouse to core: every chemical addition is isotopically and impurity certified for LWR and D_2O service.
- For sodium, operate cold traps at the lowest temperature that achieves impurity removal without excessive sodium loss; trend plugging temperature against outlet ΔT for early warning.
- For helium, operate getters to a CO/CO_2 setpoint that stabilizes graphite while minimizing alloy oxidation; verify by coupon testing.
- For molten salts, treat redox as the master variable. Calibrate probes against coupon response, not only open-circuit potential, and schedule periodic alloy autopsies to keep the model honest.

SOLAR THERMAL FLUIDS: MOLTEN SALTS AND THERMAL OILS

Concentrated solar power (CSP) plants depend on specialized heat transfer and storage fluids to move and retain thermal energy captured from solar collectors. The two dominant fluid families; molten salts and thermal oils offer contrasting performance envelopes that shape system design, operating strategies, and safety protocols. Selection between them involves trade-offs between temperature capability, stability, freeze protection, fire risk, and long-term degradation pathways (see Tables 5.14 and 5.15).

Molten nitrate salts are widely used in tower and trough-based CSP systems for their high thermal capacity, stability up to ~565 °C, and ability to provide multi-hour thermal storage. Their non-flammability and low volatility make them inherently safe, but their high freezing points and corrosivity at elevated temperature demand rigorous impurity control, freeze protection, and material selection. Advanced chloride and

TABLE 5.14
Performance and Lifecycle Comparison of Solar Thermal Fluids

Attribute	Molten Salts (Nitrate)	Molten Salts (Adv. Chloride/Fluoride)	Thermal Oils (Synthetic Aromatic)	Mineral-Based Thermal Oils
Liquid range	~220 °C to 565 °C	~300 °C to 700 °C+	~12 °C to 390 °C	~-20 °C to 300 °C
Heat capacity (kJ/kg·K)	~1.5	~1.3–1.6	~2.0	~2.0
Thermal conductivity (W/m·K)	~0.5–0.6	~0.5–0.7	~0.1–0.15	~0.1–0.15
Vapor pressure	Negligible	Negligible	Moderate above 300 °C	Low to moderate
Flammability	Non-flammable	Non-flammable	Flammable	Flammable
Freeze protection	Requires active heating	Requires active heating	Not required	Not required
Degradation products	Nitrite, NOx, solid precipitates	Metal chlorides, oxides	Low/high boilers, acids, carbon	Same as synthetics but faster oxidation
Materials compatibility	Mild steel, SS	High-nickel alloys, ceramics	Carbon steel, SS	Carbon steel, SS
Typical service life	10–30 years with control	TBD (emerging)	5–10 years with control	3–5 years

TABLE 5.15
Failure Modes and Degradation Pathways for Solar Thermal Fluids: Molten Salts and Thermal Oils

Mode	Mechanism	Triggers	Diagnostics	Consequences	Mitigation
Salt freezing	Solidification in pipes/tanks	Loss of trace heating, cold ambient	Temp mapping, ΔP rise	Blockages, rupture on thaw	Redundant heat tracing, freeze plugs, heated drains
Salt decomposition	Nitrate→nitrite→oxide, NOx	Overheat >565 °C, contamination	Nitrite/nitrate ratio, off-gas NOx	Corrosion, reduced heat capacity	Temperature control, vent management
Salt corrosion	Alloy dissolution	Chloride ingress, oxidizing conditions	ICP metals, visual	Leak, wall thinning	Chloride monitoring, redox control
Oil oxidation	Oxygen ingress	Seal leaks, poor blanketing	TAN, insolubles, RPVOT, MPC patch	Varnish, fouling, flow restriction	Nitrogen blanket, AO top-treat
Oil thermal cracking	High film temperature	Low flow, fouled HX	GC for low/high boilers, viscosity	Volatile buildup, fouling	Maintain velocity, clean HX surfaces
Fouling	Insoluble deposition	Degradation products, contamination	Patch microscopy, DP	HX ΔT loss, blockage	Filtration, bleed-and-feed

fluoride salt formulations promise even higher temperature operation, extending efficiency potential, but require further advances in corrosion management.

Thermal oils, by contrast, are synthetic aromatic blends or refined mineral oils optimized for liquid-phase operation in trough and Fresnel plants. They offer low pour points and simple startup logistics, but are flammable and limited to ~390–400 °C before rapid oxidative breakdown occurs. Their degradation generates low boilers and high boilers that alter vapor pressure and viscosity, requiring continuous monitoring and periodic fluid management.

Unlike conventional industrial heat transfer systems, CSP fluids must remain stable over decades of cycling between day and night while maintaining strict safety margins. Additives are minimal, but practices such as nitrogen blanketing, impurity scrubbing, antioxidant monitoring, and bleed-and-feed rejuvenation act as functional equivalents to additive packages in extending service life.

Molten Salts (Nitrate-based and Advanced Formulations)

Commercial CSP plants most often use binary or ternary nitrate eutectics (e.g., 60% $NaNO_3$ / 40% KNO_3 or 53% KNO_3 / 40% $NaNO_3$ / 7% $Ca(NO_3)_2$). These mixtures are non-flammable, non-volatile, and thermally stable to ~565 °C for nitrates before decomposition to nitrites and oxides begins. They solidify at 120–220 °C, demanding heat tracing and careful freeze protection. Nitrate salts are oxidizing and mildly corrosive to carbon steels at high temperature; corrosion increases with impurity oxygen, moisture, and chloride contamination. Advanced CSP designs investigate chloride or fluoride salts for higher temperature operation (≥700 °C), but these bring greater corrosion risk.

Additives and Processing

Molten nitrate salts are run without traditional "additives" but with strict impurity control. Water, chlorides, and sulfates are kept to ppm levels. Some plants employ nitrite/nitrate ratio monitoring as an indicator of thermal decomposition and oxidation state. For high-temperature work, small amounts of oxide scavengers or corrosion inhibitors (e.g., MgO for chloride salts) may be added, but these require OEM approval. Initial commissioning includes a drying phase to drive off residual moisture before charging systems. Inert gas blanketing can limit air ingress when the salt is stagnant.

Application Considerations

Molten salts excel in CSP storage because of high heat capacity and ability to store energy for hours without significant losses. The main hazards are freezing (solidification) and decomposition at high temperature. Long piping runs require uniform heating and reliable insulation. Drain tanks, expansion vessels, and freeze protection procedures are critical. Material selection is often carbon steel for cost, with stainless steel in high-velocity or high-temperature regions.

Additives and Refining

Continuous monitoring of nitrite/nitrate ratio allows operators to identify decomposition trends and schedule salt replacement or rejuvenation. Chloride monitoring identifies contamination from leaks or maintenance. For chloride-based salts, redox control using metallic reductants (e.g., magnesium) is under study to suppress corrosivity.

THERMAL OILS (SYNTHETIC AROMATIC AND MINERAL-BASED)

Thermal oils in solar trough plants are usually synthetic aromatic blends (e.g., biphenyl/diphenyl oxide mixtures such as Dowtherm™ A, Therminol™ VP-1) with liquid range from ~12 °C (pour point) to 390–400 °C (film temperature limit). Mineral-oil based fluids are used in lower temperature systems. Aromatic synthetics have high thermal stability within their range but oxidize rapidly above the design limit. They are flammable, degrade to low boilers (increasing vapor pressure) and high boilers (increasing viscosity and fouling) over time.

Additives and Processing

Thermal oils are generally supplied with an antioxidant reserve. They require oxygen exclusion (nitrogen blanketing, tight seals) to slow oxidation. No additional corrosion inhibitors are typically needed because they are non-aqueous, but filtration is used to control carbonaceous fines and insolubles. Processing includes hot-oil circulation during commissioning to remove volatiles and moisture, and venting to purge low boilers.

Application Considerations

Thermal oils offer freeze protection without heating (low pour point) and ease of startup compared to salts. They have lower maximum temperature limits, which constrains efficiency. They are suited to parabolic trough and some linear Fresnel systems, and require large heat exchangers for coupling to steam cycles. Fire protection and vapor handling are important safety aspects.

Additives and Refining

Antioxidant monitoring via RULER or voltammetry allows proactive top-treating before depletion. Low-boiler removal via venting and high-boiler removal via bleed-and-feed helps maintain viscosity and heat transfer. Seal compatibility and elastomer upgrades may be required when switching oil types.

Field Inspection and Fluid Sampling Protocol for Solar Thermal Fluids: Molten Salts and Thermal Oils

Safety and Cleanliness

For salts, work under hot conditions with PPE for burns and oxidizers; avoid contact with moisture. For oils, manage fire risk, use inerting where possible. Sampling lines must be flushed and preheated (for salts) or purged (for oils) to obtain representative samples.

Molten Salts Sampling

- **Points**: main loop hot leg, cold leg, storage tank.
- **Field checks**: temperature, ΔP, visual clarity, solids content.
- **Lab slate**: nitrite/nitrate ratio, chlorides, sulfates, moisture (KF), ICP metals for corrosion, particle size/distribution, redox potential (advanced salts).
- **Frequency**: monthly for nitrite/nitrate, quarterly full slate; after any overheat or freeze incident.

Thermal Oils Sampling

- **Points**: main expansion tank, pump discharge, HX outlet.
- **Field checks**: temperature, odor, clarity, foam tendency.
- **Lab slate**: viscosity (40/100 °C), TAN, GC for low/high boilers, RPVOT or RULER for AO reserve, insolubles/patch microscopy, MPC varnish rating, water by KF, metals by ICP.
- **Frequency**: quarterly in steady state; monthly during commissioning or after excursions.

Control Limits and Actions

- **Salts**: nitrite/nitrate ratio drift >10% from baseline or chlorides above spec triggers partial drain and reblend. Moisture above ppm limit requires drying cycle.
- **Oils**: TAN rise >0.5 over baseline or AO depletion >50% triggers antioxidant top-treat; low-boilers >5% by mass triggers venting; high-boilers >10% triggers bleed-and-feed.

Documentation
Maintain trending for key KPIs: salt chemistry ratios, oil TAN, AO reserve, viscosity, ΔP across HX, and operating temperatures. Record all heater outages and excursions with corresponding chemistry changes.

EXHAUST GAS RECOVERY AND COMBINED HEAT AND POWER (CHP) LOOP FLUIDS

Exhaust gas recovery units (HRUs) and combined heat and power (CHP) plants rely on closed-loop fluids to capture waste heat from prime movers, turbines, engines, or boilers, and repurpose it for district heating, process steam, or additional power generation. These loops typically operate in the 80–200 °C range, where water-based fluids dominate due to their high heat capacity, favorable thermophysical properties, and low cost. However, the chemistry of these fluids requires careful management: oxygen ingress, dissolved solids, and glycol degradation products can all accelerate corrosion, scale formation, and heat exchanger fouling.

Water-Based Fluids (Treated Water, Water-Glycol Blends)

Water and water-glycol blends represent the two principal fluid families. Straight water delivers maximum thermal efficiency but must be chemically treated to prevent corrosion and scaling. Water-glycol systems, based on ethylene glycol (EG) or propylene glycol (PG), are widely applied in climates requiring freeze protection or where higher boiling margins are needed under pressure. While glycols introduce viscosity penalties and reduce heat transfer efficiency, they extend operating flexibility, particularly in CHP plants operating across seasonal load cycles. The tradeoff is the need to monitor degradation, as thermal stress breaks glycols into organic acids, raising TAN and promoting under-deposit and galvanic corrosion.

Additives and Processing

Additive packages are central to CHP and HRU loop reliability, functioning as the equivalent of a protective film and chemical stabilizer:

- **Corrosion inhibitors**: Nitrites or molybdates protect ferrous alloys; azoles (e.g., tolyltriazole, benzotriazole) safeguard copper and brass; phosphates or silicates provide film-forming protection in select designs.
- **pH buffers**: Maintain loop chemistry between 8.5 and 10.0 at operating temperature, balancing corrosion protection with material compatibility.
- **Biocides**: Used during commissioning or lay-up to suppress microbial growth in stagnant water–glycol blends.
- **Make-up water control**: Demineralized or softened water avoids scale-forming calcium and magnesium salts.
- **Closed-loop design**: Pressurization and expansion vessels limit oxygen ingress, reducing inhibitor consumption and corrosion risk.

Application Considerations

Water-based CHP loops combine high thermal efficiency with relatively low cost, making them attractive for utility and industrial cogeneration. However, performance depends on maintaining clean heat exchanger surfaces: in HRUs, soot blowers and exhaust flow management are essential to minimize fouling. Where CHP heat recovery water also feeds boilers, strict condensate return quality control is necessary to prevent carryover of oils, metals, or treatment chemicals that could upset boiler chemistry.

Additives and Refining Practices

Loop chemistry is not static, it must be tuned for metallurgy, regulatory constraints, and local water conditions:

- **Aluminum-rich circuits**: Require silicate stabilization or organic acid inhibitors to prevent localized attack.
- **Hard water regions**: Benefit from side-stream filtration, ion exchange, or water softening to manage scale precursors.

- **Inhibitor balance**: Nitrites, molybdates, and azoles are replenished according to field testing; phosphates must be dosed carefully to avoid spalling or sludge deposition.
- **Oxidation control**: Regular venting and degassing, coupled with oxygen scavengers where allowed, extend inhibitor life.

Exhaust gas recovery and CHP loop fluids occupy a critical role in modern energy efficiency strategies, allowing waste heat to be captured and redeployed at high utilization factors. Their success hinges not just on the thermophysical properties of water or glycols but on the chemical stewardship of the loop. Proper inhibitor balance, oxygen exclusion, impurity control, and heat exchanger maintenance together determine whether a CHP or HRU system achieves decades of reliable operation or succumbs to corrosion, fouling, and premature failure.

THERMAL OILS (SYNTHETIC OR MINERAL-BASED) IN EXHAUST GAS HRUS AND CHP

Thermal oils are used in high-temperature exhaust heat recovery, especially in ORC preheaters and direct hot-oil HRUs, to transfer exhaust energy into secondary loops. Typical aromatic synthetics (biphenyl/diphenyl oxide blends) operate up to ~400 °C film temperature. Mineral-based paraffinic oils are used in lower-temperature loops (<300 °C). Oils degrade thermally into low-boiling fractions (raising vapor pressure and stripping pump NPSH) and high-boiling fractions (sludge, fouling), and oxidize in the presence of oxygen, forming acids and insolubles (see Tables 5.16 and 5.17).

TABLE 5.16
Performance and Lifecycle Comparison for Thermal Oils

Attribute	Water-Based CHP Fluid	Thermal Oil	ORC Working Fluid
Operating range	~0–200 °C (pressurized to prevent boiling)	~–20 to 390–400 °C (synthetic aromatics)	–20 to ~300 °C depending on fluid
Heat capacity (kJ/kg·K)	~4.0 (water) / ~3.4 (glycol mix)	~2.0	~1.2–2.0
Thermal conductivity (W/m·K)	~0.6 (water) / ~0.4 (glycol)	~0.1–0.15	~0.07–0.12
Flammability	Non-flammable (water/glycol)	Flammable	Flammable or non-flammable depending on fluid
Freeze protection	Requires glycol or heating	Good (low pour point)	Good
Degradation	Corrosion (metals), glycol oxidation	Thermal cracking, oxidation	Thermal decomposition
Typical service life	5–15 years with control	5–10 years	5–15 years with tight system

TABLE 5.17
Failure Modes and Degradation Pathways for Thermal Oils

Mode	Mechanism	Triggers	Diagnostics	Consequences	Mitigation
Corrosion (water loops)	Metal loss, leaks	Oxygen ingress, low inhibitor	ICP metals, pH, nitrite/molybdate	HX leaks, pump damage	Maintain inhibitors, eliminate O_2
Scaling (water loops)	Deposition, ΔT loss	Hard makeup water	Hardness, silica	Hot spots, fouling	Demineralize makeup, dispersants
Glycol oxidation	Acid formation	Overheat, O_2 ingress	TAN, glycolate/formate by IC	Corrosion, deposits	Control temp, maintain AO
Oil oxidation	Acid, varnish	O_2 ingress	TAN, MPC patch, AO by RULER	Fouling, pump failure	Nitrogen blanket, AO top-treat
Oil cracking	Low boilers, high boilers	Film temp above limit	GC analysis, viscosity change	NPSH issues, fouling	Maintain flow, clean HX
ORC fluid breakdown	Molecular decomposition	Overheat, contamination	GC, FTIR, TAN	Efficiency loss, corrosion	Temp control, moisture exclusion

Additives and Processing

Thermal oils are shipped with antioxidants to resist oxidative degradation. Nitrogen blanketing or sealed expansion tanks limit oxygen exposure. Inline filtration (typically 25–50 μm) removes carbonaceous fines; finer polishing may be used in varnish-prone systems. Degassing or vent condensers strip low boilers.

Application Considerations

Hot-oil loops enable stable delivery of exhaust heat to processes or ORC evaporators without risk of freezing and with good temperature control. Oils require fire-safe design, proper venting, and thermal expansion allowances. HX design must limit film temperature to avoid cracking. Pump seal cooling and alignment are critical to avoid leaks.

Additives and Refining

Maintain antioxidant levels via periodic testing; top-treat if reserves drop below ~50% of new-fluid levels. Bleed-and-feed replenishes degraded oil and controls high-boiler buildup. Review seal elastomer compatibility before changing oil chemistry or brand.

ORGANIC RANKINE CYCLE (ORC) WORKING FLUIDS IN CHP

In CHP plants, ORCs recover exhaust heat using organic refrigerants such as R245fa, R1233zd(E), cyclopentane, or toluene. These low-boiling fluids evaporate at modest exhaust HX temperatures and expand in a turbine to produce power. ORC fluids must have low ozone-depletion potential (ODP) and often low global warming potential (GWP). They are chemically stable in their designed temperature range but degrade under overheating or contamination.

Additives and Processing

Most ORC fluids are used neat, with no additives. Moisture and non-condensables are removed by dryers and degassing systems. Lubricants in ORC turbine bearings or generators are separate from the working fluid but must be compatible in case of seal leakage.

Application Considerations

Working fluid selection balances boiling point, thermal stability, turbine efficiency, and environmental constraints. Heat exchangers must prevent thermal decomposition, film temperatures are kept below fluid limits. System tightness is critical to prevent losses and air ingress.

Additives and Refining

No direct additives; refining is moisture and NCG control plus routine fluid analysis for decomposition products and contaminants.

Field Inspection and Fluid Sampling Protocol for Thermal Oils

Safety and Cleanliness

- Water/glycol: avoid burns; depressurize before sampling.
- Thermal oil: sample hot, avoid aeration; PPE for high-temp hydrocarbons.
- ORC: handle per refrigerant safety standards; sample under OEM procedures to avoid vent loss.

Water-Based Fluids

- **Points**: main loop, expansion tank, HX outlet.
- **Field checks**: pH, conductivity, inhibitor residuals.
- **Lab tests**: pH, reserve alkalinity, inhibitor residual, glycol %, TAN, metals (Fe, Cu, Al, Zn), anions (chloride, sulfate), particle count.
- **Frequency**: quarterly; monthly during commissioning.

Thermal Oils

- **Points**: expansion tank, pump discharge, HX outlet.
- **Field checks**: clarity, odor, foam tendency.
- **Lab tests**: viscosity, TAN, GC for low/high boilers, MPC varnish, RULER antioxidant, insolubles, particle count.
- **Frequency**: quarterly; after any overheating or contamination event.

ORC Fluids

- **Points**: liquid receiver, evaporator outlet.
- **Field checks**: pressure-temperature correlation (purity check), sight-glass clarity.
- **Lab tests**: GC for composition, moisture ppm, acid number, particulate load.
- **Frequency**: semi-annual to annual; after maintenance or excursions.

Control Limits and Actions

- Water-based: restore inhibitors if residuals < min; remove O_2 ingress; flush if contamination > action levels.
- Thermal oil: top-treat AO <50%; bleed-and-feed if high-boilers >10% or viscosity change >10%; vent low-boilers >5%.
- ORC: reclaim or replace if purity <95% or decomposition products > spec.

Section III

Fluid Selection and Compatibility Engineering

6 Fluid Property Selection Criteria

FLUID SELECTION AND COMPATIBILITY ENGINEERING

Effective selection starts with the duty profile: thermal load and gradients, electrical duty or insulation gaps, mechanical shear and pump regime, allowable pressure, ambient minima and maxima, and the materials stack that will see the fluid. A correct fluid solves a coupled problem: it must move heat with predictable film temperatures, maintain or enhance insulation margins where voltages are present, lubricate where mixed or boundary friction occurs, and remain chemically quiet toward metals, solders, coatings, elastomers, and plastics for the full maintenance interval. The workflow is simple but strict. Define the envelope. Screen by physics first, then by chemistry, then by operability. Physics: density, viscosity curve, thermal conductivity and specific heat, dielectric constants, vapor pressure, and flash or autoignition limits. Chemistry: oxidation stability, hydrolysis resistance, additive interactions with copper, aluminum, and rare earth magnets, and compatibility with ethylene propylene diene monomer (EPDM), fluoroelastomer (FKM), perfluoroelastomer (FFKM), nitrile butadiene rubber (NBR), polyamide (PA), polybutylene terephthalate (PBT), polyphenylene sulfide (PPS), and printed circuit board (PCB) laminates. Operability: pumpability at cold start, gas solubility and air release, foaming, corrosion inhibition, cleanliness targets, environmental profile, and reclaim or disposal path. For electrified hardware, add insulation coordination: dielectric strength at operating temperature, dissipation factor, resistivity, partial discharge inception margin, and electrochemical corrosion risks under bias. For nuclear and high-reliability assets, include radiolysis or neutron activation byproducts, redox control, and strictly qualified impurity budgets. The output of selection is not a nameplate fluid but a controlled specification with test methods, limits, and a qualification matrix against the actual metallurgy and polymers in the system.

DOI: 10.1201/9781003775201-9

FLUID PROPERTY SELECTION CRITERIA

Start with the Duty Profile

Effective fluid selection begins by defining the application's demands in detail:

- **Thermal Load and Gradients**: Heat input, maximum operating temperature, and stability under cycling.
- **Electrical Duty**: Presence of insulation gaps, voltage levels, and dielectric requirements.
- **Mechanical Regime**: Shear forces, pump design, and flow dynamics.
- **Operating Pressures**: Both maximum system pressure and pressure fluctuations.
- **Environmental Extremes**: Lowest ambient startup temperature and highest sustained operating temperature.
- **Materials Stack**: Metals, coatings, elastomers, plastics, and composites in contact with the fluid.

A correct fluid must address all of these factors simultaneously, moving heat efficiently, preserving insulation margins, providing boundary lubrication where needed, and staying chemically neutral toward system materials.

Follow the Workflow: Physics → Chemistry → Operability

Step 1: Screen by Physics
Evaluate fundamental physical properties:

- Density and viscosity curve across the operating range.
- Thermal conductivity and specific heat (heat transfer performance).
- Dielectric constants where electrical insulation is required.
- Vapor pressure (boiling tendency under service conditions).
- Flash point and autoignition temperature (safety margins).

Step 2: Screen by Chemistry
Assess stability and material compatibility:

- Oxidation stability and resistance to hydrolysis.
- Interaction with additives and alloys (e.g., copper, aluminum, rare earth magnets).
- Elastomer and polymer compatibility (EPDM, FKM, FFKM, NBR, PA, PBT, PPS, PCB laminates).
- Long-term chemical quietness (no corrosive or deposit-forming reactions).

Step 3: Screen by Operability
Focus on real-world handling and service behavior:

- Pumpability at cold start.
- Gas solubility and air release characteristics.
- Foam tendency and defoamant response.

- Corrosion inhibition and cleanliness targets.
- Environmental profile (toxicity, biodegradability, regulatory compliance).
- End-of-life considerations: reclaim, recycling, or safe disposal.

Special Considerations for Electrified Hardware

Where electrical service is involved, add insulation coordination criteria:

- Dielectric strength at operating temperature.
- Dissipation factor and resistivity.
- Partial discharge inception voltage/margin.
- Risks of electrochemical corrosion under bias conditions.

Special Considerations for Nuclear and High-Reliability Assets

For nuclear reactors, space systems, or other ultra-critical equipment, expand criteria to include the following:

- Radiolysis and neutron activation byproducts.
- Redox control requirements.
- Strict impurity budgets (water, oxygen, halides, metals) tied to qualified test methods.

Define the Output Specification

The final deliverable is not simply choosing a commercial fluid by name. Instead, it is a **controlled specification** that includes the following:

- Required test methods (American Society for Testing and Materials (ASTM), International Electrotechnical Commission (IEC), International Organization for Standardization (ISO)).
- Measurable property limits.
- Qualification matrix verifying compatibility against the actual metallurgy and polymers in the system.

This structured workflow ensures the chosen fluid is not only fit-for-purpose on paper but also validated against real-world system conditions, long-term maintenance intervals, and safety-critical requirements.

COMPATIBILITY ENGINEERING IN PRACTICE

Compatibility is engineered, not assumed. Start with coupon and soak testing that matches service temperature, dwell, and electrical bias. Run metals sets that reflect real galvanic pairs and surface finishes, not ideal lab coupons. Include solder alloys, plated copper, and aluminum heat spreaders with crevices, since under-deposit chemistry and crevice oxygen gradients are where many "compatible" fluids fail. Add elastomer swell, hardness, and compression set at low and high temperature, then repeat with antioxidant- or inhibitor-aged fluid to catch late life chemistry. Validate adhesives, conformal coatings, potting resins, and wire enamels, especially for immersion cooling and direct-to-chip systems. For water-based coolants, qualify inhibitors against your metals mix and test biocide interactions with seals. For esters and phosphate esters, screen hydrolysis and acid growth against copper and steel. For fluorinated dielectrics and hydrofluoroolefin (HFO)/hydrofluoroether (HFE) e-coolants, check extractables from plastics and the effect of trace moisture on breakdown voltage. Document a change-control rule: any component, additive, or supplier change retriggers a shortened but relevant slice of this matrix.

FLUID PROPERTY SELECTION CRITERIA

Thermal Conductivity

Thermal conductivity, k, controls how fast heat moves through the fluid film; it matters most when the thermal boundary layer is the bottleneck. Water sits at ~0.6 W/m·K at 25 C, glycols at ~0.3 to 0.4, mineral and synthetic oils at ~0.1 to 0.15, and engineered dielectrics lower still. In practice, k should be considered with specific heat, c_p, and density, ρ, because transient response and wall temperatures depend on thermal diffusivity, $\alpha = k/(\rho \cdot c_p)$, and on convective performance embodied by Reynolds and Prandtl numbers. A lower-k fluid can still meet targets if velocity, turbulence, and surface area raise the Nusselt number. Selection therefore ties k to geometry and allowable pump power. For EV inverters or immersion-cooled servers, low k can be offset by two-phase boiling if the dielectric permits nucleation at safe temperatures. In hydronics or jacket water, high k and high c_p in water dominate, making water or water-glycol the first choice where electrical insulation is not required.

Viscosity

Viscosity dictates pump power, film thickness, cavitation margin, and the transition to turbulence. It is also the master variable for frictional heating in bearings and gears. The viscosity–temperature relation is inverse and non-linear; Arrhenius or Vogel models best describe it, so you must evaluate at the coldest start and hottest film, not just at 40 °C. For lubricants, select an ISO grade that maintains elastohydrodynamic film thickness under the peak Hertzian contact while avoiding churning losses and heat rise at speed. For coolants, pick a viscosity curve that achieves target Reynolds number in the smallest channels. Watch low-temperature pumpability: cold crank simulation (CCS) and MRV for engine oils, Brookfield for thermal oils, and dynamic viscosity for glycols at expected ambients. For immersion dielectrics and PAO coolants, confirm that the chosen viscosity at operating temperature balances convection and leakage through microgaps without starving pumps.

Pour Point

Pour point defines the lowest temperature at which the fluid appears to flow under standardized conditions. It is a screening tool, not an operating limit. Real systems need margin for suction losses, filters, and control valves, so the minimum ambient or standby temperature should clear pour point by at least 10 to 20 °C, often more for long suction lines. Mineral and synthetic oils achieve low pour points through base stock selection and pour-point depressants, but depressants do not improve true pumpability at very low shear rates. PAO and certain esters provide intrinsically low pour points and are useful in exposed lines or start-stop duty. Nitrate molten salts fail this criterion entirely and require heat tracing and freeze management; that is a design choice, not a fluid deficiency. For water systems, pour point translates to freeze risk: use glycol only to the concentration that meets the worst-case ambient plus safety margin, since excess glycol penalizes viscosity and heat transfer.

Flash Point

Flash point and autoignition temperature constrain both safety and attainable wall temperatures. Thermal oils rely on flash point margin and vapor management to avoid combustible atmospheres in expansion tanks and vents. Select a fluid with a closed-cup flash point that remains comfortably above any hot surface or vent temperature, then control low-boiler formation that drags the effective flash point down over time. Hydrocarbon dielectrics used for immersion cooling should be judged by both flash point and heat of combustion, and by enclosure ventilation if a fault could spray fluid. Fluorinated dielectrics and water-based coolants are attractive where non-flammability is required, but you still manage hot-surface decomposition limits and non-condensables in two-phase rigs. In diesel and gas plants, the lubricant's flash point should exceed expected sump and local hot-spot temperatures with margin; for transformer liquids, use the fire point and sustained ignition testing as the safety benchmark, not flash point alone.

Putting the Criteria Together by Application

Thermal power and combined heat and power (CHP) hydronics favor treated water first. The physics are unbeatable: high k and high c_p with low viscosity produce high heat transfer at modest pump power. Add glycols only to the concentration needed for freeze and burst protection, and pick an inhibitor package for the actual metals in the loop. Nuclear primary and secondary circuits demand water chemistry control or, for advanced concepts, sodium, helium, or salts chosen for neutronics and temperature window rather than simple k or flash point. In solar troughs and exhaust heat recovery units (HRUs), thermal oils win on pour point and simplicity; select an aromatic synthetic if film temperatures flirt with 350 to 390 °C, and design velocity to keep film temperatures inside the stability map. For transformers, choose an insulating liquid by dielectric strength, dissipation factor, water tolerance, and fire performance. Mineral oils remain common, but natural esters raise fire point and moisture tolerance at the cost of higher viscosity and different seal behavior. For electrified drivetrains and power electronics, the coolant is also a dielectric. If using water-glycol, ensure creepage, clearance, and leakage detection are robust, and specify inhibitor and azole controls that do not raise conductivity beyond OEM limits. If using dielectrics like PAO, synthetic esters, or HFO/HFE fluids, select by breakdown strength at temperature, low tan δ, materials compatibility, and a viscosity that drives turbulent flow in microchannels without excessive pump power. For immersion cooling, low surface tension aids wetting, but validate compatibility with plastics, coatings, and adhesives under thermal cycling.

Fluid Property Selection Criteria Process

Selecting the right working fluid is not just a matter of picking by nameplate values. Each property ties directly into system design, operating limits, safety margins, and long-term monitoring. The key criteria are **thermal conductivity, viscosity, pour point, and flash point**, and how they interact with the specific application.

1. **Thermal Conductivity (*k*)**
   - **Definition and Role**:
     Thermal conductivity governs how efficiently heat moves through the fluid film at the solid–liquid interface. It matters most when the **thermal boundary layer** is the bottleneck to heat transfer.
   - **Comparative Values**:
     - Water: ~0.6 W/m·K (at 25 °C)
     - Glycols: ~0.3–0.4 W/m·K
     - Mineral/Synthetic Oils: ~0.1–0.15 W/m·K
     - Engineered Dielectrics: Lower still
   - **Engineering Considerations**:
     - Conductivity must be evaluated alongside **specific heat (c_p)** and **density (ρ)**.
     - Transient response depends on **thermal diffusivity ($\alpha = k / \rho \cdot c_p$)**.
     - Real heat transfer performance is captured by **Reynolds (Re), Prandtl (Pr), and Nusselt (Nu)** numbers.
     - Low-*k* fluids can still succeed if velocity, turbulence, and surface area drive convective transfer.
   - **Application Notes**:
     - **Hydronics/CHP water loops**: Water dominates due to high *k* and c_p.
     - **Immersion-cooled electronics**: Low-*k* dielectrics may rely on **two-phase boiling** to offset poor conductivity.
2. **Viscosity**
   - **Definition and Role**:

   Viscosity dictates **pump power, film thickness, cavitation margin, and frictional heating** in rotating or sliding parts.
   - **Behavior**:
     - The viscosity–temperature relation is **inverse and non-linear**.
     - Modeled best by **Arrhenius or Vogel equations**.
     - Must be evaluated at both **cold-start minimum** and **hottest film temperatures**.

- **Engineering Considerations**:
  - **Lubricants**: Select ISO grade to ensure elastohydrodynamic (EHD) film under peak Hertzian stress while avoiding churning losses.
  - **Coolants**: Choose viscosity to achieve the desired **Reynolds number** in small passages.
  - **Cold-start limits**:
- CCS and MRV for engine oils
- Brookfield viscosity for thermal oils
- Dynamic viscosity for glycols
  - **Immersion dielectrics/PAO coolants**: Viscosity must balance **convection efficiency vs leakage through microgaps**.

3. **Pour Point**
   - **Definition**:

   The lowest temperature at which a fluid still flows under standardized test conditions.
   - **Engineering Reality**:
     - It is a **screening tool, not an operating limit**.
     - Systems need **10–20 °C margin** below minimum ambient to account for suction losses, filters, and valves.
   - **Application Notes**:
     - **Mineral/Synthetic Oils**: Lowered with pour-point depressants, though real pumpability still degrades at very low shear rates.
     - **PAO and Esters**: Naturally low pour points, making them ideal for exposed lines or intermittent duty.
     - **Molten Salts**: Fail this criterion completely, solidify at 120–220 °C, requiring **heat tracing and freeze protection**.
     - **Water Systems**: Pour point = freeze point. Use glycol sparingly, only to cover worst-case ambient, since excess glycol hurts both viscosity and heat transfer.
4. **Flash Point and Autoignition**
   - **Definition and Role**:

   Flash point and autoignition set **safety margins** and maximum **allowable wall temperatures**.
   - **Key Considerations**:
     - **Thermal Oils**: Need flash point margin above vent/expansion tank temps; monitor low-boiler formation that reduces flash point over time.
     - **Dielectrics (PAO, Esters, Hydrocarbons)**: Judge by flash point, heat of combustion, and ventilation design.
     - **Fluorinated Dielectrics and Water-Based Coolants**: Offer non-flammability but still have **thermal decomposition** thresholds.
     - **Engines/Transformers**:
   - Engine oils: Flash point > sump and hot-spot temps with margin.
   - Transformer fluids: Use **fire point** and sustained ignition tests, not flash point alone.

5. **Application-Driven Selection**

Different systems prioritize properties differently:

- **Thermal Power and CHP Hydronics:**
  - Water is first choice (high k, high c_p, low viscosity).
  - Add glycol only as needed for freeze/burst protection.
  - Select inhibitor package for actual metallurgy.
- **Nuclear**:
  - Primary/secondary circuits driven by neutronics and safety, not just k or flash point.
  - Sodium, helium, or salts selected for temperature range and physics.
- **Solar troughs and HRUs**:
  - Thermal oils dominate (low pour point, manageable).
  - Aromatic synthetics needed if film temperatures approach 350–390 °C.
- **Transformers**:
  - Select by dielectric strength, moisture tolerance, and fire performance.
  - Mineral oil common; esters preferred for fire safety and moisture handling but bring higher viscosity.
- Electrified Drivetrains and Power Electronics:
  - Coolant doubles as dielectric.
  - Water-glycol only if creepage/clearance and leak detection are robust.
  - PAOs, esters, or engineered HFO/HFE fluids: choose by breakdown strength, tan δ, and viscosity for turbulent microchannel cooling.
  - For immersion: prioritize **low surface tension and compatibility** with plastics/coatings under thermal cycling.

6. **Lifecycle and Monitoring Integration**

Fluid property choices must feed directly into a monitoring plan:

- **Viscosity-critical designs**: Track viscosity at service temperature; watch oxidation trends.
- **Flash Point-Dependent Systems**: Monitor low-boiler formation, condenser performance, and vent composition.
- **Water-Based Systems**: Track oxygen ingress, inhibitor levels, conductivity, pH, reserve alkalinity, and metals by ICP.
- **Dielectric Coolants**: Pair moisture ppm with breakdown voltage, dissipation factor, and gas management.
- **Transformer/Ester Fluids**: Monitor TAN, interfacial tension, and dissolved gas analysis to protect electrical margins.
- **Change Control**: Any supplier, base stock, or additive change retriggers **compatibility testing** and resets baselines, small chemistry shifts can move viscosity curves, pour point, and flash point enough to break original design balance.

Fluid property selection is a multi-variable optimization problem. Each parameter, thermal conductivity, viscosity, pour point, and flash point, interacts with application geometry, duty cycle, safety margins, and material compatibility. Properly balancing them ensures not only performance but also safe, predictable, and monitorable operation over the system's lifecycle (see Table 6.1).

- **Bold** = Primary driver for that application.
- Secondary = important but not usually design-limiting.
- Other Key Notes flag unique considerations that often outweigh raw property values.

Lifecycle and Monitoring Tied to the Properties

The property decisions you make should show up in the monitoring plan. If viscosity was tight to achieve turbulence, trend viscosity at service temperature and watch oxidation numbers that push it upward. If you relied on a high flash point, measure low-boiler content and vent condenser performance so the effective flash point does not slide. If you chose water for k and c_p, make oxygen ingress and inhibitor residuals your primary sentinels; conductivity, pH, reserve alkalinity, and metals by ICP will tell you when the advantage is being eroded by corrosion or fouling. For dielectric coolants, couple moisture ppm and breakdown voltage at temperature with dissipation factor and gas management; non-condensables raise wall temperatures in two-phase systems even when bulk properties look fine. For transformer and ester fluids, pair moisture-in-oil with interfacial tension, acid number, and dissolved gas analysis so thermal and electrical margins are never guessed at. Finally, lock a change-control rule: any supplier, base stock, or additive shift retriggers compatibility checks and resets baselines, since the smallest change in chemistry can move viscosity–temperature behavior, pour point, and flash point enough to break your original design balance.

ADVANCED PROPERTY CRITERIA FOR ELECTRIFICATION AND POWER-GENERATION FLUIDS

Dielectric Strength and Electrical Stability

Dielectric strength is the hard gate for any fluid that shares space with live conductors. The useful number is not a catalog kV but the breakdown behavior of your aged, water-laden, particle-bearing fluid at service temperature. Moisture, entrained gases, and fibers collapse margins first: a 20 to 50 ppm water rise in transformer liquids can cut breakdown voltage and push dissipation factor up by an order of magnitude. Control variables are dryness, cleanliness, and chemistry. Dry the fluid to the low-ppm regime, keep particles well below the inception size for streamer growth, and avoid additives or contaminants that raise conductivity or tan δ. Prove performance with the right methods for the duty: breakdown voltage by IEC 60156 or ASTM D1816 (sensitive to water and particles), flat-plate D877 for legacy specs, volume resistivity by ASTM D1169, and dissipation factor by ASTM D924 across the temperature range. For immersion cooling and direct-to-chip systems, retest after materials soaks since

TABLE 6.1
Fluid Property Decision Matrix

Application	Thermal Conductivity (k)	Viscosity	Pour Point	Flash/Fire Point	Other Key Notes
Thermal Power and CHP Hydronics	**Primary** – water's high $k + c_p$ dominate efficiency	Secondary – pump power manageable	Secondary – freeze protection via glycol	Low concern (water-based)	Chemistry control (O_2, pH, inhibitors) more critical than k once chosen
Nuclear (PWR/BWR/ Advanced)	Secondary – chosen mainly for neutronics & safety margins	Secondary–governed by coolant type (water, sodium, helium, salts)	Not applicable (system-specific)	Context-dependent (sodium, salts)	Chemistry control, radiolysis, redox, impurity budgets dominate
Solar Troughs/HRUs (Thermal Oils)	Secondary – constrained by oil type	**Primary** – viscosity/ film temp defines stability map	**Primary** – low pour point = easy startup	**Primary** – flash/ fire point margin essential	Monitor low/high boiler fractions, antioxidants
Transformers (Mineral or Ester Liquids)	Secondary – insulation priority > heat transfer	Secondary – only as it affects flow/ cooling	Not critical (indoors), but esters higher viscosity	**Primary** – fire point and dielectric safety	Moisture tolerance, IFT, TAN, DGA trends critical
Electrified Drivetrains/Power Electronics	Secondary – immersion or microchannel flow dominates	**Primary** – must balancepumpability vs. turbulence	**Primary** in cold climates – coolant must stay mobile	**Primary** for hydrocarbons; non-flammable dielectrics favored	Dielectric strength, tan δ, material compatibility under bias crucial
Immersion Cooling (Servers/EV Inverters)	Secondary – low-k offset by nucleate boiling	**Primary** – sets convective regime and microgap flow	Secondary – pour point less critical indoors	**Primary** – flammability vs. non-flammable FK/HFE	Surface tension, polymer compatibility, void fraction monitoring
Molten Salt Systems (Solar Towers, Nuclear MSRs)	**Primary** – high $c_p + k$ advantage	Viscosity secondary but rising at low temp	**Critical bottleneck** – freeze point 120–220 °C	Not applicable (non-flammable)	Redox control, corrosion, freeze management drive design

plasticizers and adhesive extractables can quietly double tan δ even when breakdown looks fine.

Oxidation Resistance and Life Modeling

Oxidation governs viscosity drift, acid formation, sludge, and varnish. It is set by base stock chemistry, antioxidant reserve, temperature, oxygen availability, and catalytic metals. Mineral oils oxidize fastest, PAO and silicone slower, esters resist chain growth but can hydrolyze if wet, and fluorinated dielectrics are largely inert but can suffer from trace decomposition at hot spots. You manage three things: keep oxygen out, keep film temperatures below the stability knee, and keep antioxidants healthy. Validate with accelerated tests that match the family: RPVOT for turbine and circulating oils, IEC 61125 for transformer liquids, TOST for inhibited mineral oils, pressure differential scanning calorimetry for high-temperature synthetics, and Fourier transform infrared spectroscopy (FTIR) oxidation indices for trending in service. Tie those to field sentinels that move early: total acid number (TAN), membrane patch colorimetry (MPC) patch rating for varnish propensity, insolubles, color, and metals that signal catalytic surfaces. When the antioxidant reserve drops below roughly half of new-oil value, plan a top-treat or a bleed-and-feed before viscosity and deposits begin to accelerate.

Fire Resistance and Thermal Hazard Management

Fire behavior sets enclosure design and allowable hot-surface temperatures. Compare not just flash point but fire point, sustained ignition, and heat of combustion. Mineral transformer oils have high flash points but will sustain combustion when atomized; natural esters and some silicones carry much higher fire points and self-extinguish more readily, which is why they qualify as less-flammable in many transformer standards. Hydrocarbon immersion coolants in electronics can be acceptable if ventilation, leak containment, and hot-surface limits are engineered; many facilities still prefer fluorinated or water-based media where ignition risk must be near zero. For thermal oils, manage low-boilers rigorously, since they drag the effective flash point downward and raise vapor pressure in expansion tanks. Set alarms on vent condenser performance and low-boiler fraction, not only on bulk flash-point checks.

Ash Content and Deposit Risk

"Ash" is the non-combustible residue left after burning the fluid. In lubricants a higher sulfated ash value usually signals metallic detergents or AW chemistry; in dielectric or heat transfer service, ash should be minimal because metallic residues become conductive, abrasive, and deposit-forming. Trend sulfated ash for lubricants that sit near combustion products or exhaust leaks, and keep it low in any fluid that must remain electrically quiet. When switching from low-ash to ashless formulations in hydraulics or circulation systems, validate seal compatibility and deposit behavior, since detergents sometimes kept deposits in suspension that ashless packages will not.

Base/Acidity Balance: TAN, TBN, Reserve Alkalinity

Acidity builds from oxidation, hydrolysis, and contamination. For non-engine systems the key indicator is TAN, which tracks corrosivity, solvency shift, and the onset of varnish. Set absolute limits by fluid family and trend rate of change rather than chasing a single red line. For engine and some CHP loops, TBN gauges the capacity to neutralize acids from fuel combustion; once TBN falls below about one third of its starting value, corrosion and deposit risk escalate. Water systems use reserve alkalinity and pH as the control pair: enough buffer to hold pH in the protective window for mixed metallurgy without overshooting into silicate drop-out or aluminum attack. Pair these numbers with metals by ICP to catch under-deposit corrosion early.

Nitration and Sulfation Fingerprinting

Nitration and sulfation are chemical fingerprints of gas-path contamination and high-temperature surface chemistry. In gas engines and exhaust heat-recovery environments, NOx and SOx ingress or blow-by drives nitro and sulfo species that raise TAN and deposit on hot surfaces. FTIR indices for nitration and sulfation move faster than TAN and are your early warning that either combustion products are entering the loop or film temperatures are too high. If those indices climb while oxidation stays flat, you likely have contamination rather than simple oil life exhaustion. Find the leak path or rerate film temperature before adding more inhibitor to chase symptoms.

Heat Transfer Coefficient Optimization

Heat transfer performance is not a single property but the outcome of geometry and fluid physics. The target is a high convective coefficient with acceptable pump power and safe wall temperatures. That means driving Reynolds high enough for energetic turbulence while keeping Prandtl in a zone that favors thin thermal boundary layers. Low viscosity, high specific heat, and adequate thermal conductivity help, but you can often buy more performance by changing passage geometry or roughness than by swapping chemistries. In single-phase service, push velocity to where the incremental pump energy still returns more heat duty than it costs. In two-phase dielectric systems, manage nucleation site cleanliness and remove non-condensables so latent heat actually carries the load; fluid choice then revolves around saturation temperature, wetting, and dielectric safety rather than k alone. Always cross-check film temperature against the fluid's stability map, since many oils crack at the wall even when bulk temperature looks conservative.

Fouling Resistance and Cleanliness Control

Fouling throttles heat duty long before bulk properties drift. Mechanisms split into inorganic scale, organic oxidation products, polymerized "high-boilers," biofilms in water systems, and particulate loading from wear or assembly debris. Your program needs three layers. First, prevention: demineralized makeup for water loops, strong inhibitor control, oxygen exclusion, nitrogen blankets on oils, and biocide discipline.

Second, removal: side-stream filtration sized for the particle modes you actually see, venting to strip low-boilers, and bleed-and-feed to control high-boilers before they collapse to varnish. Third, surveillance: approach temperature trending, pressure drop across exchangers and filters, membrane patch microscopy to identify deposit chemistry, MPC for varnish tendency, and targeted coupon programs in the hottest zones. Treat the fouling factor you put in the design as a budget to be defended with data, not as a safety valve to be consumed.

Integration Selection and Monitoring Spec

Start with the electrical duty: set minimum breakdown voltage and maximum tan δ at the hottest operating temperature, plus a moisture limit that preserves both. Choose a base chemistry whose oxidation and fire behavior fit the enclosure and hot-spot map. Specify ash and metallic additive limits that support your insulation and wear strategy. Lock TAN and, if applicable, TBN targets tied to maintenance actions, and include FTIR nitration and sulfation trend limits wherever exhaust or combustion products are nearby. For thermal performance, state Reynolds and Prandtl targets at design conditions rather than only naming a viscosity grade, and add a maximum allowable fouling factor with corrective actions tied to approach temperature and ΔP trends. Finally, bind all of it to methods: dielectric by IEC 60156 and ASTM D924, oxidation by IEC 61125 or RPVOT as applicable, TAN by ASTM D664, TBN by ASTM D2896 or D4739, FTIR per ASTM E2412, reserve alkalinity for water by ASTM D1121, and a defined microscopy protocol for deposits. That turns abstract properties into an enforceable engineering envelope that will survive supplier changes and service life.

Fluid Specification Workflow – Step by Step

1. **Start with Electrical Duty**
   - Define **minimum breakdown voltage** at service temperature.
   - Set **maximum dissipation factor (tan δ)** at the hottest operating point.
   - Impose a **moisture limit** that ensures both dielectric strength and low tan δ are preserved.
2. **Choose Base Chemistry**
   - Match **oxidation stability** and **fire behavior** to the enclosure type and expected hot-spot map.
   - Ensure fluid is chemically compatible with insulation, metals, elastomers, and coatings.
3. **Set Additive and Ash Limits**
   - Define allowable levels of **ash and metallic additives** to protect both insulation and mechanical components.
   - Align these with your **wear strategy** and long-term reliability plan.
4. **Lock in Acid/Base Targets**
   - Specify **TAN (Total Acid Number)** limits that trigger corrective maintenance.

    - Where relevant, set **TBN (Total Base Number)** targets to preserve neutralization capacity.
    - Include **FTIR nitration and sulfation trend limits** in environments exposed to exhaust or combustion byproducts.
5. **Define Thermal-Hydraulic Performance**
    - Express requirements in terms of **Reynolds and Prandtl numbers** at design conditions, not just a viscosity grade.
    - Establish a **maximum allowable fouling factor**.
    - Tie corrective actions to **approach temperature rise** and **ΔP (pressure drop) trends**.
6. **Bind to Standardized Test Methods**
    - **Dielectric Strength**: IEC 60156, ASTM D924.
    - **Oxidation Stability**: IEC 61125, RPVOT (ASTM D2272) as applicable.
    - **TAN**: ASTM D664.
    - **TBN**: ASTM D2896 or ASTM D4739.
    - **FTIR Trend Analysis**: ASTM E2412 (nitration, sulfation, oxidation).
    - **Reserve Alkalinity (Water-Based Fluids)**: ASTM D1121.
    - **Deposit Microscopy**: Defined protocol for identifying sludge, varnish, or particulates.
7. **Close the Loop – Engineering Envelope**
    - The outcome is not just fluid selection but a **controlled engineering envelope**:
    - Enforceable specifications.
    - Tied directly to monitoring and maintenance actions.
    - Resilient to supplier changes and over the full service life of the asset.

Material Compatibility: Elastomers, Metals, and Coatings

Compatibility in lubricants, insulating fluids, and coolants for electrification and power generation is not a one-time specification; it is an ongoing engineering control process. The fluid, material, and operational envelope interact dynamically. Every selection must account for the actual chemistry over service life, not just fresh-fluid performance. Testing must simulate temperature extremes, pressure cycling, electrical bias (if dielectric service), and fluid aging by oxidation, hydrolysis, or additive depletion. Small formulation shifts in base fluid or additive package can alter swelling, hardness, corrosion, or coating adhesion, so change control is essential.

ELASTOMERS

Elastomer compatibility depends on polarity match, backbone chemistry, cure system, and filler package. Hydrocarbon oils generally have low swell in nitriles and hydrogenated nitriles but can attack certain silicones. Esters, both synthetic and natural, can cause severe swelling in nitriles, EPDM, and some silicones. Water-based fluids may shrink non-polar elastomers while hydrolyzing sulfur-cured rubbers. Perfluorinated elastomers have broad compatibility but are cost-intensive. Testing should follow ASTM D471 or ISO 1817 at both normal and peak operating temperatures (see Table 6.2).

METALS

Metals in contact with process fluids face corrosion, pitting, galvanic coupling, and stress corrosion cracking. Hydrocarbon oils rarely cause corrosion unless oxidized. Water-based systems require inhibitors to protect steel, copper, and aluminum. Esters can be more conductive and aggressive toward zinc and cadmium coatings. Molten salts and sodium demand specific high-temperature alloys. Always control dissolved oxygen and deposit buildup (see Table 6.3).

COATINGS AND PLASTICS

Coatings and plastics can degrade from solvent swelling, hydrolysis, plasticizer leaching, oxidation products, and permeation. Conformal coatings in immersed electronics must be tested for dielectric performance over time with aged fluids. Immersion and adhesion testing should follow ASTM D543, ASTM D4541, and IPC-CC-830 (see Table 6.4).

Summary on Ratings Use

- **1 = Good compatibility**, choose proper grade and still validate with representative testing.
- **2 = Conditional**, test under service conditions, consider alternative materials if risk tolerance is low.
- **3 = Poor compatibility**, avoid unless no alternatives exist and mitigation strategies are in place.

Failure Signatures That Point to Compatibility Mistakes

- Seals: rapid volume growth with softness points to solvent swell by aromatics or esters, shrinkage and cracks point to extraction by water-glycol or PAG. Compression set growth after dry-out indicates irreversible damage.
- Pumps and bearings: copper and lead metals rising with TAN growth implicate hot oil oxidation attacking bronzes, zinc release in ester dielectrics signals plated hardware risk.
- Heat exchangers: approach temperature climb with brown, adherent films indicates varnish from thermal oil high-boilers, shiny friable scale points to hardness ingress in water loops, green or black deposits near copper pairs indicate poor azole control.
- Electronics immersion: rising dissipation factor with no moisture increase suggests coating extractables, sudden BDV scatter after maintenance suggests fibrous contamination or surfactant residue.

TABLE 6.2
Elastomer Compatibility Matrix (Rating: 1–3)

Elastomer	Mineral and PAO oils	Synthetic esters, natural esters	Phosphate esters	Water, DI, water-glycol	HFO/HFE, FK dielectrics	PAG coolants	Silicone oils
NBR (nitrile)	1–2	3	3	2	1	2	2–3
HNBR	1	2	3	2	1	2	2
EPDM	3	3	3	1	1	1	3
FKM (Viton)	1	1	1	2	1	2	1
FFKM (perfluoroelastomer)	1	1	1	1	1	1	1
TFE/P (Aflas)	1	2	1	2	1	2	1
VMQ/FVMQ (silicone)	2	2	3	1	2	2	1
PTFE	1	1	1	1	1	1	1
PEEK (backup rings)	1	1	1	1	1	1	1

TABLE 6.3
Metals Compatibility Matrix (Rating: 1–3)

Metal	Mineral and PAO Oils	Synthetic/ Natural Esters	Water, DI, Water-Glycol	HFO/ HFE, FK Dielectrics	Thermal oils (300–390 °C)	Molten Nitrate Salts	Molten Sodium
Carbon steel	1	1	2	1	1	2	3
304/316 SS	1	1	1	1	1	2	2
Duplex SS	1	1	1	1	1	2	2
Copper	2	2	2	1	2	3	3
Brass/Bronze	2	2	2	1	2	3	3
Aluminum	2	2	2	1	2	3	3
Titanium	1	1	1	1	1	1	1
Nickel alloys	1	1	1	1	1	1	1

TABLE 6.4
Coatings/Plastics Compatibility Matrix (Rating: 1–3)

Material	Hydrocarbon Oils	Esters	Water, Water-Glycol	HFO/HFE, FK Dielectrics	Thermal Oils (hot)	Notes
Epoxy linings	1	2	1	1	2	Use novolac for higher temps
Phenolic linings	1	1	1	1	1	Good for hot oils
Polyurethane	2	3	1	2	3	Ether-based PU better in hydrolysis
PVDF, ETFE	1	1	1	1	1	Excellent
PTFE	1	1	1	1	1	Near universal
PEEK	1	1	1	1	1	High-temp structural
Polycarbonate	3	3	2	2	3	Prone to stress cracking
PBT, PPS	1	1	1	1	1	Good connector materials
Conformal urethane	2	3	1	2	3	Immersion risky
Conformal silicone	2	1	1	2	1	Check extractables
Conformal parylene C	1	1	1	1	1	Best barrier

Qualification and Monitoring Plan

1. **Design screening**: pick elastomer families by polarity match, pick metals by corrosion allowances and inhibitor strategy, pick coatings by cured chemistry not brand names.
2. **Lab proof**: elastomers per ASTM D471 or ISO 1817 at two temperatures and two durations, include dry-out; plastics per ASTM D543; coatings per D4541 and IPC-CC-830 for electronics; metals per ASTM G31 immersion with your inhibited fluid and per G48 for chloride pitting if water service is possible.
3. **Aged fluid round**: oxidize oils per IEC 61125 or RPVOT stress and retest critical pairs; hydrolyze esters at controlled moisture and retest; age water-glycol by thermal cycling with depleted inhibitor and retest; for dielectrics add moisture to the vendor limit and retest tan δ and breakdown.
4. **Assembly cleanliness**: forbid silicone mold release near adhesion surfaces, specify no zinc-plated wetted parts in ester circuits, require acid-free flux and complete rinse on PCB assemblies for immersion.
5. **Field surveillance**: couple chemistry sentinels to compatibility risks:
    - Elastomers: periodic dimensional spot checks during outages for long-life assets.
    - Metals: ICP for Cu, Zn, Al, Fe trend vs inhibitor residuals and TAN; chloride in water systems.
    - Coatings: approach temperature and ΔP as fouling proxies, dielectric tan δ and BDV for immersion.
    - Deposits: membrane patch microscopy to identify varnish, silica, rust, gasket crumbs.

Quick Selection Tables (see Tables 6.5–6.9)
How to Use This Chart

1. Start with your fluid family and locate the material under consideration.
2. Use the numeric rating:
    - 1: Generally good, use standard engineering grades and verify with supplier data.
    - 2: Acceptable only after testing under simulated service conditions.
    - 3: Avoid unless absolutely necessary with engineered mitigations.
3. For critical systems (e.g., nuclear, high-voltage dielectric immersion, CHP loops) confirm with ASTM/ISO compatibility testing and aged-fluid exposure.

Notes on Test Methods

- **Elastomers**:
    - **ASTM D471 / ISO 1817**: Fluid immersion to measure volume change, tensile strength, and hardness changes.
    - **SAE J200**: Classification of rubber materials.

TABLE 6.5
Pick-list: Seals for Common Fluids and Duties

Duty	Recommended Primary Seal	Back-up Ring	Notes
Water, DI, water-glycol	Peroxide-cured EPDM or FKM (steam grade)	PEEK or PTFE	Add azole and silicate or OAT as metallurgy requires
Mineral, PAO oils	FKM or HNBR	PTFE	Keep oxidation low, avoid aromatic-rich oils with VMQ
Synthetic ester dielectric	FKM high-fluorine or FFKM	PTFE	Remove zinc-plated wetted parts, watch zinc soaps
Phosphate ester hydraulics	TFE/P or FFKM	PTFE	Check base-resistance, avoid NBR, EPDM
HFO/HFE, FK dielectrics	FKM, FFKM, PTFE	PTFE	Check coating extractables, confirm tan δ stability
Thermal oil 300–390 C	FKM HT grade, graphite packing on valves	PEEK	Nitrogen blanket, keep film temp below limit

TABLE 6.6
Pick-list: Metals for Wetted Hardware

Fluid Family	Preferred Metals	Avoid
Water-glycol inhibited	316L, Cu alloys with azole, coated carbon steel	Bare Al without OAT, Zn plated fasteners
Mineral, PAO oils	Carbon steel, 316L, bronzes	High lead solders in hot zones
Synthetic or natural esters	316L, nickel alloys	Zinc plating, cadmium, soft solders
HFO/HFE, FK dielectrics	304/316, Al alloys, Cu with azole if any water present	Zinc and Mg if moisture control is weak
Molten nitrates	Low alloy steel at moderate temp, 316L, nickel alloys as temp rises	Cu, Al at high temp
Sodium	Specific ferritic/martensitic steels, selected Ni alloys	Cu, Al, water side anywhere near Na

- **Metals**:
  - **ASTM D130**: Copper strip corrosion for petroleum products.
  - **ASTM G31**: Laboratory immersion corrosion testing of metals.
  - **ASTM G59**: Polarization resistance for corrosion rate determination.
  - **ISO 8044**: Corrosion terminology and definitions.
- **Coatings/Plastics**:
  - **ASTM D543**: Chemical resistance of plastics.
  - **ASTM D4541**: Pull-off strength of coatings.

TABLE 6.7
Minimal Test Slate to Lock Compatibility

Component	Method	Why it Matters
Elastomer swell, hardness, tensile	ASTM D471, ISO 1817	Captures reversible and permanent changes, select cure system
Plastic chemical resistance	ASTM D543	Crazing, stress cracking screening
Coating adhesion, dielectric	ASTM D4541, IPC-CC-830	Underfilm loss in fluids, electrical stability
Metal corrosion rate	ASTM G31, G48 if chlorides	Baseline corrosion and pitting thresholds
Dielectric strength and tan δ	IEC 60156, ASTM D924	Electrical margin with moisture and extractables
Aged-fluid effects	IEC 61125, RPVOT, thermal cycling	End-of-life chemistry against materials

- **IPC-CC-830**: Qualification and performance of electrical insulating compounds for printed boards.

How SLRF Works (Simple, Auditable Math)
SLRF = BaseFactor(rating) × M_T × M_C × M_S × M_W
BaseFactor(rating): ties directly to the 1–3 compatibility rating

Rating 1 → 1.00
Rating 2 → 0.70
Rating 3 → 0.40

M_T (Temperature & Film Temp): 1.10 superb control; 0.60 if you are riding limits
M_C (Chemistry/Fluids aging): 1.10 if chem is tightly controlled; 0.60 if oxidation/hydrolysis likely
M_S (Mechanical stress/cycling): 1.10 if gentle; 0.70 if high ΔP, pulsation, or strain
M_W (Cleanliness & Water/O_2): 1.05 if ultra-clean/dry; 0.60 if dirty/wet reality
Interpretation:

≥0.90: low risk, life near design targets
0.60–0.89: moderate risk, shorten inspection interval or improve controls
<0.60: high risk, plan mitigations or alternate materials

Use the band that matches your control posture. If you want a single number, multiply the modifiers below.

Modifiers Cheat Sheet (Pick One Per Column)

TABLE 6.8
Fluid vs Material Compatibility Quick-Pick Chart
Rating Key:
1 = Good compatibility (with proper grade and controls)
2 = Conditional (verify under service conditions)
3 = Poor compatibility (avoid unless mitigated)

Fluid Family → / Material ↓	Mineral and PAO Oils	Synthetic/ Natural Esters	Phosphate Esters	Water/DI/ Water-Glycol	HFO/HFE, FK Dielectrics	PAG Coolants	Thermal Oils (300–390 °C)	Molten Nitrate Salts	Molten Sodium
Elastomers									
NBR (Nitrile)	1–2	3	3	2	1	2	1	3	3
HNBR	1	2	3	2	1	2	1	3	3
EPDM	3	3	3	1	1	1	3	3	3
FKM (Viton)	1	1	1	2	1	2	1	2	2
FFKM	1	1	1	1	1	1	1	1	1
TFE/P (Aflas)	1	2	1	2	1	2	1	2	2
VMQ/FVMQ (Silicone)	2	2	3	1	2	2	2	3	3
PTFE	1	1	1	1	1	1	1	1	1
Metals									
Carbon Steel	1	1	2	2	1	2	1	2	3
304/316 SS	1	1	1	1	1	1	1	2	2
Duplex SS	1	1	1	1	1	1	1	2	2
Copper	2	2	3	2	1	2	2	3	3
Brass/Bronze	2	2	3	2	1	2	2	**3**	**3**
Aluminum	2	2	3	2	1	2	2	**3**	**3**
Titanium	1	1	1	1	1	1	1	**1**	**1**
Nickel Alloys	1	1	1	1	1	1	1	**1**	**1**

(continued)

TABLE 6.8 (Continued)
Fluid vs Material Compatibility Quick-Pick Chart

Rating Key:
1 = Good compatibility (with proper grade and controls)
2 = Conditional (verify under service conditions)
3 = Poor compatibility (avoid unless mitigated)

Fluid Family → / Material ↓	Mineral and PAO Oils	Synthetic/ Natural Esters	Phosphate Esters	Water/DI/ Water-Glycol	HFO/HFE, FK Dielectrics	PAG Coolants	Thermal Oils (300–390 °C)	Molten Nitrate Salts	Molten Sodium
Coatings/Plastics									
Epoxy Linings	1	2	1	1	1	1	2	2	3
Phenolic Linings	1	1	1	1	1	1	1	1	2
Polyurethane	2	3	3	1	2	2	3	3	3
PVDF/ETFE	1	1	1	1	1	1	1	1	1
PTFE	1	1	1	1	1	1	1	1	1
PEEK	1	1	1	1	1	1	1	1	1
Polycarbonate	3	3	3	2	2	2	3	3	3
PBT/PPS	1	1	1	1	1	1	1	1	1
Conformal Urethane	2	3	3	1	2	2	3	3	3
Conformal Silicone	2	1	2	1	2	2	1	2	3
Conformal Parylene C	1	1	1	1	1	1	1	1	1

TABLE 6.9
Fluid vs Material Compatibility Quick-Pick Chart with Test Method References
Rating Key:
1 = Good compatibility (with proper grade and controls)
2 = Conditional (verify under service conditions)
3 = Poor compatibility (avoid unless mitigated)

Fluid Family → / Material ↓	Mineral & PAO Oils	Synthetic/ Natural Esters	Phosphate Esters	Water/ DI/Water-Glycol	HFO/ HFE, FK Dielectrics	PAG Coolants	Thermal Oils (300–390 °C)	Molten Nitrate Salts	Molten Sodium	Applicable Test Methods*
Elastomers										
NBR (Nitrile)	1–2	3	3	2	1	2	1	3	3	ASTM D471, ISO 1817, SAE J200
HNBR	1	2	3	2	1	2	1	3	3	ASTM D471, ISO 1817
EPDM	3	3	3	1	1	1	3	3	3	ASTM D471, ISO 1817
FKM (Viton)	1	1	1	2	1	2	1	2	2	ASTM D471, ISO 1817
FFKM	1	1	1	1	1	1	1	1	1	ASTM D471
TFE/P (Aflas)	1	2	1	2	1	2	1	2	2	ASTM D471
VMQ/FVMQ (Silicone)	2	2	3	1	2	2	2	3	3	ASTM D471
PTFE	1	1	1	1	1	1	1	1	1	ASTM D543
Metals										
Carbon Steel	1	1	2	2	1	2	1	2	3	ASTM D130, ASTM G31, ASTM G59
304/316 SS	1	1	1	1	1	1	1	2	2	ASTM G31, ASTM G59, ISO 8044

(continued)

TABLE 6.9 (Continued)
Fluid vs Material Compatibility Quick-Pick Chart with Test Method References
Rating Key:
1 = Good compatibility (with proper grade and controls)
2 = Conditional (verify under service conditions)
3 = Poor compatibility (avoid unless mitigated)

Fluid Family → / Material ↓	Mineral & PAO Oils	Synthetic/ Natural Esters	Phosphate Esters	Water/ DI/Water-Glycol	HFO/ HFE, FK Dielectrics	PAG Coolants	Thermal Oils (300–390 °C)	Molten Nitrate Salts	Molten Sodium	Applicable Test Methods*
Duplex SS	1	1	1	1	1	1	1	2	2	ASTM G31, ASTM G59
Copper	2	2	3	2	1	2	2	3	3	ASTM D130, ASTM G31
Brass/Bronze	2	2	3	2	1	2	2	3	3	ASTM D130
Aluminum	2	2	3	2	1	2	2	3	3	ASTM G31, ASTM G67
Titanium	1	1	1	1	1	1	1	1	1	ASTM G31
Nickel Alloys	1	1	1	1	1	1	1	1	1	ASTM G31
Coatings/ Plastics										
Epoxy Linings	1	2	1	1	1	1	2	2	3	ASTM D543, ASTM D4541
Phenolic Linings	1	1	1	1	1	1	1	1	2	ASTM D543, ASTM D4541
Polyurethane	2	3	3	1	2	2	3	3	3	ASTM D543

PVDF/ETFE	1	1	1	1	1	1	1	1	1	ASTM D543
PTFE	1	1	1	1	1	1	1	1	1	ASTM D543
PEEK	1	1	1	1	1	1	1	1	1	ASTM D543
Polycarbonate	3	3	3	2	2	2	3	3	3	ASTM D543
PBT/PPS	1	1	1	1	1	1	1	1	1	ASTM D543
Conformal Urethane	2	3	3	1	2	2	3	3	3	IPC-CC-830, ASTM D543
Conformal Silicone	2	1	2	1	2	2	1	2	3	IPC-CC-830
Conformal Parylene C	1	1	1	1	1	1	1	1	1	IPC-CC-830, ASTM D4541

TABLE 6.10
Quick-Pick with SLRF Bands (By Rating)
To avoid a phone-book table, here are SLRF bands you can apply to every cell in your existing chart:

Compatibility rating	BaseFactor	Typical SLRF band (realistic ops)	With strong controls	With poor controls
1 (good)	1.00	**0.85–1.05**	up to ~**1.10**	down to ~**0.70**
2 (conditional)	0.70	**0.55–0.80**	up to ~**0.90**	down to ~**0.45**
3 (poor)	0.40	**0.30–0.60**	up to ~**0.70**	down to ~**0.25**

TABLE 6.11
Add-on: Service Life Planning Table (Inspection Cadence by SLRF)

SLRF result	Action	Typical Inspection Interval Impact
≥1.00	Run program as designed; trend KPIs only	Baseline interval (100%)
0.80–0.99	Minor mitigations; add a check on the hottest zones	1.2× frequency (every ~80% of baseline period)
0.60–0.79	Add corrective controls; increase sampling; review materials	1.5× frequency
<0.60	Treat as life-limited; replace/upgrade material or fluid	2–3× frequency until corrected

Temperature (M_T)

• Excellent margin to limits; film temps conservative: **1.10**
= Near design temp; occasional peaks: **0.95**
– Frequent peaks; localized hot spots likely: **0.80**
–– Film temps at or above map; two-phase mismanaged: **0.60**

Chemistry and Aging (M_C)

• Dry, oxygen excluded, inhibitors/AO trended: **1.10**
= Routine control; minor excursions: **0.95**
– Oxidation/hydrolysis evident (TAN up, AO down): **0.80**
–– Persistent acid, water, or depleted inhibitors: **0.60**

Mechanical Stress (M_S)

• Steady flow, low cycling, low strain: **1.10**
= Normal duty; some starts/stops: **0.95**

– High ΔP, pulsation, vibration, thermal shock: **0.80**
–– Cavitation, frequent thermal cycles, misalignment: **0.70**

Cleanliness and Water/O_2 (M_W)

• ISO particle code 2 classes better than spec; water low-ppm: **1.05**
= Meets spec; occasional alerts: **0.95**
– Marginal filtration; intermittent water/O_2 ingress: **0.80**
–– Dirty loop; chronic moisture/air: **0.60**

Worked Examples (So it is Not Hand-Wavy)

1. **EPDM in water-glycol**: chart rating **1**
   - Good water treatment, solid filtration, moderate temps
   - SLRF = 1.00 × **0.95** (M_T) × **1.10** (M_C) × **0.95** (M_S) × **1.05** (M_W)
   - **SLRF ≈ 1.05** → low risk; full life expected
2. **NBR with synthetic ester dielectric**: chart rating **3**
   - You're pushing it; you keep temps moderate and fluid very dry
   - SLRF = 0.40 × **0.95** × **1.10** × **0.95** × **1.05** ≈ **0.44**
   - High risk even with controls: choose FKM/FFKM or TFE/P instead
3. **Carbon steel in water-glycol**: chart rating **2**
   - Average controls; some oxygen ingress, inhibitors dip occasionally
   - SLRF = 0.70 × **0.95** × **0.80** × **0.95** × **0.80** ≈ **0.40**
   - Risky: tighten O_2 control and inhibitor residuals or move to 316L
4. **FKM in HFO/HFE dielectric immersion**: chart rating **1**
   - Excellent dryness, clean assembly, mild temps
   - SLRF = 1.00 × **1.10** × **1.10** × **0.95** × **1.05** ≈ **1.22**
   - Outstanding; stretch inspection interval (but still verify tan δ/BDV)

How to Use With Your Existing Chart

1. **Take the 1–3 rating** from the cell → get **BaseFactor**.
2. **Pick modifiers** from the four lists that match your plant reality.
3. **Multiply** to get SLRF.
4. **Use the planning table** to adjust inspection/sampling intervals and to justify mitigations or material changes in specs.

7 Equipment-Based Fluid Matching

EQUIPMENT-BASED FLUID MATCHING

Transformers, Inverters, and Switchgear

Transformer liquids must solve three problems at once: dielectric insulation, heat removal, and paper/copper preservation. Pick the family by fire and environmental posture first, then by dielectric and water tolerance. In substation and generator step-up units where fire protection is critical, natural esters are often preferred for their high fire point and strong water solubility that slows paper aging; accept higher viscosity and cold-start penalties with larger radiators and careful winterization.

Mineral oil remains viable where containment, ventilation, and fire mitigation are strong; specify inhibited grades, low sulfur, tight oxidation stability, and a dryness plan that keeps moisture below vendor curves for your temperature. For sealed distribution transformers, oxidation stability and dissolved gas behavior dominate; specify passivators if copper sulfur staining is a risk.

For power electronics inverters and switchgear, treat the fluid as primarily a coolant and secondarily an insulator. If you choose water-glycol, set creepage and clearance with the worst-case conductivity and manage inhibitors so azoles do not foul contacts. If you choose a dielectric hydrocarbon, select by breakdown strength at operating temperature, low dissipation factor, plastics compatibility, and a viscosity that reaches turbulent flow in plates or channels without high pump power.

Two-phase dielectrics are viable for high heat flux in compact inverters when non-condensables are managed and surface cleanliness preserves nucleation. Condition monitoring aligns to duty: for transformers trend moisture in oil, breakdown voltage, dissipation factor, interfacial tension, acid number, inhibitor health, and dissolved gases tied to load history. For inverter coolants trend conductivity or resistivity, pH, inhibitor residuals, metals by inductively coupled plasma (ICP), particle counts, and for dielectrics add breakdown voltage and tan δ at temperature with periodic materials soak checks.

DOI: 10.1201/9781003775201-10

Turbines, Compressors, Pumps, and Blowers

The fluid decision begins with the hydrodynamic film you need and the thermal regime you can tolerate. Steam and gas turbines want ashless R&O turbine oils with strong oxidation life, air release, and foam control; viscosity is sized for bearing geometry and speed so that film thickness clears roughness at the hottest journal. Aeroderivative gas turbines and dry-gas seal compressors bring higher air entrainment and varnish pressure; specify high rotating pressure vessel oxidation test (RPVOT), robust antioxidant systems, and a varnish control plan using tight filtration and membrane patch colorimetry (MPC) trending. Rotary screw air compressors that use the lubricant as a sealant see high shear and oxygen exposure; polyalphaolefin (PAO) or ester synthetics extend life at temperature, but verify seal compatibility and water tolerance. Pumps and blowers that share fluids with process streams need base stocks that resist the process chemistry, so PAGs or specialty synthetics can be justified when hydrocarbons swell seals or dissolve varnish into the gas path.

Operability rules are simple: keep suction viscosity within pump maps at the coldest start, avoid elastomer pairs known to swell in your base stock, and design venting for degassing where air entrainment is inevitable. Monitoring is your early warning: viscosity at 40 and 100 °C for film control, acid number and oxidation indices for life, MPC and patch microscopy for varnish, particle counts for clearance control, foam and air release for seal stability, and metals for bearing distress. When varnish risk is high, add electrostatic or depth polishing and set action limits on MPC before servo valves stick.

Pumps, Compressors, and Blowers in Process Cooling Loops

Circulator pumps and process blowers that only handle coolants still fail due to the fluid choice. Water-glycol needs inhibitors tailored to carbon steel, copper alloys, and aluminum; choose nitrite or molybdate for steel, azoles for copper, and organic acid technology (OAT) or controlled silicate for aluminum. Keep glycol only to the percentage required for freeze and burst margins, since extra glycol costs pump power and heat transfer.

In high head pumps verify net positive suction head with viscosity at winter ambient. For hydrocarbon or PAO coolants, ensure low-temperature pumpability, seal material compatibility, and air-release performance. Add side-stream filtration and magnetic traps sized for your particle modes so the pump does not become the solids concentrator. Monitor pH, inhibitor residuals, conductivity, glycol percent by refractometer, particle counts, and wear metals; rise in iron or copper usually points to oxygen ingress or depleted inhibitors rather than pump design limits.

Batteries: Liquid-Cooled and Immersion-Cooled

Battery thermal management is as much a materials program as a thermal one. For cold plates and jacketed liquid cooling, water-glycol remains first choice due to heat capacity and low viscosity; protect aluminum manifolds with OAT inhibitors or controlled silicate, protect copper with azoles, and keep conductivity inside the pack builder's creepage and isolation test window. Choose elastomers accordingly: peroxide-cured EPDM or FKM handle hot water and azoles, while NBR is a poor fit for long hot-glycol duty.

For immersion-cooled packs, the fluid must be a quiet electrical neighbor. Hydrocarbon dielectrics and selected synthetics work if breakdown strength at temperature, low tan δ, wetting of cell surfaces, and plastics compatibility all clear; verify interaction with cell vent chemistry and adhesives, and test for polymer extractables that raise dissipation factor over time. Two-phase concepts can lift performance but demand tight gas management and materials that do not poison nucleation.

Monitoring is tailored to risk: for liquid cooling trend conductivity, pH, inhibitor health, metals, and leak detection current; for immersion dielectrics trend moisture ppm, breakdown voltage, tan δ at the hottest pack temperature, and inspect for foaming or polymer haze that signals extractables.

Engine Lubrication Circuits, Valve Trains, and Turbos

Engine circuits need shear-stable viscosity, detergency to suspend soot, dispersancy to prevent agglomeration, antiwear and extreme pressure capability where mixed or boundary regimes exist, and oxidation resistance for long thermal soaks. Choose viscosity grade by bearing clearance and target elastohydrodynamic film under the worst speed and temperature.

For high specific power diesels, monitor high-temperature high-shear viscosity, base number retention for acid control, oxidation, nitration and sulfation by Fourier transform infrared spectroscopy (FTIR) near exhaust gas recirculation duty, and soot by thermogravimetry. Turbos challenge the oil with thin passages and very high skin temperatures at shutdown; select base stocks and antioxidants that resist coking, design feed and drain geometry to avoid oil pooling, and consider auxiliary scavenge or cooldown routines to suppress heat soak. Valve train contact wants zinc dialkyldithiophosphate (ZDDP) or modern antiwear systems balanced to emissions controls. Rising iron and lead alongside soot signals film collapse and abrasive wear; rising total acid number (TAN) and insolubles signal oxidation and overtemperature. Keep air release, foaming, and volatility within bounds so pumps do not cavitate and vapors do not stress crankcase ventilation.

Nuclear Actuator Assemblies, Rod Drives, and Shielded Electronics

Fluids in nuclear auxiliaries trade performance for predictability and radiological cleanliness. Hydraulic fluids for rod drives and actuator circuits should be phosphate ester or synthetic hydrocarbon families with documented radiation tolerance, low outgassing, strict flash and fire performance, and elastomers that do not generate conductive debris after dose. Where phosphate esters are used, avoid nitrile butadiene rubber (NBR) or ethylene propylene diene monomer (EPDM) and favor tetrafluoroethylene/propylene copolymer (TFE/P), fluoroelastomer (FKM), or perfluoroelastomer (FFKM) with polytetrafluoroethylene (PTFE) back-ups; design filtration to capture fine acidic decomposition products and specify acid acceptance and water tolerance tests. Shielded electronics require coolants that do not activate problematically and that do not attack conformal coatings or potting compounds under dose and heat; parylene C and high-temperature silicone coatings perform well, but confirm tan δ and adhesion after soaking in aged fluid at temperature. Monitoring focuses on acid number, resistivity, particle counts at fine codes, metals that indicate valve cage or spool wear, and for ester systems water content plus neutralization capacity. Any supplier or additive change should retrigger radiation and thermal aging so dose-by-chemistry interactions are caught on the bench rather than during an outage.

Turbines Again: Combined Cycle Interfaces and Generator Hydrogen

At the combined cycle boundary the coolant and lubricant choices must respect cross-contamination consequences. Keep turbine oil and water-glycol loops physically segregated with double seals or leak alarms; a small leak of glycol into turbine oil accelerates oxidation and varnish, while oil into glycol ruins heat transfer and fouls plates.

Generator hydrogen reduces windage and improves cooling but raises demands on seal-oil chemistry; specify low volatility, strong air release, and antifoam that does not break into conductive fragments. Maintain seal-oil differential to the hydrogen space and treat CO_2 purge periods as chemical upsets for the oil, since carbonic acid and foam appear after transitions. Trend hydrogen purity, dew point, seal-oil TAN and water, air release, foam tendency, and dissolved gas content.

Solar Thermal Loop Piping and Storage Tanks

Molten salt storage and solar heat transport reward meticulous thermal housekeeping and materials discipline. Nitrate salts are non-flammable and stable in their window but freeze hard at 120 to 220 °C and decompose at the top end into nitrites and oxides. Design uniform heat tracing, expansion accommodation, and drain-down that avoids cold traps in dead legs. Use steels proven for the window and switch to stainless or nickel alloys in hot or high velocity regions. Keep moisture and chlorides out at ppm levels and watch nitrite to nitrate ratio as a decomposition proxy.

Thermal oils for parabolic trough circuits offer easier starts at low temperature but cannot run near molten salt temperatures. Choose aromatic synthetics with sufficient film stability, ensure velocity holds film temperature below the cracking knee, and design vents and condensers to strip low boilers while preventing oxygen ingress. Monitor salts for nitrite ratio, moisture, chlorides, and corrosion metals. Monitor oils for viscosity drift, TAN, low and high boiler fractions by gas chromatography, varnish tendency by MPC, and particle contamination at filterable sizes that map to exchanger fouling. Coatings and linings in hot storage tanks should be novolac epoxy or phenolic that are post-cured and proven in the fluid, with adhesion and blister resistance verified after fluid soak and thermal cycles.

ESTABLISHING SPECIFICATIONS

Start with the equipment physics and failure mechanisms, then bind the fluid to those risks. For any asset, write the selection in three parts: the fluid family and grade justified by heat flux, dielectric duty, and mechanical regime; the materials and seals justified by polarity, temperature, and expected aging chemistry; and the monitoring plan that tracks the exact properties that defend the choice.

For transformers and inverters this means dielectric strength at temperature, moisture control, and tan δ. For turbines and compressors it means viscosity control, oxidation life, varnish propensity, and air release. For batteries it means conductivity, inhibitor health, or dielectric stability plus materials extractables. For nuclear auxiliaries it means acid control, radiation tolerance, and ultra-fine cleanliness. For solar loops it means freeze protection or film temperature and decomposition control. If any of these pillars relies on a narrow margin, raise the monitoring cadence, add side-stream polishing, or step up one materials grade so the program remains stable over supplier and operating variability (see Tables 7.1 and 7.2).

Step-Wise Fluid Selection and Assurance Process

(1) **Define equipment physics and credible failures**
- Duty: heat flux, surface and film temperatures, duty cycle, flow regime, pressure, cavitation risk.
- Electrical: voltage class, creepage and clearance, partial-discharge risk, electric field hotspots.
- Mechanical: contact stress, shear, speed, pump curve and NPSH, seal type.
- Environment: ambient minima and maxima, contamination vectors, fire zone, radiation (if applicable).
- Failure mechanisms to consider: dielectric breakdown, PD, oxidation and varnish, hydrolysis, sludge, corrosion, aeration and foaming, erosion, radiolysis, freeze and fire.

(2) **Translate risks into functional envelopes**
- Thermal envelope: target wall temperature, allowable film temperature, required heat flux, ΔT, fouling factor.
- Dielectric envelope: minimum breakdown at hottest operating temperature, maximum tan δ at temperature, moisture limit that preserves both, PD inception margin.
- Tribology envelope: viscosity window across cold-start to hottest film, air release, demulsibility, antiwear and EP needs.
- Chemical compatibility envelope: metals, solders, coatings, elastomers, plastics; polarity fit and aging chemistry.
- Operability envelope: pumpability, gas solubility and air release, cleanliness class, environmental and disposal profile.

(3) **Write the selection in three parts for every asset**

A. Fluid family and grade justified by physics
- Choose base chemistry and grade using heat flux, dielectric duty, and mechanical regime.

TABLE 7.1
Summary of Equipment-Based Fluid Matching

Equipment/System	Common Fluid Families	Primary Selection Criteria	Key Compatibility Concerns	Core Monitoring Parameters
Transformers, Inverters, Switchgear	Mineral oil (inhibited), natural ester, synthetic ester, silicone, dielectric hydrocarbon, water-glycol	Dielectric strength, fire point, oxidation stability, moisture tolerance, viscosity	Paper insulation, copper, elastomers (NBR, EPDM, FKM), plastics in inverters	Breakdown voltage, moisture ppm, tan δ at temperature, IFT, acid number, inhibitor residual, DGA (transformers), pH/conductivity (water-glycol)
Turbines (Steam/Gas)	Ashless R&O turbine oil (mineral or synthetic), synthetic esters (high temp), PAG (special cases)	Viscosity at service temp, oxidation life (RPVOT), air release, varnish resistance	Seal elastomers, varnish on servo valves, copper/bronze alloys	Viscosity, MPC varnish index, acid number, RPVOT, foam/air release, particle counts, wear metals
Compressors, Pumps, Blowers	R&O oils, PAO synthetics, PAGs, esters	Lubricity, oxidation stability,[12] water tolerance, viscosity under load	Seal elastomers (PAG vs NBR), paint/coatings	Viscosity, acid number, moisture, particle counts, wear metals
Liquid-Cooled Batteries	Water-glycol (OAT, hybrid inhibitors)	Heat capacity, corrosion protection (Al, Cu), low conductivity	Aluminum, copper, EPDM/FKM elastomers, polymer manifolds	pH, conductivity, inhibitor residuals, glycol %, particle counts, metals
Immersion-Cooled Batteries	Dielectric hydrocarbon, synthetic ester, fluorinated dielectric (HFE/HFO)	Breakdown voltage, low tan δ, thermal stability, materials compatibility	Cell adhesives, plastics, potting compounds, elastomers	Moisture ppm, BDV at temperature, tan δ, foaming, polymer extractables
Engine Lubrication Circuits, Valve Trains, Turbos	Engine oils (API/ACEA, mineral or synthetic)	Viscosity grade (H/T/H-S), detergency/dispersancy, oxidation resistance, volatility	Elastomers, turbo bearing seals, catalysts (ZDDP phosphorus)	Viscosity, TBN/TAN, oxidation/nitration/sulfation (FTIR), soot %, metals, insolubles

(continued)

TABLE 7.1 (Continued)
Summary of Equipment-Based Fluid Matching

Equipment/System	Common Fluid Families	Primary Selection Criteria	Key Compatibility Concerns	Core Monitoring Parameters
Nuclear Actuator Assemblies, Rod Drives	Phosphate ester, synthetic hydrocarbon, silicone	Radiation tolerance, fire resistance, oxidation stability	Seal elastomers (avoid NBR, EPDM with esters), metals under radiation	Acid number, resistivity, fine particle counts, water, neutralization capacity
Shielded Nuclear Electronics Cooling	Dielectric hydrocarbon, fluorinated dielectric	Low activation, breakdown voltage, thermal stability	Conformal coatings, potting compounds	Moisture, BDV, tan δ, extractables
Generator Hydrogen Seal Oil	Mineral seal oil (low volatility R&O)	Volatility, air release, oxidation stability	Seal elastomers, hydrogen permeation	TAN, water, air release, foam tendency, DGA
Solar Thermal Loops – Molten Salts	Nitrate/nitrite molten salts	Melting point, decomposition temperature, corrosion resistance	Carbon steel, stainless, nickel alloys	Nitrite/nitrate ratio, moisture ppm, chlorides, metals
Solar Thermal Loops – Thermal Oils	Aromatic synthetic thermal oils	Film temp vs cracking point, oxidation stability	Elastomers, coatings, oxygen ingress	Viscosity, TAN, low/high boiler %, varnish index, particles
CHP / Heat Recovery Fluids	Water-glycol, thermal oil	Heat capacity, corrosion inhibition, thermal stability	Carbon steel, copper alloys, elastomers	pH, inhibitor residuals, glycol %, TAN, viscosity, particles

TABLE 7.2
Failure Modes and Corresponding Monitoring Tests

Equipment/ System	Common Critical Failure Modes	Corresponding Monitoring Tests
Transformers, Inverters, Switchgear	Insulation breakdown from moisture or contamination, oxidation sludge formation, copper sulfur corrosion, dielectric loss increase	Breakdown voltage (ASTM D1816/ D877), moisture ppm (Karl Fischer), tan δ at operating temp (IEC 60247), interfacial tension (ASTM D971), acid number (ASTM D664), inhibitor content, dissolved gas analysis (ASTM D3612)
Turbines (Steam/ Gas)	Varnish-induced servo/control valve sticking, oxidation thickening, air entrainment, foam-induced pump cavitation	Viscosity at 40/100 °C (ASTM D445), MPC varnish index (ASTM D7843), RPVOT (ASTM D2272), foam tendency/stability (ASTM D892), air release (ASTM D3427), particle counts (ISO 4406), patch microscopy
Compressors, Pumps, Blowers	Oxidation sludge, seal swelling/ leakage, corrosion pitting, abrasive wear	Viscosity, acid number, moisture, ICP metals, particle counts, FTIR oxidation/nitration, elastomer soak testing
Liquid-Cooled Batteries	Corrosion of aluminum/copper, inhibitor depletion, electrical leakage via high conductivity, pump wear from solids	pH, conductivity, inhibitor residual (nitrite/molybdate/azoles), glycol %, particle counts, ICP metals
Immersion-Cooled Batteries	Dielectric breakdown, polymer swelling or leaching, moisture ingress, foaming	Moisture ppm, breakdown voltage, tan δ, foaming tendency, GC-MS for extractables, visual inspection for haze or deposits
Engine Lubrication Circuits, Valve Trains, Turbos	Film collapse, oxidation/nitration sludge, soot thickening, turbo coking, wear of bearings and cams	Viscosity, TBN/TAN, FTIR oxidation/nitration/sulfation, soot %, insolubles, ICP wear metals, particle counts
Nuclear Actuator Assemblies, Rod Drives	Acidic decomposition of esters, fine particle generation, elastomer embrittlement or swelling, loss of fire resistance	Acid number, resistivity, Karl Fischer water, fine particle counts (ISO 4406), ICP metals, acid acceptance testing
Shielded Nuclear Electronics Cooling	Dielectric loss increase, moisture ingress, coating delamination, chemical attack on potting	Breakdown voltage, tan δ, moisture, extractables, visual coating inspection
Generator Hydrogen Seal Oil	Foaming, oxidation sludge, hydrogen purity loss via seal leaks, corrosion from CO_2 purge acidification	TAN, water, foam tendency, air release, hydrogen purity, DGA of seal oil

(continued)

TABLE 7.2 (Continued)
Failure Modes and Corresponding Monitoring Tests

Equipment/ System	Common Critical Failure Modes	Corresponding Monitoring Tests
Solar Thermal Loops – Molten Salts	Freeze blockage, nitrate-to-nitrite decomposition, chloride-induced corrosion, moisture ingress	Nitrite/nitrate ratio, chlorides, moisture ppm, ICP metals
Solar Thermal Loops – Thermal Oils	Oxidation thickening, low-boiler loss, high-boiler formation, fouling on heat surfaces	Viscosity, TAN, low/high boiler fractions (GC), MPC varnish index, particle counts
CHP/Heat Recovery Fluids	Corrosion, inhibitor loss, fouling from degradation, viscosity drift	pH, inhibitor residual, glycol %, TAN, viscosity, particle counts

- Declare property targets at operating temperature: viscosity curve, breakdown strength, tan δ, moisture ppm, k and c_p if heat transfer is limiting, flash or fire point, pour point, air release, foaming, oxidation life.

B. **Materials and seals justified by chemistry**
   - Specify elastomers and plastics by polarity and aging chemistry: EPDM, FKM, FFKM, NBR, polyamide (PA), polybutylene terephthalate (PBT), polyphenylene sulfide (PPS), printed circuit board (PCB) laminates.
   - Include acceptance tests: volume change, hardness and tensile retention, mass change after soak at temperature, adhesive and coating compatibility.

C. **Monitoring plan that defends the choice**
   - Name each property to trend, the method, the sampling point, the frequency, the alarm and action.
   - Include conditioning: breathers, dryers, side-stream filtration or polishing, sludge and varnish control, water removal.

(4) **Asset-specific checklists**
Transformers and inverters
- Dielectric strength at temperature; tan δ at temperature; resistivity; moisture ppm and drying method.
- Partial discharge (PD) inception margin; interfacial tension for insulating liquids; dissolved gas analysis (DGA) or gas management for immersion systems.
- Actions: dehydration, reconditioning, antioxidant top-treat if approved, tighten seals and breathers.

Turbines and compressors

- Viscosity at operating temperature; VI; air release and foam tendency; water separation; oxidation life and varnish propensity.
- Trends and tests: RPVOT or International Electrotechnical Commission (IEC) oxidation, MPC or patch rating for varnish, ISO 4406 cleanliness, TAN.
- Actions: side-stream polishing, antioxidant management, filter upgrade, load and temperature derate if needed.

Batteries and power electronics

- If coolant: conductivity window, inhibitor health in water-glycol, materials extractables and leachables.
- If dielectric immersion: breakdown strength at temperature, tan δ, moisture ppm, polymer and adhesive compatibility, void fraction monitoring if two-phase.
- Actions: inhibitor top-ups, leak detection and isolation, drier sizing, component material upgrade.

Nuclear auxiliaries

- Acid control and TAN; radiation tolerance; radiolysis management; ultrafine cleanliness targets.
- Impurities: halides, sulfur, metals; dryness; redox where applicable.
- Actions: cold traps or getters, sub-micron filtration, qualified materials only, strict sampling and chain of custody.

Solar thermal loops

- Freeze protection for water-glycol or freeze management for salts; film temperature and decomposition control.
- For thermal oils: low and high boiler control, antioxidant health; for salts: nitrite to nitrate ratio, chloride limits, dryness.
- Actions: heat tracing and procedures, bleed-and-feed or venting, salt rejuvenation or replacement, corrosion surveillance.

(5) Tie performance to measurable targets and methods

- Thermal performance: specify Reynolds and Prandtl at design conditions; set maximum fouling factor with actions linked to approach temperature and ΔP trends.
- Dielectric and chemistry methods: IEC 60156 and ASTM D924 for dielectric; IEC 61125 or ASTM D2272 for oxidation; ASTM D664 TAN; ASTM D2896 or D4739 TBN where relevant; ASTM E2412 FTIR trends; ASTM D1121 reserve alkalinity for water; defined microscopy protocol for deposits.

(6) Narrow-margin rule and escalation

- If any pillar runs on a narrow margin, increase monitoring cadence, add side-stream polishing or dryers, lower load or temperature, or step up one materials grade to restore margin over supplier and operating variability.

(7) Qualification, baseline, and change control

- Screen in the lab: property verification, compatibility soaks, deposit tendency, dielectric at temperature and moisture steps.
- Prove in a loop: pilot rig under representative heat flux and flow, record start-up and transient behavior.
- Baseline: record initial values for all trended properties; store retain samples.
- Change control: any supplier, base stock, or additive change retriggers compatibility checks and resets baselines; document and approve before use.

8 Regulatory and Environmental Considerations

REGULATORY AND ENVIRONMENTAL CONSIDERATIONS (IEC • IEEE • ASTM • ISO • SAE)

Regulatory and environmental compliance for lubricants, insulating fluids, and coolants is not one monolithic rulebook; it is a patchwork that depends on the asset (transformer vs inverter vs turbine), the facility type (thermal, nuclear, renewable, industrial), and where you operate. The five standards ecosystems below shape *how* you specify, test, label, operate, and retire fluids. A good program harmonizes them into one specification that procurement can enforce and operations can audit (see Table 8.1).

Governing Bodies

- **IEC (International Electrotechnical Commission):** Electrical properties and apparatus performance. You will use IEC for dielectric testing (breakdown, dissipation factor, resistivity), transformer guides, and liquid/fluid behavior inside energized equipment.
- **IEEE (Institute of Electrical and Electronics Engineers):** North American-centric equipment application/maintenance guides, especially transformers and monitoring (e.g., dissolved gas analysis, acceptance criteria, condition assessment).
- **ASTM (American Society for Testing and Materials):** The lab "how-to." Methods for viscosity, acidity, oxidation, water, metals, dielectric strength (legacy methods), corrosion tests for coolants, and fuel/coolant property checks.
- **ISO (International Organization for Standardization):** Management systems (ISO 14001 for environmental management), lubricant classifications (e.g., environmentally acceptable lubricant (EAL)/biodegradable hydraulics), cleanliness (particle counting), sampling, and test method twins to ASTM.
- **SAE (Society of Automotive Engineers):** Engine/gear/transmission oil performance categories, viscosity grade systems, and coolant specifications commonly applied to CHP, engines, and traction inverters in mobility-adjacent power assets.

DOI: 10.1201/9781003775201-11

Cross-cutting Regulatory Themes

- **Fire and life safety:** Flash/fire point and sustained ignition drive enclosure design and siting. Align fluids with *NFPA 30* (flammable and combustible liquids), *NFPA 70/70E* (NEC/electrical safety), and *NFPA 850* (power-plant fire protection). For transformers in dense settings, *high-fire-point* liquids (e.g., natural esters, some silicones) often simplify fire protection and secondary containment.
- **Environmental releases and storage:** Secondary containment, drainage isolation, and spill kits per local law; in the U.S., think *SPCC* planning. Waste classification hinges on halogen content, metals, and acidity; legacy checks for PCB contamination (<2 ppm) remain routine for older transformer systems.
- **Chemical policy trends:**
  - **Per- and poly-fluoroalkyl substances (PFAS) phase-downs/restrictions** affect fluorinated dielectrics and some surfactants; expect procurement disclosure and substitution plans.
  - **Refrigerant/hydrofluorocarbon (HFC) phasedown and global warming potential (GWP) targets** (Kigali-aligned rules) influence two-phase e-coolants and organic Rankine cycle (ORC) working fluids.
  - **Registration, Evaluation, Authorisation and Restriction of Chemicals (REACH)/Restriction of Hazardous Substances (RoHS)/Toxic Substances Control Act (TSCA)** push on substances of very high concerns (SVHCs), certain phthalates, and additive chemistries (e.g., some amines, phenols, and phosphorus compounds).
- **Worker exposure and product stewardship:** Harmonized labeling/ safety data sheet (SDS) under **globally harmonized system (GHS)/ classification, labeling and packaging (CLP)**, ventilation for volatile fractions, and hot-work permitting for thermal oil maintenance.
- **End-of-life and circularity:** Rerefining of hydrocarbon oils, reclaim/dry-out of esters, strict destruction for certain fluorinated materials, and salt reclamation/drying programs.

Standards Touchpoints

Transformers, Inverters, Switchgear

- **Property and performance:** IEC 60156 (BDV), IEC 60247 (tan δ, resistivity), IEC/IEEE transformer guides (insulating liquids, condition limits), IEEE C57 series (acceptance, DGA, maintenance).
- **Environmental levers:** Prefer high-fire-point liquids where siting demands it; set moisture/tan δ limits at *operating temperature*, not just 25 °C; require PCB-free declarations and additive disclosure (passivators, metal deactivators).
- **Monitoring framework:** ASTM D1816/D877 (legacy breakdown voltage (BDV)), ASTM D664 (total acid number (TAN)), ASTM D971 (interfacial

tension (IFT)), IEC 61125 (oxidation stability), ASTM D3612 (dissolved gas analysis (DGA)).

Turbines, Compressors, Pumps, Combined Heat and Power (CHP) Engines

- **Property and performance:** ASTM D445 (viscosity), D2272 (rotating pressure vessel oxidation test (RPVOT)), D892 (foam), D3427 (air release), ISO 4406 (cleanliness), SAE J300 (engine oil viscosity), applicable original equipment manufacturer (OEM) performance sequences (American Petroleum Institute (API)/European Automobile Manufacturers' Association (ACEA)/Society of Automotive Engineers (SAE)).
- **Environmental levers:** Oxidation control = fewer varnish cleanouts; choose EALs (ISO 15380) where discharge risk exists; document nitrogen blanketing for thermal oils (fire safety + oxidation).
- **Monitoring:** TAN/TBN, MPC varnish (ASTM D7843), Fourier transform infrared spectroscopy (FTIR) oxidation/nitration/sulfation (ASTM E2412), water by Karl Fischer (ASTM D6304).

Batteries (Liquid-Cooled and Immersion-Cooled)

- **Property and performance:** For water-glycol, ASTM D1384 (corrosion), D3306/D6210-style coolant expectations, conductivity windows tied to creepage/clearance; for dielectrics, IEC 60156/60247 applied at pack temperature.
- **Environmental levers:** Low toxicity glycols (PG) in people-dense zones; dielectric fluids with documented plastics/elastomer compatibility and low volatility; plan reclaim and moisture management.
- **Monitoring:** pH, inhibitor residuals, conductivity (coolant); BDV, tan δ, moisture ppm (immersion).

Nuclear Auxiliaries and Shielded Electronics

- **Property and performance:** IEC/IEEE equipment guides plus ASTM/ISO corrosion and dielectric methods; radiation tolerance protocols (dose conditioning) added to standard compatibility tests.
- **Environmental levers:** Phosphate esters (fire) vs synthetic hydrocarbons (materials); particle-cleanliness and resistivity limits; radiolysis/byproduct management.
- **Monitoring:** Resistivity, TAN, ultrafine particle codes, moisture, acid acceptance for esters.

Solar Thermal (Molten Salts and Thermal Oils)

- **Property and performance:** ASTM/ISO methods for moisture, chloride, nitrite/nitrate ratio (programmatic, often internal method); thermal oils per ASTM (viscosity, TAN, GC low/high boilers).

- **Environmental levers:** Heat tracing (freeze prevention) is environmental protection, blockage equals spill risk; oxygen exclusion lowers frequency of oil changeouts and waste volumes.
- **Monitoring:** Nitrite/nitrate trend, chlorides, moisture (salts); TAN, GC speciation, MPC, particle counts (oils).

Add jurisdictional fire/environment codes (e.g., NFPA 30, spill/containment rules) in the same section so engineering + EHS read one spec.

Environmental Profiling

- **Biodegradability & ecotoxicity (where discharge risk exists):** OECD 301/302 (ready/inherent biodegradation), OECD 201/202/203 (algae/daphnia/fish), acute aquatic toxicity thresholds; request data packages for candidate EALs (ISO 15380).
- **Volatility and emissions:** Vapor pressure and VOC content for worker exposure and permit accounting; for two-phase coolants or ORC fluids, demand **GWP/ODP** declarations and end-of-life reclaim plans.
- **Hazard classification and labeling:** Current **SDS** per GHS, transport classes (UN/DOT/ADR/IATA), and any country-specific disclosure (REACH registration or TSCA inventory).

TABLE 8.1
Standards for Which Check

Decision/Check	Typical Standard(s) to Cite
Dielectric strength and losses	IEC 60156 (BDV), IEC 60247 (tan δ, resistivity); ASTM D1816/D877 (legacy)
Oxidation stability	IEC 61125 (insulating liquids), ASTM D2272 (RPVOT) for turbine/circulating oils, RULER
Acidity/neutralization	ASTM D664 (TAN), ASTM D974 (colorimetric TAN/TBN), ASTM D2896/D4739 (TBN)
Water content	ASTM D6304 (Karl Fischer), ISO 12937 (water in petroleum)
Cleanliness/particles	ISO 4406 (coding), ISO 11171 (calibration), membrane patch microscopy (internal/ASTM practices)
Coolant corrosion	ASTM D1384 (simulated service), ASTM D4340 (Al hot surface), ASTM D3306/D6210 guidance
Foam and air release	ASTM D892 (foam), ASTM D3427 (air release)
Viscosity	ASTM D445 (kinematic), SAE J300 (engine oil grades)
Transformer condition	IEEE C57 series (acceptance/maintenance), ASTM D3612 (DGA), ASTM D971 (IFT)
Environmental management	ISO 14001 (EMS) + local spill/permit rules; ISO 15380 (EAL criteria for hydraulics)

- **PFAS/Halogen statements:** Written disclosure of intentionally added PFAS/halogens, plus mass-balance and pathway for end-of-life (incineration vs reclaim).
- **Recyclability/reuse:** Rerefining routes for hydrocarbon oils; reclaim/dry-out for esters; closed-loop reclaim for special dielectrics; salt dehydration and reuse criteria.

Compliance Workflow

1. **Pre-bid spec:** Lock *properties + tests + environmental declarations*. Cite IEC/IEEE for electrical, ASTM/ISO for lab methods, NFPA/local codes for fire/containment, and require SDS + PFAS/GWP/biodeg statements.
2. **First-article qualification:** Verify electrical (BDV/tan δ at **operating temp**), chemistry (TAN/water/oxidation), compatibility (ASTM D471/ISO 1817 elastomers; D543 plastics; coating adhesion), and coolant corrosion panels (D1384/D4340). Include *aged-fluid retests*.
3. **Operational controls:** Moisture/oxidation/cleanliness limits tied to alarm/action levels; for coolants, inhibitor residuals and conductivity/pH windows; for dielectrics, moisture + tan δ trending and DGA (transformers).
4. **Change control:** Any supplier/base-stock/additive change triggers a delta-qualification (key IEC/ASTM tests + compatibility slices).
5. **End-of-life:** Define reclaim vs disposal, transport codes, and certificates of destruction/recycling. Document spill plans and training.

Procurement language

- "Fluid shall meet **IEC 60156** and **IEC 60247** limits **at 80–90 °C** and after **IEC 61125** oxidation aging."
- "Turbine oil shall achieve **ASTM D2272 RPVOT ≥ X min**, **ASTM D892** foam stability within limits, **ISO 4406 ≤ 16/14/11** in service, and **MPC ≤ Y**."
- "Coolant shall pass **ASTM D1384** corrosion limits for our metals set and maintain **conductivity ≤ Z µS/cm** at 25 °C."
- "Supplier must disclose **PFAS**, **GWP**, **VOC**, and provide **REACH/TSCA** status, **SDS (GHS)**, and reclaim/disposal pathway."
- "Transformer liquid must be **PCB-free**, meet **IEEE/IEC insulating liquid guides**, and be compatible with cellulose and copper at operating moisture."

Use **IEC/IEEE** to define *electrical fitness*, **ASTM/ISO/SAE** to define the *lab numbers and durability*, and bolt on **fire, spill, chemical policy, and end-of-life** requirements that match your jurisdiction. Put the whole bundle into procurement up front and tie it to acceptance testing and operating alarms. That way, environmental and regulatory compliance becomes a property of the *spec* and not a scramble after the fluid is already in your equipment.

Fire Code and Approval Framework (FM Global • UL)
Fire safety for fluids in power infrastructure is dictated not only by **NFPA** and local building codes but also by **third-party approval bodies** like FM Global and Underwriters Laboratories (UL). For transformers, turbines, inverters, and high-energy electrical systems, fluid fire performance directly influences enclosure design, containment requirements, and insurance premiums. Both FM Global and UL maintain their own **flammability and fire point performance standards**, which overlay the electrical and chemical performance rules of IEC, IEEE, ASTM, ISO, and SAE.

FM GLOBAL – PROPERTY LOSS PREVENTION STANDARDS

FM Global's *Property Loss Prevention Data Sheets* define the insurance-accepted minimums for flammability, containment, and protection systems. Key FM documents relevant to power fluids include:

- **FM Global DS 5-4 – Transformers**
  Requires transformer insulating liquids to meet minimum fire point thresholds (typically >300 °C for "less flammable" classification) and pass FM Global's full-scale fire exposure tests. Fluids meeting criteria are listed in the **FM Approved – Less Flammable Transformer Fluid** database.
- **FM Global DS 7-99 – Turbines and Lubrication Systems**
  Guidance on oil system layout, secondary containment, and nitrogen blanketing for high-volume lube reservoirs to prevent pool fires.
- **FM Approvals Standard 6930** – Certification protocol for "Less Flammable Electrical Insulating Liquids," including fire point verification, sustained combustion testing, and material compatibility.

Practical impacts:

- Meeting FM Approved status often allows reduced spacing from buildings or between transformers.
- Failure to meet FM criteria may force physical fire barriers, deluge systems, or expanded containment, raising both capex and opex.

UL (UNDERWRITERS LABORATORIES) – SAFETY CERTIFICATION

UL's **Safety Standards** and **Component Recognition Programs** provide third-party validation for fluid flammability and compatibility in electrical equipment:

- **UL 340 – Outline of Investigation for Insulating Liquids for Transformers and Similar Electrical Equipment**
  Covers flash point, fire point, and sustained combustion tests, as well as dielectric breakdown, corrosivity, and compatibility with copper and insulation.
- **UL 94 – Flammability of Plastic Materials for Parts in Devices and Appliances**

Applied to polymer housings, seals, and tanks that contact the fluid, important for immersion cooling and battery packs.

- **UL 1978 – Grease Duct Construction** (applicable by analogy in oil piping) for materials exposed to hot oil and fire conditions.
- **UL Environmental Claim Validation** programs can verify sustainability claims (e.g., biodegradability, bio-based content) relevant for "green" transformer fluids.

Practical impacts:

- UL recognition of a fluid or fluid-contact component can be a procurement requirement for OEM supply.
- For battery immersion fluids and e-coolants, UL listing may be a precondition for product safety certification of the whole assembly.

Fire Safety Property Criteria

Across National Fire Protection Association (NFPA), FM approvals/factory mutual (FM), and underwriters laboratories (UL), the defining fluid flammability parameters are:

Property	Typical Standard Method	Typical "Less Flammable" Threshold
Flash Point	ASTM D92 (Cleveland Open Cup)	>275 °C
Fire Point	ASTM D92	>300 °C
Sustained Combustibility	ASTM D5306 / FM Proprietary Method	No sustained burn under specified exposure
Self-Extinguishing Time	FM / UL Method	<1–2 seconds after ignition source removal
Heat Release Rate	Cone calorimetry or FM full-scale	Below FM's max limit for class

Integration with IEC / IEEE / ASTM

- IEC/IEEE specify **electrical performance**, ASTM provides **property test methods**, and FM/UL validate **fire safety and real-world hazard mitigation**.
- Example: A natural ester transformer oil might pass IEC 60156 and ASTM D1816 for BDV, ASTM D92 for fire point, and then achieve FM Approved listing, unlocking reduced fire protection requirements under NFPA 850 and FM DS 5-4.

Environmental and Insurance Considerations

- **FM Approved** or **UL Listed** fluids can reduce insurance surcharges and improve insurability of high-value assets.

- Fire point classification can affect **secondary containment design** (e.g., oil-water separators for hydrocarbons vs sealed sumps for esters).
- For renewable and nuclear assets, using FM Approved fluids can lower permitting complexity where environmental and fire codes overlap.

Specification Language Example

Insulating liquid shall have a flash point ≥275 °C and fire point ≥300 °C per ASTM D92, pass IEC 61125 oxidation stability, and be FM Approved as a 'Less Flammable Electrical Insulating Liquid' per FM Standard 6930. All fluid-contact polymers shall meet UL 94 V-0 flammability classification.

REACH (EU REGULATION 1907/2006)

REACH is about **chemicals placed on the EU market**: registration of substances, communication of hazardous properties down the supply chain, and controls on substances of very high concern (SVHC). For fluids, three parts matter:

- **Registration** (your supplier/importer's job): any **substance** ≥1 t/y placed on the EU market must be registered with ECHA (polymers largely exempt; monomers are not). Formulated fluids are **mixtures** of substances; each substance must be covered by a registration somewhere up the chain.
- **Communication and SDS**: you must receive a **CLP-compliant SDS** (16 sections) in the local language; if a substance is registered with exposure scenarios, an **extended SDS** is required. You must pass safety info to your users.
- **SVHC/SCIP**: if any **SVHC (Candidate List)** is present ≥0.1% w/w in your **mixture**, you must **provide an SDS upon request** and proactively inform professional recipients. (SCIP database notifications are for **articles**, but your fluid-wetted hardware will still need article-level SVHC handling.) Watch also **Annex XIV (Authorisation)** and **Annex XVII (Restrictions)**: if your fluid contains a substance on Annex XIV past its sunset date (without an authorisation), it can't be used; Annex XVII bans/restricts uses (e.g., certain PAHs, some phthalates, NMP, etc.).
- **Newer hazard classes** under CLP (the EU's GHS) include **ED/EDa (endocrine disruptors)** and **PBT/vPvB/PMT/vPvM**; suppliers are starting to relabel accordingly. Expect more candidate listings (e.g., **PFAS** proposals).

GHS / CLP (GLOBALLY HARMONIZED SYSTEM; EU CLP REG. 1272/2008)

GHS harmonizes **classification, labels, and SDS**. In the EU, GHS is implemented via **CLP**; in the U.S., via **OSHA HCS**. For fluids, this defines:

- **Classification** (e.g., flammable liquid, aspiration hazard, skin sensitizer, aquatic toxicity).

- **Label elements** (pictograms, signal word, H- and P-statements) on every container.
- **SDS** format and content (16 sections) and language/local poison-center requirements. In the EU, **UFI codes + Poison Centre Notification (PCN)** are required for most hazardous **mixtures** (your coolants/lubes) placed on the market.

ROHS (EU 2011/65/EU AS AMENDED "ROHS 3" 2015/863)

RoHS restricts **ten substances in electrical/electronic equipment (EEE)** at the **homogeneous material** level (Pb, Hg, Cd, Cr^{6+}, PBB, PBDE, DEHP, BBP, DBP, DIBP). **Fluids themselves aren't EEE**, so RoHS doesn't directly regulate a bottle of oil or coolant. It applies **if the fluid is an integral part inside EEE sold as a product** or if your **system-level Declaration of Conformity** demands RoHS across *all* supplied materials. In practice, many OEMs still collect **RoHS declarations for fluids** to avoid introducing restricted flame retardants or phthalates via maintenance kits or pre-fills.

What This Means in Real Purchasing/Spec Language

- **Always require**: (1) **REACH compliance statement** naming any SVHC ≥0.1% w/w; (2) **current CLP/GHS SDS** in the local language; (3) for EU: **UFI** and confirmation of **PCN submission** for hazardous mixtures; (4) **Annex XVII** conformance; (5) disclosure of any **Annex XIV** substance and authorisation status.
- **Often require**: **RoHS 3 declaration** (yes/no for the 10 substances), even though fluids are mixtures, this keeps restricted phthalates and brominated FRs out of your electrical asset BOM.
- **If global**: ask for **REACH-like compliance** (UK REACH, Turkey **KKDIK**, Korea **K-REACH**, China **MEE Order 12**) and **GHS country labels**.
- **PFAS and halogen policy**: many utilities now require a **PFAS disclosure** (intentionally added?) and total organic fluorine screening thresholds for certain applications (immersion coolants).

Two Quick Tables You Can Drop into a Spec (see Tables 8.2 and 8.3)

Implementation Tips That Save Rework

- **Use "latest Candidate List" language** in every PO/spec so your supplier remains obligated when ECHA adds SVHCs (happens ~2x/year).
- **Mixtures vs articles**: Fluids are mixtures; RoHS/SCIP mainly hit **articles**. Still, require declarations to protect your *system* compliance.
- **UFI/PCN** (EU): if you privately rebrand/repackage a coolant, *you* become the duty holder for **UFI and PCN**, plan for it.
- **Labels must match SDS** (same classification, same languages). In multi-country projects, collect country-specific GHS labels up front.

TABLE 8.2
Scope and Trigger Matrix

Regime	What it Covers for Fluids	Triggers You Must Check	Your Evidence to File
REACH	Substances in mixtures; SVHC, Restrictions/ Authorisation	Any SVHC ≥0.1% w/w; Annex XVII bans; Annex XIV authorisation	Supplier **REACH statement**, **SDS**, SVHC disclosure; (EU) confirm **registrations** exist upstream
CLP / GHS	Classification, labels, SDS format/content	Mixtures with hazardous classification placed on market	**SDS (16-section)**, compliant **label**, (EU) **UFI + PCN** proof
RoHS 3	EEE homogeneous materials; fluids only if part of EEE or by OEM policy	Ten substances at limit values (0.1%/ 0.01%)	**RoHS declaration** (yes/ no) for fluid; supplier test data if high risk
PFAS (policy)	Not (yet) one law; multiple proposals/restrictions	Intentionally added PFAS; total fluorine screening	**PFAS disclosure**, tech data; substitution plan if required

- **Test when needed**: if a supplier won't give a clean RoHS/SVHC declaration for a "legacy" fluid, spot-test high-risk analytes: **ICP-MS** for Pb/Cd/Cr, **GC-MS** for phthalates/PBB/PBDE, **ion chromatography** for halides; consider **TOF (total organic fluorine)** screening where PFAS risk exists.
- **Document substitutes**: If you swap from mineral to natural ester (transformers) or from HFC to HFO/HFE (two-phase cooling), update **SDS libraries, labels, maintenance instructions**, and **spill plans** (viscosity/fire behavior differ).

Sample Compliance Clauses

- "Supplier warrants REACH compliance for all substances in the mixture and shall notify the purchaser within 15 days of any SVHC addition ≥0.1% w/w."
- "Mixture shall be classified, labeled, and supplied with an SDS compliant with **CLP/GHS**; for EU, a **UFI** shall be affixed and **PCN** completed prior to delivery."
- "Supplier shall provide a **RoHS 2011/65/EU (2015/863)** declaration covering Pb, Hg, Cd, Cr^{6+}, PBB, PBDE, DEHP, BBP, DBP, DIBP."
- "Supplier shall disclose intentionally added **PFAS**; fluids containing such substances may be rejected or require an approved substitution plan."

TABLE 8.3
Supplier Questionnaire (attach to RFQ/PO)

Topic	Ask the Supplier...	Why it Matters
REACH registration	"Are all **substances** in this mixture registered for EU use? Provide registration numbers or Only Representative details."	Legal placement on EU market
SVHC	"List any **SVHC** ≥0.1% w/w (latest Candidate List date). If none, state 'none'."	Customer communication duty
Annex XVII/XIV	"Confirm **no Annex XVII** restriction is breached; declare any **Annex XIV** substances and authorisation status."	Use bans / sunset control
SDS and CLP	"Provide a current **SDS** (language of destination), **label artwork**, and (EU) **UFI** + **PCN** status."	Safety, transport, poison centre
RoHS	"Provide a **RoHS 2011/65/EU (2015/863)** declaration for the 10 substances."	Keep restricted phthalates/FRs out
Halogens/PFAS	"Declare intentionally added **PFAS/halogens**; provide total halogen/fluorine if available."	Future bans; compatibility
Biodegradability	"Provide **OECD 301/302** and ecotox (201/202/203) if marketing as 'environmentally acceptable'."	EAL claims, permits
Change control	"Commit to **90-day notice** of formulation/base-stock/additive changes."	Requalification trigger

Global Warming Potential (GWP)

Definition and Relevance

- **GWP** measures the climate impact of a greenhouse gas over a time horizon (usually 100 years, GWP_{100}), expressed relative to CO_2 (GWP = 1).
- In insulation and cooling applications, GWP is relevant for:
  - **Two-phase dielectric cooling fluids** (fluoroketones, fluoronitriles, HFCs, HFOs, HFEs).
  - **Gas insulation** (SF_6, N_2/SF_6 blends, Novec 4710/5110).
  - **Refrigeration loops** in liquid-cooling systems.

Engineering Trade-Offs

- Many legacy fluids (e.g., SF_6, HFC-245fa) have extremely high GWP (>7,000 for SF_6) but excellent dielectric or thermal performance.
- Low-GWP replacements (e.g., 3M Novec 649, Chemours Opteon series) may have:
  - Lower boiling points → reduced high-temperature stability.
  - Higher vapor pressure → stricter containment requirements.

- Different material compatibility profiles (e.g., elastomer swelling in F-Ketones).

Regulatory and Market Pressures

- **EU F-gas Regulation**: GWP thresholds and phasedown schedule for placing fluorinated gases/mixtures on the market.
- **Kigali Amendment to the Montreal Protocol**: global HFC phasedown, affects OEM selection for high-GWP coolants.
- Many utilities and data center operators now include **GWP < 10 or < 150** as a procurement spec.

Best Practices

- Specify **GWP limit** at the product level (mixture-weighted value).
- Require **lifecycle leakage rate modeling**: GWP impact = fluid GWP × annual leakage × asset life.
- Evaluate **recovery/reclamation pathways** for end-of-life fluids.

PFAS (Per- and Polyfluoroalkyl Substances)

Definition and Scope

- PFAS is a broad chemical class containing a **perfluorinated carbon backbone**. This includes many fluorinated dielectric fluids (e.g., perfluorocarbons, some fluoroketones, fluoronitriles) and additives used for thermal stability or surfactancy.
- PFAS are persistent, bioaccumulative, and increasingly regulated due to toxicity and environmental persistence.

PFAS in Insulating and Cooling Fluids

- **Present** in many fluoroketone and fluoronitrile fluids (Novec 4710, 649) and some HFEs, though short-chain PFAS have shorter half-lives than legacy PFOA/PFOS.
- **Potentially present** as processing aids or degradation products in non-fluorinated coolants if manufactured in PFAS-containing plants.
- **Absent** in hydrocarbon-based fluids, esters, glycol-water mixes, unless contaminated in manufacturing.

Regulatory and Policy Developments

- **EU REACH PFAS restriction proposal** (2023): would ban or severely restrict manufacture/use of most PFAS with a few time-limited derogations (some for high-reliability electrical equipment).

- **US EPA**: PFAS reporting under TSCA, potential designations as hazardous substances under CERCLA, and drinking water MCLs for specific PFAS.
- **OEM & data center PFAS policies**: voluntary phase-outs, especially where non-PFAS alternatives exist.

Engineering Risks

- If PFAS-containing fluids leak, they create long-term remediation liability; in some jurisdictions, costs are strict-liability.
- Disposal routes are narrowing: incineration requirements are tightening, and landfill bans are emerging.
- In high-heat duty, some PFAS fluids can thermally degrade to form toxic perfluoroisobutylene (PFIB) or HF gas, requires engineering controls.

Selection and Procurement Guidance

Condition Monitoring for GWP and PFAS-Critical Fluids

Even though GWP and PFAS are environmental/compliance metrics (see Table 8.4), the same **fluid health testing** that prevents equipment failure also prevents leaks and unplanned releases:

TABLE 8.4
GWP and PFAS Decision Matrix

Fluid Category	Typical GWP	PFAS Content Risk	Notes
SF_6	~23,500	High (perfluorinated)	Outstanding dielectric; near-total phaseout pressure
HFCs (R-245fa, R-134a)	1,300–1,430	Low PFAS	Kigali phaseout; climate liability
HFOs (R-1234ze, R-1234yf)	<10	Low PFAS	Low-GWP refrigerants; may require higher system charge pressures
Fluoroketones (Novec 649, 4710, 5110)	1–2	Yes (short-chain PFAS)	Marketed as low-GWP SF_6 alternatives; PFAS scrutiny growing
Hydrocarbon synthetics (PAO, alkylbenzenes)	0	No	Non-dielectric; use in thermal loops only
Esters (natural/synthetic)	0	No	Biodegradable options for transformer insulation
Water-glycol	0	No	High heat capacity; dielectric limitations
Molten salts	0	No	For high-temp thermal storage; corrosion concerns

- **Tightness Testing**: Pressure decay or helium leak detection to prevent fugitive GHG/PFAS emissions.
- **Headspace GC/MS**: Can confirm composition stability for refrigerants/fluorochemicals.
- **Fluoride Ion Analysis**: Early indicator of PFAS fluid thermal decomposition.
- **Annual Mass Balance Audits**: Tracks leakage rate vs allowable limits.

Specification Language Example

Fluid shall have a 100-year GWP ≤ 10 per IPCC AR6, verified by supplier declaration, and contain no PFAS substances as defined by OECD 2021 classification unless specifically approved by the purchaser. If PFAS are present, supplier shall provide a phase-out roadmap and disposal/recycling plan meeting applicable REACH, TSCA, and local regulations.

Biodegradability

Definitions and Test Standards

- **Readily biodegradable**: ≥60% mineralization within 28 days (OECD 301 series; ISO 9439, 9408).
- **Inherently biodegradable**: >20% but <60% within 28 days (OECD 302 series).
- **Not biodegradable**: <20% degradation in standard tests.
- **Marine/aqueous-specific testing**: OECD 306, ASTM D6691 for seawater conditions.
- **Eco-label thresholds** (EU Ecolabel, Vessel General Permit [VGP]): also include aquatic toxicity (OECD 201/202/203) and bioaccumulation (BCF/LogKow limits).

Typical Fluid Profiles

- **Natural esters (transformer oils)**: readily biodegradable; pass OECD 301B; often >95% degradation in 21 days.
- **Synthetic esters (polyol esters, phosphate esters)**: high biodegradation for many POEs; phosphate esters variable due to hydrolysis resistance.
- **PAOs and mineral oils**: generally not readily biodegradable; can meet inherent biodegradability with highly isoparaffinic base stocks.
- **Water-glycol coolants**: glycol fraction readily biodegradable; inhibitors may slow degradation.
- **Fluoroketones/fluoronitriles**: persistent; low biodegradation; environmental persistence concerns.
- **Molten salts**: non-biodegradable but inert in natural systems; dissolution creates ionic load rather than organic pollution.

Disposal and End-of-Life Requirements

Regulatory Frameworks

- **Basel Convention**: controls cross-border movement of hazardous waste (includes waste oils, halogenated coolants).
- **OECD Council Decision C(2001)107**: aligns hazardous waste codes between member countries.
- **EU Waste Framework Directive (2008/98/EC)**: classifies waste oil as hazardous; mandates priority for regeneration/rerefining over incineration.
- **US RCRA (40 CFR 279)**: defines used oil management standards; prohibits disposal in landfills; requires tracking of hazardous components (halogens, heavy metals).
- **Local permits**: Many jurisdictions require generator permits for storage >90 days or above certain volumes.

Preferred End-of-Life Routes

- **Mineral/PAO lubricants**:
  - *Best*: rerefining to base oil.
  - *Acceptable*: energy recovery via high-temp incineration (>1100 °C) with scrubbing.
- **Natural/synthetic esters**:
  - *Best*: biodegradation in controlled compost or wastewater treatment if contamination is minimal.
  - *Acceptable*: energy recovery if heavily contaminated.
- **Water-based coolants**:
  - Treat for pH and inhibitor content; discharge to sewer only if meeting local effluent criteria.
  - Remove glycols if >25%, many plants use vacuum distillation.
- **Halogenated fluids (SF_6, fluoroketones)**:
  - Capture/recover in closed-loop systems.
  - Contract with certified reclaimers; destruction by plasma arc or specialized thermal oxidation.
- **Molten salts**:
  - Allow to solidify; manage as inorganic industrial waste.
  - Possible recovery via purification and reuse if contamination is low.

Engineering and Monitoring for Biodegradability and Disposal Planning

- **Baseline testing**: Require biodegradability and ecotox data in supplier SDS or separate test reports (OECD 301/302 + aquatic toxicity) (see Table 8.5).

TABLE 8.5
Biodegradability and Disposal Summary Table

Fluid Type	Biodegradability	Typical Waste Code (EU/US)	Disposal Route	Notes
Natural ester transformer oil	Readily biodegradable	13 03 07* / Used Oil	Composting, energy recovery	Sensitive to oxidation
Synthetic ester	Readily to inherently biodegradable	13 03 08*	Energy recovery, rerefining	Hydrolysis risk
Mineral oil (non-halogenated)	Inherently biodegradable	13 02 05*	Rerefining, energy recovery	Persistent in soil
PAO	Inherently biodegradable	13 02 08*	Energy recovery	Non-polar, low water solubility
Water-glycol	Readily biodegradable (glycol)	16 01 14 / 16 10 01*	Wastewater treatment	pH adjustment needed
Fluoroketones	Persistent	14 06 03*	Reclamation, high-temp incineration	PFAS policy risk
SF_6	Persistent	N/A (gas)	Capture/reclaim	Extreme GWP
Molten salts	Non-biodegradable	06 03 13*	Solid waste management	Corrosive potential

- **Segregated waste streams**: Prevent mixing biodegradable fluids with persistent or halogenated fluids, keeps disposal cost and hazard classification down.
- **Condition monitoring link**: Early fluid change-out (before severe contamination) increases rerefining feasibility and avoids reclassification as hazardous waste.
- **Spill planning**: For biodegradable fluids, biodegradation rates still drop in cold/anaerobic soils, containment and recovery are still required by law.

Biodegradability vs Performance Trade-offs

- **Natural esters**: High biodegradability but higher viscosity, lower oxidation stability at elevated temps, requires nitrogen-blanketed storage and sealed transformer design.
- **Synthetic esters**: Balance of biodegradability and high-temp stability; more expensive; ester hydrolysis risk with water ingress.
- **Non-biodegradable synthetics**: PAOs, silicone oils, stable but persistent; disposal cost and compliance burden higher.

Example Specification Clause

Fluid shall meet at least 60% degradation in 28 days per OECD 301B (readily biodegradable) or >20% per OECD 302B (inherently biodegradable) unless a functional requirement dictates otherwise. Waste management shall prioritize regeneration or recovery over disposal. Supplier shall provide full end-of-life handling instructions, including waste codes (EWC or RCRA) and recommended disposal contractors.

Section IV

Condition Monitoring and Diagnostics

9 Oil and Grease Condition Monitoring

CONDITION MONITORING AND DIAGNOSTICS FOR OIL AND GREASE

Effective condition monitoring of lubricants and greases in electrification and power generation assets, turbines, compressors, generators, gearboxes, transformers, inverters, and high-speed rotating equipment, relies on correlating **chemical, physical, and wear-related indicators** to both **fluid health** and **machine health**. By combining standard laboratory analysis with in situ measurements, operators can optimize drain intervals, detect early failures, and ensure compliance with equipment and environmental requirements (see Table 9.1).

Viscosity

- **Purpose**: Primary performance property for lubricants; affects hydrodynamic film thickness, cooling, and sealing.
- **Test Methods**:
  - ASTM D445 (kinematic viscosity at 40 °C and 100 °C)
  - ASTM D7279 (automatic viscometry)
- **Interpretation**:
  - ↑ Viscosity: oxidation, polymerization, contamination with soot or solids.
  - ↓ Viscosity: fuel dilution, shear of viscosity index improvers.
- **Application**: For turbine oils, ±5% deviation from baseline is actionable; for engine oils, ±10–15% may be acceptable depending on duty.

Viscosity is the single most critical property of a lubricant because it governs the thickness and stability of the fluid film that separates moving surfaces, directly determining whether hydrodynamic, elastohydrodynamic, or boundary lubrication regimes are maintained. If viscosity is too low, the lubricant film thins, allowing asperities on mating surfaces to contact and accelerate adhesive and abrasive wear; if it is too high, fluid flow becomes restricted, increasing internal friction, energy loss, and operating temperatures. In cooling and insulating fluids, viscosity also controls

DOI: 10.1201/9781003775201-13

convective heat transfer rates and pumpability, particularly in low-temperature startup conditions.

Changes in viscosity from the new-fluid baseline often signal chemical or mechanical distress; oxidation and polymerization increase viscosity, while fuel dilution, base oil shear, or thermal cracking reduce it. Monitoring viscosity through standardized methods such as American Society for Testing and Materials (ASTM) D445 or ASTM D7279 allows operators to detect these shifts early, correlate them with other condition monitoring data, and take corrective action before loss of lubrication integrity leads to equipment failure. In practice, maintaining viscosity within tight limits is fundamental to extending component life, optimizing energy efficiency, and ensuring that lubricants and coolants perform their intended mechanical, thermal, and protective functions under all operating conditions.

Total Base Number (TBN)/Total Acid Number (TAN)

- **Purpose**:
  - **TBN** (ASTM D2896 / D4739): Indicates reserve alkalinity for neutralizing acids in engine oils.
  - **TAN** (ASTM D664): Measures acidity from oxidation or additive depletion in industrial oils.
- **Interpretation**:
  - Declining TBN → loss of acid-neutralizing capacity.
  - Rising TAN → oxidation, contamination, acidic degradation products.
- **Application**: TBN should not drop below 50% of new oil value; TAN increase of 1–2 mg KOH/g above baseline for turbines/transformers triggers change.

TBN and TAN are key chemical indicators of a lubricant's remaining service life and its ability to protect equipment from corrosive wear. TBN measures the reserve alkalinity in a lubricant, typically provided by detergent and dispersant additives, which neutralizes acids formed during fuel combustion, oxidation, or contamination. In engines, a declining TBN signals depletion of this neutralizing capacity, leaving surfaces vulnerable to acid attack and corrosion, particularly in high-sulfur or high-nitrogen fuel environments. TAN, on the other hand, measures the concentration of acidic compounds in the lubricant, which can arise from oxidation of the base oil, breakdown of additives, or ingress of acidic contaminants. A rising TAN is often an early sign of lubricant degradation, varnish or sludge formation, and increased risk of corrosive wear in bearings, gears, and other load-bearing surfaces. For transformer and turbine oils, an increase in TAN of 1–2 mg KOH/g above the new-oil value is generally considered actionable. Monitoring TBN and TAN together provides a more complete picture of lubricant health, TBN reveals the remaining defense against acids, while TAN shows the degree of acidic degradation already occurring. By trending these values over time and comparing them to original equipment manufacturer (OEM) limits, operators can accurately determine change-out intervals, avoid premature lubricant disposal, and prevent corrosion-related failures.

Oxidation/Nitration

- **Purpose**: Tracks chemical degradation of oil base stock.
- **Test Methods**: as Fourier transform infrared (FTIR) spectroscopy (ASTM E2412, D7414).
- **Interpretation**:
  - ↑ Oxidation → loss of antioxidant package, increased TAN, sludge formation.
  - ↑ Nitration (combustion engines) → exposure to NOx, high-temperature stress.
- **Application**: Used alongside rotating pressure vessel oxidation test (RPVOT) (ASTM D2272) for turbine and hydraulic fluids to forecast antioxidant exhaustion.

Oxidation and nitration are chemical degradation processes that progressively compromise a lubricant's performance and service life, making their monitoring essential for reliable operation of engines, turbines, compressors, and other lubricated systems.

Oxidation occurs when the lubricant reacts with oxygen, typically accelerated by high temperatures, catalytic metal surfaces, and the depletion of antioxidant additives. This reaction produces acidic compounds, sludge, and varnish precursors, which increase viscosity, reduce heat transfer efficiency, and promote deposit formation on critical surfaces.

Nitration results from the reaction of lubricant molecules with nitrogen oxides (NOx) formed during combustion or high-temperature operation in gas-fired equipment. It leads to the formation of nitro-compounds that promote oil thickening, elastomer degradation, and additive depletion. Both processes can occur simultaneously, particularly in high-load, high-temperature conditions, and their effects are cumulative, accelerating wear, restricting oil flow, and increasing operating temperatures.

Monitoring oxidation and nitration using techniques such as FTIR spectroscopy (ASTM E2412, ASTM D7414) allows for early detection of chemical stress before physical properties like viscosity and TAN shift beyond acceptable limits. By trending oxidation and nitration values against baseline and OEM thresholds, maintenance teams can intervene with oil reclamation, additive replenishment, or timely replacement, preventing varnish-induced sticking of valves, bearing failures, or thermal runaway conditions.

Wear Metals and Elemental Analysis

- **Purpose**: Identifies wear trends, contamination, and additive levels.
- **Test Methods**:
  - Inductively coupled plasma optical emission spectroscopy (ICP-OES ASTM D5185): detects metals in <5–8 µm range.
  - Rotating disk electrode spectroscopy (RDE) spectroscopy (ASTM D6595): good for trending larger wear particles.

    - X-ray fluorescence (XRF): Non-destructive, field-usable for some metals.
- **Key Wear Metals**: Fe (steel), Cu (bearings, windings), Al (pistons, housings), Sn/Pb (babbitt), Cr (plating), Ag (contacts).
- **Application**: Trending more critical than single-point readings; use control charts.

Wear metals and elemental analysis provide a direct window into both the health of the lubricant and the condition of the machine components it protects. By detecting trace metals suspended in the oil or grease, these tests identify active wear, contamination, and additive depletion before failure symptoms appear.

Using techniques such as ICP-OES (ASTM D5185), RDE (ASTM D6595), or XRF, elemental analysis can quantify wear metals like iron (steel components), copper (bearings, windings, and coolers), aluminum (housings and pistons), lead and tin (babbitt bearings), chromium (plating and hard surfaces), and silver (contacts and relays).

In addition to wear debris, these methods detect elements from environmental contamination such as silicon from dirt ingress or from the lubricant's additive package, enabling tracking of additive depletion over time. Trending wear metal concentrations is more important than single-sample results, as gradual increases can reveal early-stage problems like misalignment, inadequate lubrication, or electrical discharge damage. For electrified and high-value power generation equipment, pairing elemental analysis with particle count and ferrography enhances root cause accuracy, distinguishing between normal break-in wear, corrosion, and catastrophic failure modes. Regular wear metal and elemental trending form a cornerstone of predictive maintenance, reducing unplanned downtime and extending asset life.

FTIR Spectroscopy

- **Purpose**: Multiplex test detecting oxidation, nitration, sulfation, glycol contamination, soot, water.
- **Advantages**: Rapid, minimal sample prep; can be embedded in online sensors.
- **Application**: Establish baseline spectral fingerprints for each lubricant grade to enhance anomaly detection.

FTIR spectroscopy is a versatile condition monitoring tool that detects a wide range of lubricant degradation products and contaminants by measuring the infrared absorption spectrum of a fluid sample. Each chemical bond in the lubricant whether in the base oil, additives, or contaminants absorbs infrared light at characteristic wavelengths, creating a "fingerprint" that can be compared against baseline spectra from new oil.

In lubrication and coolant analysis, FTIR is commonly used to quantify oxidation, nitration, and sulfation products, as well as detect fuel dilution, glycol contamination, water, and soot loading. Methods such as ASTM E2412 and ASTM D7414 provide standardized approaches for interpretation, enabling consistent trending across service intervals. FTIR is particularly valuable because it is rapid, requires minimal sample preparation, and can analyze multiple degradation markers in a single scan.

In high-value power generation, electrification, and transformer systems, FTIR can be deployed both in the laboratory and in online sensor platforms for near-real-time monitoring, providing early warning of chemical stress before physical properties like viscosity or TAN drift significantly.

By integrating FTIR trending with other condition monitoring data; elemental analysis, particle count, and water content, maintenance teams can detect emerging problems sooner, extend fluid life, and avoid equipment damage from varnish formation, acid buildup, or loss of dielectric integrity.

Particle Count – ISO 4406

- **Purpose**: Quantifies particulate contamination by size range.
- **Test Methods**: ISO 4406:2017 code format (e.g., 18/15/12 for >4/6/14 μm).
- **Application**:
  - Critical for hydraulics, turbines, gearboxes.
  - Online laser particle counters enable real-time cleanliness control.
- **Interpretation**: Deviation from target cleanliness class can be directly linked to filter efficiency and fluid life.

Particle counting per ISO 4406 is a critical cleanliness assessment method for lubricants, hydraulic fluids, and insulating oils, providing a quantitative measure of solid contamination that can accelerate wear, clog filters, and impair fluid performance. The ISO 4406:2017 standard expresses contamination as a three-part code (e.g., 18/15/12) representing the number of particles larger than 4, 6, and 14 microns per milliliter, respectively.

Counts are typically obtained using optical laser particle counters, pore-blockage instruments, or in some cases microscopic analysis. In high-value power generation and electrification systems, turbines, transformers, switchgear, and precision hydraulic actuators maintaining fluid cleanliness within OEM-specified ISO codes is essential to preserving component life and preventing premature failures. Excessive particle levels can lead to abrasive wear, surface fatigue, and in electrical systems, tracking or arcing across insulating surfaces.

Trending ISO 4406 codes over time, particularly in conjunction with filter performance data and wear metal analysis, allows maintenance teams to identify ingression events, filtration deficiencies, or system wear before they escalate. For critical assets, online particle counters can provide continuous cleanliness monitoring, enabling immediate corrective actions like filtration, flushing, or seal replacement to maintain optimal operating reliability.

Water Content

- **Purpose**: Prevents corrosion, additive depletion, dielectric breakdown (transformer oils), and micropitting.
- **Test Methods**:
  - Karl Fischer titration (ASTM D6304) for ppm precision.
  - Relative humidity sensors for online monitoring.

- **Application**: Maintain below OEM or IEC limits (e.g., <20 ppm for transformers; <0.1% for turbines).

Water content in lubricants, insulating fluids, and coolants is a critical parameter because even small amounts of moisture can dramatically reduce fluid performance, accelerate degradation, and cause equipment damage. In lubricating oils, excess water promotes corrosion of ferrous and non-ferrous metals, depletes additives, and fosters microbial growth, particularly in storage or low-circulation zones. In insulating fluids such as transformer oil, water drastically lowers dielectric strength, increasing the risk of partial discharge, arcing, and catastrophic electrical failure; for example, moisture above 20 ppm in high-voltage transformer oil can be problematic. In heat transfer fluids, water can cause localized boiling, cavitation in pumps, and sludge formation from chemical reactions with fluid constituents.

Water content is most accurately measured by Karl Fischer titration (ASTM D6304), which detects moisture down to parts-per-million levels; other methods include relative humidity sensors and crackle tests for field screening. Continuous or periodic monitoring is essential for high-reliability assets, as water ingress can result from condensation, seal failure, cooler leaks, or improper storage. Trending water content alongside particle counts and acid number enables early detection of contamination pathways and supports timely corrective actions such as filtration, dehydration, or component repair. Maintaining moisture within OEM or IEC-specified limits is one of the most effective ways to extend fluid life, preserve dielectric integrity, and prevent corrosion-related failures.

GREASE-SPECIFIC TESTS

- **Bleed Rate**: ASTM D6184 – measures oil separation under static conditions; high bleed can cause starvation or leakage.
- **Penetration (Consistency)**: ASTM D217 – monitors mechanical stability.
- **Oxidative Stability**: ASTM D942 – predicts shelf and service life.
- **Ferrography**: Analytical and direct reading to visualize wear particle morphology.

Grease-specific condition monitoring tests focus on the unique physical and chemical stability requirements of semi-solid lubricants, where both the base oil and the thickener system must work together to provide long-term protection.

Bleed rate, measured per ASTM D6184, quantifies the separation of base oil from the thickener under static conditions. Controlled oil release is necessary for lubrication at the contact surface, but excessive bleed can lead to grease starvation, leakage, and contamination of surrounding components, while too little bleed can prevent adequate film formation.

Penetration (consistency), determined by ASTM D217, measures the depth a standardized cone penetrates the grease and reflects its mechanical stability and firmness. Changes in penetration over time or after mechanical working indicate structural breakdown or stiffening due to contamination, oxidation, or base oil loss.

Oxidative stability, evaluated by ASTM D942, assesses a grease's resistance to chemical degradation when exposed to oxygen at elevated temperatures, predicting both shelf life and in-service performance. Loss of oxidative stability can result in hardening, additive depletion, and deposit formation in bearings and seals.

Ferrography; both analytical and direct reading examines the wear particle content within the grease, providing visual evidence of wear mode and severity, such as cutting wear, fatigue spalling, or corrosive attack. Because grease in service is not always continuously replenished, these tests collectively provide a comprehensive understanding of its ability to release oil, maintain structure, resist oxidation, and protect against wear over extended intervals, especially in sealed-for-life applications and critical electrification or power generation components.

Ferrograms

- **Purpose**: Qualitative wear particle analysis to identify fatigue, adhesive, abrasive wear modes.
- **Application**: Links lubricant health directly to component wear root cause analysis.

Ferrograms are a qualitative wear particle analysis technique used to identify and classify wear mechanisms occurring within lubricated equipment, providing a direct link between lubricant condition and component health. In this method, ferrous and non-ferrous particles suspended in the lubricant are separated often magnetically in the case of ferrous debris and deposited onto a slide in order of size. The prepared slide is then examined under a microscope to determine particle morphology, composition, and surface characteristics. This allows the analyst to distinguish between **fatigue wear** (smooth, flat particles from rolling contact surfaces), **adhesive wear** (scuffed, elongated particles from metal-to-metal contact), and **abrasive wear** (angular particles from hard contaminants cutting into surfaces), as well as identify evidence of corrosion, fretting, or electrical discharge damage.

By correlating particle type, size distribution, and quantity with operating history and other oil analysis results, ferrograms enable accurate root cause analysis and early fault detection before vibration or temperature changes occur. This makes them particularly valuable in high-value electrification and power generation assets where failure consequences are severe. When combined with trending data from elemental analysis, particle counting, and FTIR, ferrographic analysis provides a powerful tool for predictive maintenance, enabling timely intervention to prevent progressive damage.

In-Service Interval Tracking and Drain Optimization

- **Approach**:
    - Combine all above properties into a **fluid health index** (FHI) with weighting for critical parameters per equipment type.
    - Implement **condition-based oil changes** rather than fixed-hour changes, extends drain intervals safely.

- **Example**: Gas turbine lube oil life extended from 8,000 to 20,000 hours via varnish potential monitoring, RPVOT trending, and particle count control.

In-service interval tracking and drain optimization involve integrating multiple condition monitoring parameters into a structured decision-making framework to determine the precise timing for lubricant or coolant replacement. The most effective approach is to develop a **FHI**, which combines key indicators such as viscosity stability, TBN/TAN balance, oxidation/nitration levels, wear metal concentration, particle count (ISO 4406), water content, and for greases, bleed rate and penetration into a single weighted score. The weighting factors are tailored to the equipment type and criticality: for example, dielectric strength and moisture content may be heavily weighted for transformer oils, while viscosity, oxidation, and soot loading carry more influence for engine lubricants. By trending the FHI over time and comparing it to pre-established service thresholds, maintenance teams can confidently shift from fixed-hour or calendar-based oil changes to **condition-based maintenance**. This not only extends drain intervals safely but also reduces waste oil generation, lowers operating costs, and minimizes environmental impact. In high-value electrification and power generation assets, integrating real-time sensor data with laboratory analysis results further refines the process, allowing for predictive drain scheduling that accounts for actual load, temperature, contamination events, and additive depletion rates. When implemented correctly, in-service interval tracking and drain optimization preserve equipment reliability while maximizing fluid service life, a core best practice in modern asset management.

FHI: What it is and How to Build One That Actually Works

The FHI is a single, decision-ready score that condenses multiple condition-monitoring signals into a normalized 0–1 (or 0–100) scale so you can trigger actions keep running, correct, reclaim, or change out, without arguing over six different test reports. Done well, it is objective, portable across assets, and auditable. Done poorly, it hides problems. Below is a practical, end-to-end process you can lift into your program.

Define Scope and Baselines

Start by locking the **asset class**, **fluid family**, and **operating envelope**. Capture the following:

- **New-oil baseline** values (viscosity @ 40/100 °C, TAN/TBN, FTIR oxidation/nitration, additive metals, dielectric metrics for insulating fluids, etc.).
- **Operating limits** (OEM limits, site limits, regulatory or safety limits).
- **Criticality** (consequence of failure) and **mission profile** (load, temperature, duty cycles).
- **Alarm philosophy** (warn vs action) and **sampling cadence**.

The baselines and limits anchor normalization. Without them, the math is theater.

Choose Indicators by Asset Type (Do Not Overfit)

Pick **5–10 signals** that move first and matter most. Examples are as follows:

- **Transformers/insulating liquids:** breakdown voltage (BDV), moisture (ppm), dissipation factor (tan δ @ temp), TAN, interfacial tension (IFT), DGA severity (Rogers/Dornenburg), particle count.
- **Turbines/hydraulics:** viscosity shift, MPC varnish index, RPVOT (or anti-oxidant health), TAN, particle count (ISO 4406), water (ppm), air release/foam.
- **Engines:** viscosity, TBN retention, TAN rise, FTIR oxidation/nitration/sulfation, soot, fuel dilution, wear metals (Fe/Cu/Al/Pb), particle count.
- **Water-glycol loops:** pH, conductivity, inhibitor residuals (nitrite/molybdate/azole), glycol %, metals (Fe/Cu/Al), particle count.
- **Greases:** penetration change, bleed (D6184), oxidation stability (D942), wear debris (ferrography), water.

If a metric rarely influences decisions, drop it or park it as an advisory KPI.

Normalize Each Metric to a 0–1 Health Score

Convert raw measurements to **unitless health subscores** HiH_iHi where 1.0 is "new/ideal" and 0.0 is "failed." Use monotonic mappings that reflect physics and risk.

Common Mappings

- **Two-sided limit (e.g., viscosity):**

$$H = 1 - \min\left(1, \frac{|x - x_0|}{\Delta x \max}\right)$$

where x_0 is baseline and Δx_{max} is allowable deviation to action.

- **One-sided increasing (e.g., TAN, oxidation, water, particles):**

$$H = \max\left(0, 1 - \frac{x - x_0}{L_{act} - x_0}\right)$$

- **One-sided decreasing (e.g., TBN, RPVOT):**

$H = \max\left(0, \frac{x_0}{L_{act}}\right)$ if expressing remaining vs action level.

- **Ordinal/severity (e.g., DGA diagnosis)**: map severity classes to fixed H values (e.g., Normal = 0.9, Caution 0.6, Alarm = 0.2).

Use **temperature-corrected** or **at-temperature** values (tan δ, BDV, viscosity) so the mapping reflects reality.

Weight by Consequence and Observability
Assign weights wiw_iwi summing to 1.0. Heavier weight for **leading indicators** of catastrophic failure (e.g., moisture in transformer oil), and for metrics with **good signal-to-noise**. Typical weight ranges:

- Life-limiting chemistry (oxidation, TAN/TBN): 0.10–0.25 each.
- Catastrophic risk (BDV, moisture, particle surge): 0.20–0.35 collectively.
- Advisory/confirmatory (IFT, color, foam): 0.05–0.10 each.

Keep the total number of high-weight metrics to **three or fewer** to avoid dilution.

Combine into the FHI and Enforce "No-Masking" Logic
Basic aggregation (weighted mean):

$$\text{FHI} = \sum w_i \cdot H_i$$

But prevent a good average from hiding a red flag:

- **Hard stops:** If any Hi < HtripH (e.g., 0.25), **cap FHI** at that level or trigger an **immediate action** regardless of the average.
- **Penalty for divergence:** Subtract a small penalty if the **worst** sub-score is far from the mean (spikiness indicates emerging failure).

Grade Bands → Decisions
Create three or four bands tied to actions and computerized maintenance management system (CMMS) work orders:

- **FHI ≥ 0.85 (Green):** Run; standard sampling.
- **0.70–0.84 (Yellow):** Corrective action window (drying, filtration, AO top-treat); shorten interval.
- **0.50–0.69 (Orange):** Plan change/reclaim; execute corrective actions now; management of change review.
- **< 0.50 (Red):** Remove from service/replace or reclaim; investigate root cause.

Make the **band thresholds asset-specific** and include a **hysteresis** (do not bounce across boundaries on noise).

Example (Transformer Insulating Liquid)
Inputs and limits (illustrative):

- Moisture: 12 ppm (baseline 5, action 25) → $H = 1–(12–5)/(25–5) = 0.65$
- BDV (D1816): 32 kv (action 25; best 40) → scale linearly: $H = (32–25)/(40–25) = 0.47$
- tan δ @ 90 °C : 0.5% (action 1.0%, baseline 0.2%) → $H = 1–(0.5–0.2)/(1.0–0.2) = 0.62$
- TAN: 0.10 (action 0.03, baseline 0.02) → $H = 1–(0.10–0.02)/(0.30–0.02) = 0.71$
- IFT: 24 mN/m (action 18, baseline 36) decreasing-is-bad → $H = (24–18)/(36–18) = 0.33$
- Particles (ISO code to scalar): trending adverse → assing $H = 0.70$
- DGA severity : Caution → $H = 0.06$

Weights: moisture 0.20, BDV 0.20, tan δ 0.15, TAN 0.15, IFT 0.15, particles 0.10, DGA 0.05.

FHI = 0.20 · 0.65 + 0.20 · 0.47 + 0.15 · 0.62 + 0.15 · 0.71 + 0.15 · 0.33 + 0.10 · 0.70 + 0.05 · 0.60

FHI ≈ **0.58 (Orange)**. Hard stop check: worst HHH=0.33 (IFT) > 0.25 → no trip, but **plan reclaim/dry-out now**.

Data Quality, Confidence, and Trend Behavior

- **Confidence score:** Reduce FHI confidence if sample chain-of-custody is weak, bottle sat >48 h unmixed, or repeatability flags (e.g., particle counter coincidence). Display FHI ± CI.
- **Rate-of-change:** Track $\Delta FHI/\Delta t$. A slow decline may be manageable; a steep drop triggers immediate action even if still "Yellow."
- **Sensor fusion:** Where online sensors exist (moisture, particles, conductivity, viscosity), use them for **inter-sample interpolation**, but keep the lab values as the **authoritative anchors**.

Grease Variant (No Easy BDV, Limited Sampling)
Use a **lighter stack**: penetration change, bleed, oxidation stability, ferrogram severity, water. Normalize similarly; weight penetration/bleed higher for bearings that starve or sling. Tie FHI bands to **regrease intervals** and **bearing inspections**, not drains.

Governance: Make it Auditable

- **Spec annex:** Publish mappings, limits, and weights by asset class.
- **Change control:** Any change to fluid brand, additive package, or test method → **reestablish baseline** and review mappings.
- **Automate:** Build FHI in your CMMS/PI historian; auto-generate work orders when thresholds breach with prescribed actions (dry, filter, top-treat, reclaim, change).
- **Post-mortems:** When a unit is changed/reclaimed, capture pre-change FHI and actual findings (varnish, sludge, moisture) to **retrain weights** yearly.

TABLE 9.1
Condition Monitoring Parameters

Parameter	Method(s)	Typical Limit(s)	Equipment Impact
Viscosity	ASTM D445/ D7279	±5–15% from baseline	Film strength, heat control
TBN/TAN	ASTM D2896/ D664	TBN ≥50% new; TAN +1–2	Neutralization, corrosion
Oxidation/Nitration	FTIR	OEM/trend limit	Sludge, varnish
Wear Metals	ICP, RDE, XRF	OEM/trend limit	Component wear detection
Particle Count	ISO 4406	Class ≤18/15/12	Filter life, wear rate
Water Content	Karl Fischer	<20 ppm–0.1%	Corrosion, dielectric loss
Grease Bleed	ASTM D6184	OEM spec	Starvation risk
Penetration	ASTM D217	OEM spec	Application fit
Oxidative Stability	ASTM D942	≥ baseline	Shelf life, high-temp use
Ferrograms	Analytical ferrography	N/A (qualitative)	Wear mode identification

Using FHI for Drain Optimization

Tie **drain/reclaim** to bands and rate-of-change:

- If **FHI ≥ 0.85** and flat: **extend** interval (up to policy cap).
- If **0.70–0.84** or trending down: **correct** (dry/filtration/AO top-treat) and retest; do not drain yet.
- If **0.50–0.69**: **plan** reclaim or change at next outage; raise sampling cadence.
- **<0.50**: **execute** now; root-cause the driver (heat, O_2, contamination).

Document savings (oil avoided, downtime avoided) to keep the program funded.

Bottom Line

An effective FHI is **simple enough to use, strict enough to trust**, and tailored by asset class. Normalize to physics, weight by consequence, block masking, and wire it into your work management. Do that, and you will extend fluid life safely while catching the failures that actually matter.

10 Dielectric Fluid Monitoring

DIELECTRIC FLUID MONITORING

In electrification and power generation systems, it is essential to ensure the continued reliability and safety of transformers, switchgear, and other oil-filled high-voltage equipment. The goal is to detect early signs of insulation breakdown, contamination, and thermal or electrical faults so corrective actions can be taken before catastrophic failure occurs (see Tables 10.2 and 10.3).

Dissolved Gas Analysis (DGA) and Fault Gas Interpretation

Purpose

To detect and quantify gases generated by thermal or electrical faults within oil-filled electrical equipment. Gases are byproducts of oil or cellulose degradation and indicate arcing, overheating, partial discharges, or insulation breakdown. Trending gas generation over time enables early fault detection before catastrophic failure (see Table 10.1).

Test Methods

- **ASTM D3612** (Methods A, B, or C – gas extraction and GC analysis)
- **IEC 60567** (Sampling and analysis procedures)
- Interpretation tools: **Rogers Ratios**, **Duval Triangle**, **Dornenburg Ratios**, and rate-of-change analysis

Applications

- Continuous monitoring of power transformers, shunt reactors, and oil-filled breakers
- Differentiating between electrical faults (PD, arcing) and thermal faults (low- or high-temperature overheating)
- Establishing transformer maintenance priorities and outage planning

DOI: 10.1201/9781003775201-14

This test is the primary diagnostic tool, measuring concentrations of fault gases such as hydrogen (H_2), methane (CH_4), ethane (C_2H_6), ethylene (C_2H_4), acetylene (C_2H_2), carbon monoxide (CO), and carbon dioxide (CO_2) in transformer oil per ASTM D3612 or IEC 60567. Each gas is associated with specific fault modes: hydrogen with partial discharges, acetylene with arcing, ethylene with high-temperature faults, and carbon monoxide/dioxide with cellulose insulation degradation. Interpretation methods such as the **Rogers Ratio**, **Duval Triangle**, or **Dornenburg ratios** provide structured fault classification. Trending over time is critical, as rates of change are often more diagnostic than absolute concentrations.

Rogers Ratios

Purpose

Rogers Ratios is a diagnostic method developed by R.R. Rogers (IEC 60599, ASTM D3612 interpretive framework) to determine fault type and severity by analyzing **ratios between specific dissolved fault gases** in transformer oil. The underlying premise is that different failure modes (partial discharge, thermal overheating at various temperatures, arcing) generate different proportions of gases such as methane (CH_4), ethylene (C_2H_4), acetylene (C_2H_2), and ethane (C_2H_6).

Method

Four primary ratios are calculated from DGA results:

1. CH_4 / H_2
2. C_2H_6 / CH_4
3. C_2H_4 / C_2H_6
4. C_2H_2 / C_2H_4

Each ratio is compared against diagnostic ranges in a look-up table. A code (0, 1, or 2) is assigned based on the range into which each ratio falls. The resulting four-digit code is then matched to fault categories such as

- PD (Partial Discharge)
- D1 (Low Energy Discharge)
- D2 (High Energy Discharge)
- T1, T2, T3 (Thermal faults at <300 °C, 300–700 °C, >700 °C)

Application Notes

- Works well for clearly defined faults with strong gas signatures.
- Less accurate when multiple faults exist simultaneously or when oil has been degassed.
- Often used in combination with trending and Duval Triangle for confirmation.

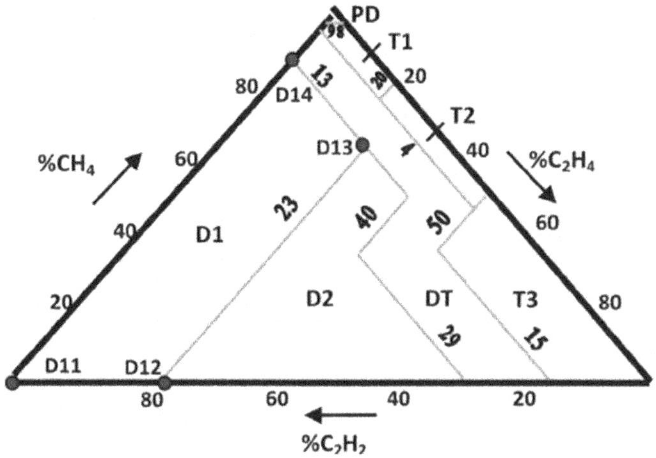

FIGURE 10.1 Duval triangle.

Duval Triangle

Purpose

The Duval Triangle (see Figure 10.1) is a graphical method developed by Michel Duval (IEC 60599 Annex A) that uses **three key hydrocarbon gases, methane (CH_4), ethylene (C_2H_4), and acetylene** (C_2H_2), to classify transformer faults. Unlike Rogers Ratios, which use multiple ratio boundaries, the Duval Triangle uses relative percentages plotted within a triangular diagram.

Method

- Calculate the percentage contribution of each gas relative to the total of the three:
  $\%CH_4 = (CH_4 / \text{Total}) \times 100$
  $\%C_2H_4 = (C_2H_4 / \text{Total}) \times 100$
  $\%C_2H_2 = (C_2H_2 / \text{Total}) \times 100$
- Plot the point inside the equilateral triangle diagram, which is subdivided into regions corresponding to fault types:
  - PD (Partial Discharge)
  - D1 (Low Energy Discharge)
  - D2 (High Energy Discharge)
  - T1, T2, T3 (Thermal faults at increasing temperatures)
  - Mixed zones for combined electrical and thermal faults.

Application Notes

- Very effective for single, dominant fault conditions.
- Less sensitive to minor gases not included in the three-gas set.

- Widely used in IEC and CIGRÉ guidelines and considered one of the most visual and intuitive DGA tools.

Dornenburg Ratios

Purpose
The Dornenburg Ratios method is an older approach that also uses gas ratios but with **slightly different ratio selections and interpretation rules** compared to Rogers Ratios. It was developed by E. Dornenburg in the 1970s for early transformer fault classification.

Method
Four key ratios are calculated:

1. CH_4 / H_2
2. C_2H_2 / C_2H_4
3. C_2H_2 / CH_4
4. C_2H_6 / C_2H_2

Each ratio is evaluated against fixed limits to determine the presence or absence of faults such as arcing, corona, and various overheating conditions. Dornenburg used "yes/no" decision rules instead of a multi-level code.

Application Notes

- Historically important but now considered less accurate than Rogers or Duval due to limited dataset origin and higher susceptibility to misinterpretation with mixed faults.
- Still used for historical trending in legacy fleets where long-term ratio history exists.

Rate-of-Change Analysis

Purpose
Rate-of-change analysis focuses not only on **absolute gas concentrations** but also on **how quickly they are increasing**. The assumption is that rapid increases in certain gases indicate active or accelerating fault conditions.

Method

- Monitor periodic DGA test results and calculate the change in ppm per day, week, or month for each gas.
- Compare against alarm thresholds, such as the following:
  - Hydrogen >50 ppm/month increase
  - Acetylene >5 ppm/month increase (especially critical for arcing)
  - Total combustible gas (TCG) >10% rise per month

Some utilities adopt IEEE C57.104 or IEC 60599 "rate alarms" alongside absolute concentration alarms.

Application Notes

- Highly valuable for early detection of worsening conditions.
- Detects active faults even if absolute concentrations are still below absolute alarm limits.
- Particularly important in online monitoring systems where rapid gas accumulation can be caught between offline sampling intervals.

Best Practice Integration
In modern transformer diagnostics:

- **Duval Triangle** is used as a primary visualization and classification tool.
- **Rogers Ratios** confirm and refine the fault type, especially in mixed conditions.
- **Rate-of-change analysis** provides the time-critical dimension, allowing operators to determine urgency.
- **Dornenburg Ratios** are mainly referenced for legacy correlation or cross-checks in fleets with decades of historical ratio data.

Moisture content (Karl Fischer)

Purpose
To measure dissolved water in dielectric fluids, which reduces dielectric strength, promotes corrosion, and accelerates insulation aging.

Test Methods

- **ASTM D6304** (Coulometric or volumetric Karl Fischer titration)
- **IEC 60814** (Water content in insulating liquids)

Applications

- Determining the need for oil dehydration or filtration
- Correlating moisture levels with breakdown voltage trends
- Assessing risk of bubble formation under load

Moisture is determined by Karl Fischer titration (ASTM D6304), directly affects dielectric strength and accelerates aging of both fluid and paper insulation. Values above OEM or IEC thresholds (often 10–20 ppm for power transformers) can reduce breakdown voltage dramatically.

TABLE 10.1
Comparative Summary of DGA Interpretation Methods

Method	Required Gases	Calculation Process	Diagnostic Output	Strengths	Limitations
Rogers Ratios	H_2, CH_4, C_2H_6, C_2H_4, C_2H_2	Calculate four gas ratios: (CH_4/H_2), (C_2H_6/CH_4), (C_2H_4/C_2H_6), (C_2H_2/C_2H_4). Assign codes per range tables (0,1,2). Match code set to fault type.	Fault type classification: PD, D1, D2, T1, T2, T3.	Structured, IEC-recognized, good for well-developed faults.	Less accurate for multiple concurrent faults; affected by degassing.
Duval Triangle	CH_4, C_2H_4, C_2H_2	Convert each gas to % of total of the three, plot inside triangular diagram with predefined fault zones.	Graphical fault classification: PD, D1, D2, T1, T2, T3, mixed faults.	Intuitive visual output, robust for single dominant faults, widely used in IEC/CIGRÉ.	Ignores gases outside the three; less accurate for subtle/mixed faults.
Dornenburg Ratios	H_2, CH_4, C_2H_2, C_2H_4, C_2H_6	Calculate 4 ratios: (CH_4/H_2), (C_2H_2/C_2H_4), (C_2H_2/CH_4), (C_2H_6/C_2H_2). Compare to fixed limits with yes/no rules.	Fault indications for corona, arcing, overheating.	Historically important, simple yes/no interpretation.	Lower accuracy; obsolete for modern multi-fault cases; based on limited early datasets.
Rate-of-Change Analysis	All combustible gases (esp. H_2, C_2H_2, C_2H_4)	Calculate ppm change per day/week/month from serial DGA results; compare to trend thresholds.	Activity status: stable, increasing, rapidly escalating.	Detects developing faults before absolute limits reached; essential for online monitoring.	Requires consistent sampling intervals; influenced by load/temperature fluctuations.

Interfacial Tension (IFT)

Purpose
To assess contamination by polar compounds, oxidation products, and degradation residues. Lower IFT values indicate oil aging or contamination.

Test Methods

- **ASTM D971** (Du Noüy ring method)

Applications

- Trending oxidation and sludge-forming tendencies
- Determining the need for reclamation or replacement
- Correlating with acidity and color changes in oil

This is measured per ASTM D971, is a sensitive indicator of polar contamination and oxidation products declining IFT typically signals advancing oil deterioration.

Furan Analysis

Purpose
To quantify degradation products of cellulose insulation dissolved in the oil. Indicates paper insulation aging and remaining life.

Test Methods

- **ASTM D5837** (high-performance liquid chromatography – HPLC)

Applications

- Estimating transformer paper degree of polymerization (DP) indirectly
- Life assessment for asset replacement planning
- Confirming suspected high-temperature faults affecting solid insulation

This analysis detects degradation products from cellulose paper insulation, providing a direct measure of solid insulation condition and remaining life.

Breakdown Voltage (BDV)

Purpose
To measure dielectric fluid's ability to withstand electrical stress without failure.

Test Methods

- **ASTM D1816** (electrically stressed under liquid – VDE electrodes, oil circulation optional)
- **ASTM D877** (flat disk electrodes – less sensitive to small contaminants)
- **IEC 60156** (AC breakdown voltage test)

Applications

- Acceptance testing for new fluids
- In-service quality assessment
- Guiding filtration or dehydration requirements

This testing (ASTM D1816/D877) evaluates the oil's ability to resist electrical stress under controlled conditions; declining BDV often correlates with high moisture, dissolved gases, or particulate contamination.

Color (ASTM D1500)

Purpose
To monitor oil darkening due to oxidation, contamination, or thermal stress.

Test Methods

- **ASTM D1500** (color comparator scale)

Applications

- Quick screening for oxidation or contamination
- Supplementary indicator to acidity and IFT trends

Neutralization (Acid Number)

Purpose
To determine concentration of acidic compounds formed by oxidation or contamination. High acidity accelerates sludge formation and corrodes metals.

Test Methods

- **ASTM D974** (potentiometric or color-indicator titration)

Applications

- Identifying need for oil reclamation or change-out
- Correlating with sludge presence and IFT decline

This test serves as secondary indicators of oxidation and sludge formation.

TABLE 10.2
Dielectric Fluid Monitoring – Purpose, Test Methods, and Applications

Parameter	Purpose	Test Method(s)	Applications
Dissolved Gas Analysis (DGA)	Detect and quantify fault gases from thermal/electrical degradation; classify fault type and severity.	ASTM D3612 (A/B/C), IEC 60567; Interpretation via Rogers Ratios, Duval Triangle, Dornenburg Ratios.	Early fault detection in transformers/reactors; distinguish PD, arcing, overheating; prioritize maintenance.
Moisture (Karl Fischer)	Measure dissolved water to assess risk of dielectric breakdown, corrosion, and accelerated aging.	ASTM D6304 (coulometric/volumetric), IEC 60814.	Determine need for dehydration/filtration; correlate with BDV decline; assess bubble formation risk.
Interfacial Tension (IFT)	Detect polar contamination and oxidation products; trend oil degradation.	ASTM D971 (Du Noüy ring).	Identify sludge-forming tendencies; guide reclamation/replacement; correlate with acidity and color changes.
Furan Analysis	Quantify cellulose degradation products; estimate paper insulation condition.	ASTM D5837 (HPLC).	Transformer life assessment; plan asset replacement; confirm overheating of paper insulation.
Breakdown Voltage (BDV)	Measure ability to withstand electrical stress without failure.	ASTM D1816, ASTM D877, IEC 60156.	Acceptance testing; in-service dielectric quality check; trigger filtration/dehydration actions.
Color	Monitor oil darkening from oxidation, contamination, or overheating.	ASTM D1500.	Quick oxidation screening; supplementary indicator with acidity/IFT data.
Acidity (Neutralization Number)	Measure acidic compounds from oxidation/contamination; high acidity leads to sludge and corrosion.	ASTM D974 (potentiometric or indicator titration).	Identify need for reclamation; correlate with sludge presence; assess aging stage.
Sludge Presence	Detect precipitated oxidation products that block cooling or insulate hot spots.	Visual inspection, ASTM D1698.	Diagnose cooling restrictions; determine end-of-life; plan reclamation/cleaning.

TABLE 10.3
Oil and Grease Condition Monitoring – Purpose, Test Methods, and Applications

Parameter	Purpose	Test Method(s)	Applications
Viscosity	Assess fluid film thickness and load-carrying ability; detect shear thinning, fuel dilution, or thickening from oxidation/contamination.	ASTM D445 (kinematic), ASTM D7279 (rotational).	Predict lubrication performance; schedule oil changes; identify contamination events.
Total Base Number (TBN)	Measure reserve alkalinity to neutralize acids in combustion environments.	ASTM D2896, ASTM D4739.	Monitor detergent/dispersant additive health; determine oil change intervals for engines.
Total Acid Number (TAN)	Quantify acidic compounds from oxidation or contamination.	ASTM D664.	Detect oxidation, additive depletion; trigger reclamation or oil change.
Oxidation / Nitration	Identify chemical degradation from oxygen/NOx exposure; linked to high temperature and combustion.	FTIR per ASTM E2412.	Predict oil life; correlate with TAN rise; identify abnormal thermal loads.
Wear Metals & Elemental Analysis	Detect metallic particles from wear, corrosion, or additive depletion.	ICP (ASTM D5185), RDE (ASTM D6595), XRF.	Diagnose component wear mode/location; trend severity; root cause analysis.
FTIR Spectroscopy	Detect oxidation, nitration, sulfation, additive depletion, fuel dilution, and glycol contamination.	ASTM E2412, ASTM D7414.	Rapid screening for multiple degradation pathways; confirm contamination events.
Particle Count – ISO 4406	Quantify solid contamination >4, 6, 14 μm; assess cleanliness level.	ISO 4406:2017; optical laser or pore-blockage counting.	Maintain cleanliness targets; assess filtration; detect ingression.
Water Content	Measure dissolved/emulsified water that accelerates corrosion and additive depletion.	ASTM D6304 (Karl Fischer), Crackle Test.	Trigger dehydration/filtration; correlate with BDV or corrosion issues.

Grease Bleed Rate	Measure base oil separation; excessive bleed leads to starvation, low bleed to poor lubrication.	ASTM D6184.	Select suitable grease for application; monitor in-service stability.
Grease Penetration (Consistency)	Monitor mechanical stability and firmness.	ASTM D217.	Detect hardening or softening from contamination, oxidation, or oil loss.
Grease Oxidative Stability	Predict grease shelf/service life under oxygen exposure.	ASTM D942.	Determine relubrication intervals; evaluate storage conditions.
Ferrography	Visual analysis of wear particles to determine wear mechanism.	Analytical/direct reading ferrography.	Identify fatigue, adhesive, abrasive wear; correlate to failure mode.
Ferrograms	Qualitative wear particle morphology identification.	Microscopic analysis after slide preparation.	Direct root cause linkage between lubricant condition and component wear.
Fluid Health Index (FHI)	Weighted composite score combining multiple parameters for condition-based maintenance.	Custom algorithms using lab and/or online sensor inputs.	Optimize drain intervals; shift from fixed to predictive maintenance.

Sludge Presence

Purpose

To detect advanced oxidation products precipitated in the oil or deposited on solid insulation surfaces.

Test Methods

- Visual inspection during sampling or internal inspection
- Sediment and sludge quantification per **ASTM D1698**

Applications

- Root cause analysis for cooling blockage and hot spots
- End-of-life determination for transformer fluids
- Planning reclamation and cleaning procedures

11 Coolant and Thermal Fluid Monitoring

Coolant and thermal fluids only deliver reliable service when their chemistry stays inside a tight operating window. A monitoring program translates that window into measurable targets: sample on a fixed cadence, test with validated methods, trend results against limits, and trigger corrective actions before corrosion, fouling, loss of dielectric margin, or biofouling erode performance. The goal is simple: keep heat transfer efficient, materials protected, and safety margins intact across seasons, load swings, and supplier variability.

This section organizes monitoring by the dominant failure drivers in real systems. Chemistry control comes first: pH, conductivity, and inhibitor health determine whether metals passivate or corrode. The freeze and boil envelope follows: glycol fraction, freeze point, and boil point set mechanical safety and thermal efficiency. Ionic contamination is next: trace chlorides, sulfates, and hardness ions detected by ion chromatography are early signals of leaks, makeup water issues, or cross-contamination. For nuclear and other high-reliability services, radiolytic decomposition products and suspended "crud" must be quantified to protect both materials and dose rates. Finally, biofouling is addressed: biocide residuals, microbial activity, and biofilm growth are monitored so heat exchangers stay clean and pumping power remains low.

Use the guidance below as a practical framework: what to measure, why it matters, which methods to use, and what actions to take when trends drift. Tying each parameter to a specific failure mechanism ensures the program is not just compliant but predictive, turning routine sampling into an early-warning system that preserves efficiency, asset life, and safety (see Table 11.1).

PH, CONDUCTIVITY, AND CORROSION INHIBITORS

pH is a primary coolant health indicator because even small shifts from the design range can significantly accelerate corrosion rates or cause precipitation of protective films. American Society for Testing and Materials (ASTM) D1287 provides both colorimetric and electrometric methods to measure pH accurately in coolant and thermal fluid samples. A stable, manufacturer-recommended pH ensures the corrosion inhibitors remain in their active form, protecting system metals from acid or alkaline

DOI: 10.1201/9781003775201-15

TABLE 11.1
Overview Table: Purpose, Test Methods, and Applications

Parameter	Purpose	Test Method(s)	Applications
pH	Assess acid/alkaline balance affecting corrosion rate and inhibitor performance.	**ASTM D1287 (colorimetric/electrometric).**	Ensure corrosion control; monitor additive depletion; adjust chemistry in closed loops.
Conductivity	Measure ionic contamination and dissolved solids that influence corrosion and scaling.	**ASTM D1125 (electrometric).**	Detect leakage, contamination, or degradation; assess deionization system performance.
Corrosion Inhibitors	Quantify active inhibitor concentration (e.g., nitrite, molybdate, phosphate).	**ASTM D8045 (alkalinity), ASTM D1126 (chloride), specific colorimetric assays.**	Maintain protective film integrity; avoid under- or over-treatment.
Glycol Content	Verify concentration for proper freeze/boil protection and thermal efficiency.	**ASTM D3321 (refractometry), ASTM E202 (specific gravity), FTIR.**	Maintain target performance window; avoid pump cavitation, thermal inefficiency, or additive precipitation.
Freeze/Boil Point	Determine temperature limits of protection.	**ASTM D1177 (freezing point depression), ASTM D1120 (boiling point).**	Ensure suitability for site climate; prevent burst damage or vapor lock.
Ion Chromatography (Anion/Cation Contamination)	Identify corrosive species (chloride, sulfate, sodium, potassium, calcium).	**ASTM D4327, ASTM D6919.**	Pinpoint contamination source; prevent stress corrosion cracking and pitting.
Radiolytic Decomposition Products (nuclear systems)	Detect breakdown of coolant molecules under radiation producing corrosive or conductive byproducts.	**HPLC, ion chromatography, GC-MS.**	Protect reactor cooling circuits; plan resin replacement; minimize dose rates.
Coolant Crud	Identify suspended oxide and corrosion debris.	**Gravimetric analysis, filtration and particle characterization, XRD/SEM-EDS.**	Maintain heat transfer efficiency; reduce localized corrosion risk.
Biocide Monitoring	Verify active biocide residual levels for microbial control.	**DPD colorimetric method, HPLC, ASTM E645 (microbial count).**	Prevent biofilm formation; maintain heat exchanger performance; control microbiologically influenced corrosion (MIC).
Biofouling Control	Monitor microbial growth and biofilm development.	**Biofilm thickness probes, ATP luminometry, culture methods.**	Enable proactive cleaning schedules; reduce heat transfer losses; extend system service life.

attack. Conductivity, measured per ASTM D1125, reflects the total ionic content of the coolant. Elevated conductivity often indicates leakage, contamination ingress, or additive depletion, all of which can increase galvanic corrosion potential. Regular monitoring allows operators to adjust deionization systems or perform partial drains and refills to restore quality. Corrosion inhibitors, such as nitrites, molybdates, or phosphates, must be present at specified levels to maintain a passivating film on metal surfaces. Field or lab testing (via titration or colorimetry) ensures these concentrations remain within target ranges, preventing both under-treatment (leading to corrosion) and over-treatment (leading to scaling or deposits).

GLYCOL CONTENT, FREEZE POINT, AND BOIL POINT

The glycol fraction in water-based coolants is critical for both freeze protection in cold climates and boil-over prevention in high-heat environments. ASTM D3321 (refractometry) and ASTM E202 (specific gravity) offer quick, accurate determination of glycol content, while Fourier transform infrared spectroscopy (FTIR) can identify the specific glycol type (ethylene, propylene) and degradation products. Inadequate glycol concentration can cause freezing and mechanical rupture of components, whereas excessive glycol increases viscosity, reduces thermal conductivity, and may promote additive drop-out. Freeze point determination via ASTM D1177 ensures coolant remains liquid at the lowest expected operating temperature, while boiling point testing per ASTM D1120 confirms vapor-lock and cavitation risks are minimized. Together, these parameters allow optimization of thermal performance while avoiding mechanical stress from phase changes.

ION CHROMATOGRAPHY FOR ANION/CATION CONTAMINATION

Ion chromatography (IC) is a high-sensitivity method for detecting corrosive ions such as chloride, sulfate, and nitrate, as well as cationic metals like sodium, potassium, calcium, and magnesium. These species, even in low ppm levels, can cause pitting, crevice corrosion, and scaling in both ferrous and non-ferrous system metals. ASTM D4327 and ASTM D6919 define procedures for anion and cation analysis, respectively. IC results not only indicate contamination but also help pinpoint its source, such as condenser leaks, makeup water impurities, or cross-contamination from other process fluids. In high-reliability systems, IC is integrated into routine monitoring programs to pre-empt catastrophic corrosion events.

RADIOLYTIC DECOMPOSITION PRODUCTS (NUCLEAR SYSTEMS) AND COOLANT CRUD

In nuclear applications, coolant fluids, whether water, heavy water, or molten salts are subjected to intense radiation fields that cause radiolysis, breaking molecular bonds and generating reactive species such as hydrogen, oxygen, peroxides, and organic acids. These products can increase coolant conductivity, attack structural alloys, and degrade ion-exchange resins. Monitoring via high-performance liquid chromatography (HPLC), ion chromatography, or gas chromatography-mass spectrometry (GC-MS) is critical to ensure radiolysis byproducts remain below operational limits. In addition, nuclear coolants often carry "crud," a term for suspended corrosion products (oxides, particulates) originating from fuel cladding, piping, or structural components. Crud can deposit in heat exchangers, reducing thermal performance, and in extreme cases, elevate local radiation fields. Gravimetric analysis, particle filtration, and characterization by X-ray diffraction (XRD) or scanning electron microscopy with energy dispersive spectroscopy (SEM-EDS) are standard practices to track and control crud accumulation.

BIOCIDE MONITORING AND BIOFOULING CONTROL

Biological growth in cooling water and closed-loop systems can lead to biofilm formation, which reduces heat transfer efficiency, increases energy pumping costs, and creates under-deposit corrosion sites (microbiologically influenced corrosion, MIC). Biocide monitoring ensures that oxidizing (e.g., chlorine, bromine) or non-oxidizing agents (e.g., isothiazolinones, glutaraldehyde) are maintained at effective concentrations without exceeding regulatory discharge limits. Methods include DPD colorimetry for oxidizing biocides, HPLC for organics, and ASTM E645 microbial count for culture-based enumeration. For proactive biofouling control, ATP (adenosine triphosphate) luminometry offers rapid total microbial activity measurement, while biofilm thickness probes installed in representative flow areas can detect early accumulation. Continuous monitoring allows corrective actions, such as targeted biocide dosing, mechanical cleaning, or system flushing, before fouling impacts operational performance.

Section V

Failure Forensic Case Studies

12 Lubrication Failures in Power Equipment

FAILURE FORENSICS AND CASE STUDIES: LUBRICATION FAILURES IN POWER EQUIPMENT

Lubrication failures in power generation equipment present recurring causes of forced outages, accelerated wear, and unplanned maintenance. Forensic analysis identifies the initiating mechanism, contributing conditions, and degradation pathways by combining lubricant condition data, component inspection, and operational records. The following cases illustrate failure modes observed in thermal, nuclear, renewable, and reciprocating engine-based power assets.

CASE STUDIES

Varnish in Turbine Bearings

Gas and steam turbine bearings operate under high bulk oil temperatures with localized thermal gradients at journal surfaces. Oxidative polymerization of base oil, catalyzed by metallic wear debris and hot spots, produces insoluble high molecular weight deposits. Varnish formation reduces oil film clearance, increases bearing metal temperatures, and can lead to oil wedge collapse. Infrared spectroscopy shows carbonyl and carboxylic functional group growth, while MPC (Membrane Patch Colorimetry) numbers above 30 indicate active deposition risk (American Society for Testing and Materials (ASTM) D7843). Root cause in many cases is insufficient antioxidant reserve (low RULER® or ASTM D6971 values), inadequate offline filtration for soft contaminants, or prolonged operation with oil above 90 °C without varnish mitigation equipment. Corrective measures include chemical cleaning, charge of oil with higher oxidative stability per ASTM D943 results, and installation of electrostatic or depth filtration units (see Table 12.1).

DOI: 10.1201/9781003775201-17

TABLE 12.1
Turbine Bearing – Varnish Formation FMEA

Failure Mode	Potential Causes	Effects	Severity (S)	Occurrence (O)	Detection (D)	RPN	Detection Methods
Varnish deposits on bearing surfaces	High oil temperature, oxidation, additive depletion, poor filtration	Increased friction, reduced clearance, bearing failure	9	5	4	180	MPC (ASTM D7843), FTIR, RULER antioxidant testing, visual inspection during outage

Varnish in Turbine Bearings – Diagnostic Checklist

Lab Analysis

- Obtain recent oil sample from active system, following ASTM D4057.
- Run membrane patch colorimetry (MPC) per ASTM D7843; record ΔE values.
- Conduct Fourier transform infrared spectroscopy (FTIR) analysis for carbonyl/carboxyl group peaks.
- Test antioxidant level with RULER® or ASTM D6971.
- Measure total acid number (TAN) per ASTM D664.
- Review historical oxidative stability (ASTM D943) trend.

Field Inspection

- Inspect bearing surfaces for brown or amber deposits.
- Check oil film clearance against original equipment manufacturer (OEM) tolerances.
- Measure localized bearing metal temperatures during operation.
- Examine filters and varnish removal units for saturation.

Root Cause Confirmation

- Verify operating bulk oil temperature history.
- Cross-check filtration and varnish mitigation equipment performance.
- Review oil change and flush history for past varnish events.

GREASE INCOMPATIBILITY IN WIND TURBINES

Yaw and pitch bearings, along with generator bearings in wind turbines, are often relubricated with different thickener chemistries due to supply changes or contractor practices. Incompatibility between polyurea, lithium complex, and calcium sulfonate complex thickeners can cause structural collapse of the grease matrix. The mixture releases base oil prematurely, leading to channeling and starvation in rolling elements. ASTM D6185 grease compatibility testing and worked penetration changes beyond ±30 points (ASTM D217) confirm incompatibility. Microscopic analysis shows soap fiber breakdown and oil bleed exceeding OEM maximums. Contributing factors include unverified substitution of OEM-specified grease and cross-contamination from multi-use grease guns. Remediation requires purge and purge-volume verification, followed by a controlled re-lubrication program using a single approved grease chemistry (see Table 12.2).

Grease Incompatibility in Wind Turbines – Diagnostic Checklist

Lab Analysis

- Collect representative grease samples from bearings, purge points, and cartridges.

TABLE 12.2
Wind Turbine Main Bearing – Grease Incompatibility FMEA

Failure Mode	Potential Causes	Effects	S	O	D	RPN	Detection Methods
Thickener collapse, oil separation	Mixing incompatible greases, contamination during relube	Loss of lubrication film, bearing overheating, failure	8	6	5	240	Grease compatibility testing (ASTM D6185), penetration change (ASTM D217), oil bleed (ASTM D6184)

- Perform ASTM D6185 compatibility testing between greases in use.
- Conduct worked penetration (ASTM D217) before/after mixing.
- Test oil bleed per ASTM D6184.
- Check dropping point for thermal stability changes.

Field Inspection

- Inspect bearing cavities for hardened or channelled grease.
- Look for evidence of oil leakage and runout from bearing housings.
- Disassemble one bearing to inspect thickener structure under microscope.

Root Cause Confirmation

- Verify grease type, thickener chemistry, and supplier batch records.
- Review lubrication PM records for cross-contamination risk.
- Identify uncontrolled substitution of OEM-specified grease.

DIESEL ENGINE WEAR FROM SOOT AND FUEL DILUTION

Standby and prime diesel generators suffer accelerated ring and liner wear when incomplete combustion increases soot generation and when fuel dilution from injector leakage lowers oil viscosity. Inductive coupled plasma (ICP) elemental analysis shows elevated Fe and Cr, while thermogravimetric analysis quantifies soot loading. Fuel dilution is measured by gas chromatography per ASTM D3524 or flash point depression. Soot agglomeration raises abrasive wear rates and can block hydraulic lifters. TAN/TBN imbalance occurs as soot and fuel accelerate oxidation and acid formation. Root causes include extended idling, low load operation, poor injector spray patterns, and cold starts without sufficient warm-up. Corrective actions involve injector overhaul, periodic high-load operation to burn off deposits, oil formulation with higher dispersant capacity, and shortening oil change intervals based on condition monitoring results (see Table 12.3).

TABLE 12.3
Diesel Engine – Wear from Soot and Fuel Dilution FMEA

Failure Mode	Potential Causes	Effects	S	O	D	RPN	Detection Methods
Abrasive and corrosive wear	Poor combustion, injector leaks, excessive idling, late oil change	Cylinder/piston scoring, reduced compression, early overhaul	9	7	3	189	ICP wear metals, soot measurement (TGA), fuel dilution by GC, flash point drop

Diesel Engine Wear from Soot and Fuel Dilution – Diagnostic Checklist

Lab Analysis

- Collect oil sample hot and running, mid-drain.
- Test wear metals by ICP spectroscopy (Fe, Cr focus).
- Quantify soot loading by TGA or ASTM D5967 insolubles test.
- Measure fuel dilution via GC (ASTM D3524) or flash point drop.
- Record TBN and TAN for balance assessment.

Field Inspection

- Inspect exhaust color and opacity.
- Check crankcase breather and blow-by rates.
- Listen for injector knock or irregular combustion noise.
- Inspect injectors and fuel spray pattern.

Root Cause Confirmation

- Verify load profile (percentage of rated load) from log data.
- Review start/stop and idle time records.
- Correlate injector condition with dilution levels.

DRY SUMP STARVATION IN GAS ENGINES

Gas turbine starter engines and high-performance reciprocating units using dry sump lubrication rely on scavenge and pressure pump coordination. Aeration or vapor lock in scavenge lines can cause the oil reservoir level to drop below pickup, producing momentary starvation. Wear pattern analysis reveals scoring on cam lobes and crankshaft journals consistent with boundary lubrication. Viscosity testing shows no major deviation, indicating an acute rather than chronic lubrication failure. Factors include

TABLE 12.4
Gas Engine – Dry Sump Starvation FMEA

Failure Mode	Potential Causes	Effects	S	O	D	RPN	Detection Methods
Insufficient oil delivery to bearings	Scavenge pump failure, aeration, blocked return line	Rapid bearing wear, seizure, catastrophic engine failure	10	4	4	160	Oil pressure monitoring, return line inspection, flow rate testing

improper oil tank venting, blocked scavenge screens, or incorrect oil return line routing. Prevention requires sump return flow verification, vent pressure checks, and installation of sight-glass monitoring in return lines during commissioning (see Table 12.4).

Dry Sump Starvation in Gas Engines – Diagnostic Checklist

Lab Analysis

- Confirm oil viscosity within spec (ASTM D445).
- Check sample for entrained air or foam visually.
- Review used oil elemental analysis for wear debris spikes.

Field Inspection

- Monitor oil return flow during operation; use sight-glass if available.
- Inspect scavenge pump and suction screens.
- Verify oil tank vent function and pressure.
- Check for return line restrictions or improper routing.

Root Cause Confirmation

- Cross-reference event logs with oil pressure alarm history.
- Document tank fill level during operation.
- Test scavenge pump flow rate against OEM spec.

Cavitation Pitting and Aeration

Hydraulic and lubrication circuits with low inlet pressure at pump suction can experience vapor bubble formation. Implosion of these bubbles on metal surfaces generates microjets causing surface fatigue and pitting. Aeration from suction-side air leaks or return-line agitation reduces pump efficiency and accelerates oil oxidation. Field evidence includes spongy oil appearance, entrained air bubbles visible under strobe inspection, and acoustic signatures detected via ultrasound. Ferrography reveals spherical wear particles typical of cavitation erosion. Corrective measures

TABLE 12.5
Hydraulic Pump – Cavitation Pitting and Aeration FMEA

Failure Mode	Potential Causes	Effects	S	O	D	RPN	Detection Methods
Metal surface pitting, erosion	Low inlet pressure, suction leaks, high fluid temperature, poor air release	Reduced pump efficiency, catastrophic failure	8	6	5	240	Air release test (ASTM D3427), foam test (ASTM D892), acoustic monitoring, inlet pressure measurement

include ensuring net positive suction head (NPSH) margins, sealing suction joints, and installing anti-vortex plates in reservoirs. ASTM D3427 air release testing and ASTM D892 foam testing confirm susceptibility of the oil to entrainment issues (see Tables 12.5 and 12.6).

Cavitation Pitting and Aeration – Diagnostic Checklist

Lab Analysis

- Test air release properties per ASTM D3427.
- Conduct foam testing per ASTM D892.
- Examine wear particles with analytical ferrography for spherical morphology.

Field Inspection

- Listen for high-frequency cavitation noise near pump inlets.
- Inspect pump vanes and housings for pitting.
- Check oil for milky/cloudy appearance under light.
- Look for vortexing in reservoir return zones.

Root Cause Confirmation

- Measure inlet pressure to confirm adequate NPSH margin.
- Pressure test suction side for leaks.
- Verify antivortex plate presence and positioning.

ROOT CAUSE DETERMINATION PROCESS

The following is a **Root Cause Determination Process** tailored for lubrication-related failures in power equipment, structured for reliability engineering field use.

TABLE 12.6
Lubrication Failure Forensics in Power Equipment

Failure Mode	Lubricant Test Indicators	Physical Inspection Findings	Likely Root Causes
Varnish in Turbine Bearings	• High MPC value (>30, ASTM D7843) • FTIR carbonyl/carboxyl peak growth • Low antioxidant reserve (RULER® / ASTM D6971) • TAN increase	• Brown/gold deposits on bearing surfaces • Reduced oil film clearance • Localized hot spots	• Prolonged high oil temperature (>90 °C) • Antioxidant depletion • Inadequate varnish removal/filtration
Grease Incompatibility in Wind Turbines	• Worked penetration change >±30 points (ASTM D217) • Excessive oil bleed (ASTM D6184) • Dropping point reduction	• Grease channeling and hardening • Base oil leakage from bearings • Collapsed thickener structure under microscopy	• Mixing incompatible thickener types • Cross-contamination from grease guns • Substitution without compatibility testing
Diesel Engine Wear from Soot and Fuel Dilution	• High Fe, Cr in ICP results • Soot loading via TGA or insolubles test • Fuel dilution by GC or flash point drop • TAN/TBN imbalance	• Ring and liner scoring • Hydraulic lifter sticking • Piston crown deposits	• Extended idling/low load operation • Injector leakage or poor spray pattern • Cold starts without proper warm-up
Dry Sump Starvation in Gas Engines	• No major change in viscosity or additive levels • Oil aeration evident in visual sample check	• Scoring on cam lobes, crank journals • Bluing/discoloration from heat • Low oil level in reservoir during operation	• Scavenge pump vapor lock or blockage • Incorrect return line routing • Tank venting issues
Cavitation Pitting and Aeration	• High air entrainment in ASTM D3427 • Foam tendency/failure in ASTM D892 • Elevated wear particles on ferrography (spherical morphology)	• Pitting on pump vanes or housing • Cloudy/spongy oil appearance • Audible cavitation noise	• Low NPSH at pump suction • Suction-side air leaks • Reservoir design faults (vortexing)

This process is designed to move from symptom identification through lab confirmation to verified cause, avoiding premature conclusions.

Lubrication Failure Root Cause Determination Process

1. **Initial Symptom Documentation**
   - Record equipment ID, location, and operating status at the time of fault.
   - Note abnormal sounds, temperatures, vibration levels, odors, or performance changes.
   - Capture oil pressure, temperature, and flow readings if applicable.
   - Photograph visible lubricant condition (color, clarity, foaming, deposits).
2. **Fluid Sampling and Preservation**
   - Use proper sampling procedures (ASTM D4057 for oils, ASTM D7718 for greases).
   - Label samples with time, date, equipment ID, and load condition.
   - Take comparative samples from a reference or control unit if available.
   - Store samples under conditions preventing contamination or property change.
3. **Laboratory Testing**
   - Select targeted tests based on failure symptoms:
     - **Viscosity (ASTM D445)** for flow characteristics.
     - **Wear metals (ICP, ASTM D5185)** for component wear.
     - **FTIR** for oxidation, nitration, contamination.
     - **Particle count (ISO 4406)** for cleanliness level.
     - **Specific property tests** such as MPC for varnish, Karl Fischer for moisture, grease penetration/bleed rate.
   - Compare results to OEM, ASTM, or in-house alarm limits.
   - Trend against historical data from the same equipment.
4. **Physical Inspection**
   - Disassemble or open inspection ports as permitted.
   - Examine lubricated surfaces for scoring, discoloration, varnish, or deposits.
   - Inspect filters, strainers, seals, and breathers for debris or restriction.
   - For grease-lubricated components, assess thickener integrity and distribution.
5. **Correlation of Lab and Field Findings**
   - Match laboratory indicators with physical evidence.
     - High Fe/Cr with visible scuffing.
     - TAN increase with acidic deposits.
     - High MPC with surface varnish.
   - Rule out test anomalies by confirming with repeat sampling if necessary.
6. **Operating Condition Review**
   - Retrieve operating logs for load, speed, temperature, and pressure.
   - Review recent maintenance activities (lubricant changes, filter replacement, alignment work).
   - Identify any changes in lubricant type, supplier, or additive package.
   - Check for environmental influences (ambient temperature swings, contamination sources).

7. **Hypothesis Development**
    - List all plausible root causes based on evidence:
        - Wrong lubricant selection.
        - Contamination ingress.
        - Overheating or inadequate cooling.
        - Mechanical misalignment or overload.
        - Incompatible grease mixing.
    - Eliminate causes inconsistent with test and inspection results.
8. **Verification Testing**
    - Conduct targeted confirmatory tests:
        - Compatibility testing for suspected grease mixing.
        - DGA for electrical faults in dielectric oils.
        - Foam and air release tests for suspected aeration.
    - If necessary, run short-term monitoring with sensors for pressure, temperature, or vibration to validate the fault progression.
9. **Root Cause Confirmation**
    - Select the cause supported by the greatest weight of converging evidence from:
        - Lab data.
        - Physical inspection.
        - Operating history.
    - Document evidence chains for traceability.
10. **Corrective and Preventive Actions**
    - Define immediate corrective action to restore function (fluid change, flushing, bearing replacement).
    - Define preventive measures:
        - Change lubricant specification or supplier.
        - Install improved filtration or moisture control.
        - Adjust maintenance intervals.
        - Implement contamination control procedures.
    - Schedule follow-up verification testing after corrective action.

EXAMPLE

Lubrication Failure Field Diagnostic Form

Equipment ID:
Location:
Date:
Technician/Engineer:
Oil/Grease Sample ID:

Section A – Varnish in Turbine Bearings

Lab Analysis

☐ Oil sample collected per ASTM D4057
☐ MPC (ASTM D7843) – Result: ______ ΔE
☐ FTIR carbonyl/carboxyl peaks – Observed? ☐ Yes ☐ No

- ☐ Antioxidant reserve (RULER/ASTM D6971) – % remaining: ______
- ☐ TAN (ASTM D664): ______ mg KOH/g
- ☐ Oxidative stability trend reviewed (ASTM D943)

Field Inspection

- ☐ Brown/amber deposits on bearings
- ☐ Oil film clearance measured – Value: ______ mm
- ☐ Localized hot spots measured – Temp: ______ °C
- ☐ Varnish removal/filter system inspected

Root Cause Notes: ____

Section B – Grease Incompatibility in Wind Turbines

Lab Analysis

- ☐ Grease samples collected from multiple points
- ☐ Compatibility test (ASTM D6185) result: ___________
- ☐ Worked penetration (ASTM D217) change: ______ pts
- ☐ Oil bleed (ASTM D6184) result: ______ %
- ☐ Dropping point: ______ °C

Field Inspection

- ☐ Hardened/channelled grease observed
- ☐ Oil leakage/runout present
- ☐ Thickener collapse confirmed by microscopy

Root Cause Notes: ____

Section C – Diesel Engine Wear from Soot and Fuel Dilution

Lab Analysis

- ☐ Hot oil sample collected mid-drain
- ☐ ICP wear metals – Fe: ______ ppm, Cr: ______ ppm
- ☐ Soot loading (TGA/ASTM D5967): ______ %
- ☐ Fuel dilution (GC/ASTM D3524): ______ %
- ☐ Flash point drop from baseline: ______ °C
- ☐ TBN: ______ mg KOH/g, TAN: ______ mg KOH/g

Field Inspection

- ☐ Exhaust smoke observed – Color: ______
- ☐ Crankcase breather/blow-by checked
- ☐ Injector noise/spray pattern checked

Root Cause Notes: ____

Section D – Dry Sump Starvation in Gas Engines

Lab Analysis

- ☐ Viscosity within spec (ASTM D445) – cSt: ______ at 40 °C
- ☐ Visual air entrainment/foam present
- ☐ Wear debris spike in elemental analysis

Field Inspection

☐ Return flow verified during operation
☐ Scavenge pump inspected
☐ Tank vent functional
☐ Return lines free of restriction

Root Cause Notes: ____

Section E – Cavitation Pitting & Aeration

Lab Analysis

☐ Air release test (ASTM D3427): ______ min
☐ Foam test (ASTM D892) – Sequence I: ______ ml
☐ Ferrography – spherical wear particles present? ☐ Yes ☐ No

Field Inspection

☐ Cavitation noise audible at pump inlet
☐ Pitting on pump vanes/housing observed
☐ Oil appears milky/cloudy
☐ Vortexing in reservoir return zones

Root Cause Notes: ____

Technician Summary and Recommendations:

Signature: ____________________ **Date:** ___________________

Failure Modes and Effects Analysis (FMEA) Process – Step-by-Step

1. **Define Scope and Boundaries**
   - Identify the system, subsystem, or component to be analyzed.
   - Clarify operating context: load, duty cycle, environmental conditions, regulatory requirements.
   - Determine whether the FMEA is for design (DFMEA) or maintenance/operation (PFMEA).
2. **Assemble a Cross-Functional Team**
   - Include engineering, maintenance, operations, reliability, quality, and supplier representatives.
   - Ensure participants have access to equipment history, drawings, and failure records.
3. **Break Down the System**
   - Decompose into functional blocks or process steps.
   - Use P&IDs, schematics, equipment BOMs, or process flow diagrams.
   - For each item, define its intended function and performance requirements.

4. **Identify Potential Failure Modes**
    - List all credible ways each function could fail to meet its requirement.
    - Use field history, OEM data, and subject matter expertise.
    - Examples: wear, corrosion, contamination, fatigue, incorrect lubrication, overheating.
5. **Determine Failure Effects (see Tables 12.7–12.14)**
    - Describe the immediate, downstream, and end-user consequences of each failure mode.
    - Consider safety, environmental, production, and asset damage impacts.
6. **Identify Potential Causes**
    - Determine root-level physical, human, or process factors.
    - Link causes to operating conditions, material incompatibility, poor maintenance, design flaws, or process variation.
7. **List Current Controls**
    - Document inspections, preventive maintenance tasks, alarms, condition monitoring, design safeguards, or procedural controls already in place.
8. **Assign Severity (S), Occurrence (O), and Detection (D) Ratings**
    - **Severity:** Rate the consequence of the effect (1 = negligible, 10 = catastrophic).
    - **Occurrence:** Rate the likelihood of the cause/failure mode happening (1 = rare, 10 = frequent).
    - **Detection:** Rate the likelihood that current controls will detect the failure before impact (1 = almost certain detection, 10 = no detection).
9. **Calculate the Risk Priority Number (RPN)**
    - **RPN = S × O × D**.
    - Rank the failure modes from highest to lowest RPN to prioritize mitigation efforts.
10. **Recommend Actions**
    - Identify actions that will reduce Severity, Occurrence, or improve Detection.
    - Assign responsible owners and due dates.
    - Examples: design changes, upgraded materials, lubrication changes, improved monitoring, procedural training.
11. **Reassess After Actions**
    - Update S, O, and D ratings after mitigation.
    - Recalculate RPN to verify improvement.
    - Confirm that the residual risk is acceptable for safety, compliance, and business objectives.
12. **Document and Control**
    - Keep the FMEA as a living document in a controlled system.
    - Link updates to incident reports, inspection data, and maintenance history.
    - Review periodically or when changes in design, operation, or environment occur.

TABLE 12.7
FMEA for Transformers

Item	Failure Mode	Cause	Effect	S	O	D	RPN	Current Controls	Detection Methods	Recommended Actions	Owner	Interval
T1	Loss of dielectric strength	Moisture ingress	Trip from insulation breakdown	10	4	3	120	Sealed tank. Breather maintenance.	Karl Fischer per ASTM D6304. BDV per IEC 60156.	Hot oil dry out. Seal inspection. Online moisture alarm.	Substation	Quarterly
T2	Copper sulfur corrosion	Reactive sulfur in oil	Winding shorts	9	3	5	135	Passivator. Approved oil spec IEC 60296.	Copper strip corrosion per ASTM D130.	Oil change to non corrosive grade. Add passivator.	Engineering	Annual
T3	Oxidation sludge	High temp. Low antioxidant reserve	Blocked cooling. Hot spots	8	5	4	160	Nitrogen blanket. Antioxidant spec.	TAN per ASTM D664. IFT per ASTM D971. MPC per ASTM D7843.	Reclaim. Add polish filtration. Adjust loading.	Ops	Semiannual
T4	Contact wear in OLTC oil	Carbon and metal debris	Switching failure	8	4	4	128	OLTC filtration skid.	Particle count ISO 4406. Gas in oil for OLTC compartment.	Offline cleaning. Filter upgrade.	Maintenance	Annual

TABLE 12.8
FMEA for Steam and Gas Turbines

Item	Failure Mode	Cause	Effect	S	O	D	RPN	Current Controls	Detection Methods	Recommended Actions	Owner	Interval
TG1	Varnish deposits	Oxidation. Soft contaminants	Servo sticking. Bearing heat	9	6	4	216	Depth filters. Reservoir cooling.	MPC ASTM D7843. RPVOT ASTM D2272. FTIR oxidation.	Electrostatic polishing. Top treat AO. Reservoir cleanout.	Reliability	Monthly
TG2	Air entrainment	Return agitation. Low air release oil	Pump cavitation	8	5	5	200	Baffles in tank. Antifoam control.	ASTM D3427 air release. ASTM D892 foam. Ultrasound.	Rework returns. Lower return velocity. Antifoam check.	Maintenance	Quarterly
TG3	Water contamination	Cooler leak. Breather fault	Bearing corrosion	8	4	4	128	Leak tests. Desiccant breathers.	Karl Fischer. Patch microscopy for rust.	Isolate cooler. Dehydrate oil. Replace breather.	Ops	Monthly

TABLE 12.9
FMEA for Hydraulic Systems and Compressors

Item	Failure Mode	Cause	Effect	S	O	D	RPN	Current Controls	Detection Methods	Recommended Actions	Owner	Interval
HY1	Cavitation pitting	Low NPSH. Suction leak	Pump failure	8	6	5	240	Strainers. Correct hose routing.	Inlet pressure log. Ultrasound. Ferrogram spheres.	Seal suction joints. Raise supply head. Larger inlet line.	Maintenance	Monthly
HY2	Wear from solids	Poor filtration. Ingress	Valve sticking	7	6	4	168	Beta-rated filters.	ISO 4406 counter. Patch microscopy.	Finer filters. Offline kidney loop. Seal upgrades.	Reliability	Monthly
HY3	Zinc soap deposits	Ester or AW depletion	Spool stiction	7	4	6	168	Approved additive chemistry.	FTIR for phosphate ester. Valve ΔP trend.	Flush. Change to compatible fluid.	Engineering	As needed

TABLE 12.10
FMEA for Wind Turbines

Item	Failure Mode	Cause	Effect	S	O	D	RPN	Current Controls	Detection Methods	Recommended Actions	Owner	Interval
WT1	Grease incompatibility	Mixed thickener chemistries	Main bearing overheating	8	6	5	240	Lube IDs. Color coded guns.	ASTM D6185. D217 penetration shift.	Purge old grease. Single approved product. Training.	Site lead	Semiannual
WT2	Micropitting gearbox	Low film due to viscosity loss	Gear noise. Early overhaul	9	5	4	180	Correct ISO grade. Cooling control.	Viscosity ASTM D445. PQ index. Wear metals.	Raise grade in cold season. Heater control. Filter upgrade.	Reliability	Quarterly
WT3	Water in generator bearings	Breather failure. Condensation	Corrosion. Spalling	7	5	4	140	Sealed housings.	Karl Fischer on purge grease. Visual rust.	Replace breathers. Add purge schedule.	Maintenance	Quarterly

TABLE 12.11
FMEA for Diesel and Gas Engines

Item	Failure Mode	Cause	Effect	S	O	D	RPN	Current Controls	Detection Methods	Recommended Actions	Owner	Interval
EN1	Fuel dilution	Injector leak. Cold starts	Low viscosity. Wear	9	6	3	162	Injector PM.	GC ASTM D3524. Flash point drop. Viscosity shift.	Injector service. Raise load profile. Shorten drain by condition.	Ops	Every sample
EN2	Soot overload	Low load. EGR faults	Abrasive wear	9	7	3	189	API category match.	TGA soot. ICP Fe Cr. FTIR nitration.	Correct EGR. Periodic high load run. Higher dispersant oil.	Reliability	Monthly
EN3	Turbo coking	Hot shutdowns	Oil coke in bearings	8	4	5	160	Cooldown run.	Boroscope. Differential temp on housing.	Add post-lube pump. Improve drain routing.	Maintenance	Quarterly
EN4	Dry sump starvation	Scavenge fault	Seizure risk	10	4	4	160	Low level alarms.	Return flow check. Pressure trend.	Vent fix. Screen clean. Line reroute.	Maintenance	After event

TABLE 12.12
FMEA for Cooling Loops and Thermal Oils

Item	Failure Mode	Cause	Effect	S	O	D	RPN	Current Controls	Detection Methods	Recommended Actions	Owner	Interval
CL1	Corrosion in water glycol	Low inhibitors. Low pH	Leaks. Fouling	7	6	4	168	OAT program.	pH ASTM D1287. Nitrite or molybdate residual. IC anions.	Re-treat inhibitors. Fix makeup water.	Water chem	Monthly
CL2	Glycol misblend	Wrong percent	Freeze or cavitation	8	5	4	160	Refractometer checks.	Refractometer ASTM D3321. D1120 boil point.	Drain and refill to spec. Label tanks.	Ops	Seasonal
CL3	Thermal oil cracking	High film temperature	Varnish. Viscosity rise	8	5	5	200	Vent condenser. Nitrogen blanket.	GC low and high boilers. TAN. MPC.	Lower film temp. Add polish filtration. Partial bleed and feed.	Reliability	Quarterly

TABLE 12.13
FMEA for Nuclear Auxiliaries

Item	Failure Mode	Cause	Effect	S	O	D	RPN	Current Controls	Detection Methods	Recommended Actions	Owner	Interval
NA1	Ester acid build	Hydrolysis. Dose effects	Valve stiction	9	3	5	135	Filter. Dry control.	TAN. Resistivity. Water ppm.	Dehydrate. Top treat with acid scavenger.	Engineering	Quarterly
NA2	Particulate fines	Wear in actuators	Spool jam	8	4	4	128	Fine filtration.	ISO 4406 code. Patch microscopy.	Upgrade filter rating. Add bypass polishing.	Maintenance	Monthly

TABLE 12.14
FMEA Switchgear and Inverters with Dielectric Coolant

Item	Failure Mode	Cause	Effect	S	O	D	RPN	Current Controls	Detection Methods	Recommended Actions	Owner	Interval
SI1	Rise in dissipation factor	Polymer extractables. Moisture	Partial discharge risk	9	4	4	144	Approved BOM. Dry controls.	IEC 60247 tan delta at temp. Karl Fischer.	Materials review. Fluid change if persistent.	Engineering	Semiannual
SI2	Conductivity rise in water coolant	Ion ingress. Inhibitor loss	Tracking and leakage	8	6	4	192	DI bed. Closed system.	Conductivity ASTM D1125. pH. IC anions.	Resin changeout. Leak repair. Rebalance inhibitors.	Water chem	Monthly

Master Lubrication FMEA Sheet for Power Generation Assets
Use a 1–10 scale for Severity, Occurrence, and Detection.
1 is best. 10 is worst.
RPN = S × O × D.
Re-calculate after each corrective action.

Severity (S), Occurrence (O), and Detection (D) Ratings

- Severity: Rate the consequence of the effect (1 = negligible, 10 = catastrophic).
- Occurrence: Rate the likelihood of the cause/failure mode happening (1 = rare, 10 = frequent).
- Detection: Rate the likelihood that current controls will detect the failure before impact (1 = almost certain detection, 10 = no detection).

Rating Guidance

- Severity
- 10. Personnel or large asset loss
- 7 to 9. Forced outage or major overhaul
- 4 to 6. Degraded output or accelerated wear
- 1 to 3. Minor effect
- Occurrence
- 8 to 10. Frequent in fleet history
- 5 to 7. Periodic
- 2 to 4. Rare with controls

1. Remote

- Detection
- 8 to 10. Low chance to detect before failure
- 5 to 7. Moderate chance
- 2 to 4. High chance with current program

1. Certain detection

The following is a **Field-Ready FMEA Execution Checklist** for lubrication, insulating fluid, and coolant failures in power generation and electrification systems.

It is structured so a reliability team can work through it in a single session or during a structured failure review.

FMEA Execution Checklist – Field Application

1. **Preparation**
   - Define equipment or system under review.
   - Identify whether this is a **Design FMEA (DFMEA)** or **Process/Operational FMEA (PFMEA)**.

- Gather system drawings, lubrication diagrams, process flow diagrams, and P&IDs.
- Retrieve historical failure data, maintenance logs, and oil/grease analysis results.
- Assign facilitator and note-taker.

2. **Team Assembly**
    - Include representatives from engineering, operations, maintenance, reliability, quality, and suppliers if relevant.
    - Confirm subject matter experts are present for lubricant, coolant, or dielectric fluid specifics.
3. **System Breakdown**
    - Divide system into subsystems and components.
    - Define each component's intended function and performance requirements.
    - List relevant lubrication, cooling, or insulation interfaces for each.
4. **Failure Mode Identification**
    - For each component, list all possible ways it can fail to perform its intended function.
    - Use historical cases and OEM data for reference.
    - Consider lubrication-specific modes: oxidation, varnish, contamination ingress, grease hardening, additive depletion, phase separation.
5. **Failure Effect Definition**
    - Describe immediate effect on the component.
    - Describe effect on the subsystem.
    - Describe effect on the full system, including safety, environmental, and production impacts.
6. **Cause Identification**
    - List mechanical, operational, chemical, and procedural causes.
    - Link to specific root causes where known (wrong viscosity grade, seal failure, excessive temperature, incompatible fluids, etc.).
7. **Current Control Review**
    - List inspections, oil analysis programs, online sensors, and PM tasks currently in place.
    - Note detection limitations for each control.
8. **S–O–D Rating Assignment**
    - **Severity (S):** Rate from 1 (no effect) to 10 (safety/environmental catastrophe).
    - **Occurrence (O):** Rate from 1 (remote) to 10 (frequent).
    - **Detection (D):** Rate from 1 (almost certain to detect) to 10 (no detection).
    - Document rationale for each rating to ensure transparency.
9. **RPN Calculation**
    - Multiply S × O × D for each failure mode.
    - Rank failure modes from highest to lowest RPN.
    - Flag high-priority items for immediate mitigation.
10. **Action Plan Development**
    - Identify actions to reduce severity, occurrence, or improve detection.

- Assign responsible owner and target completion date.
- Define verification method for action effectiveness.

11. **Post-Action Review**
    - Update S, O, D ratings after mitigation.
    - Recalculate RPN to confirm risk reduction.
    - Decide if additional action is required or residual risk is acceptable.
12. **Documentation and Control**
    - Save FMEA in controlled document repository.
    - Link to CMMS, oil analysis database, and incident reports.
    - Schedule FMEA review when design, operating conditions, or lubricants change.

Integrated Master FMEA Sheet for Lubrication, Insulating Fluids, and Coolants
Use this sheet during FMEA sessions. Fill ratings on a 1 to 10 scale. Compute RPN as S × O × D. Update after each corrective action (see Tables 12.15–12.21).

Header

- Asset or system ID
- Location
- Date and revision
- Facilitator
- Team members and roles

Rating guidance

- Severity: 10 personnel or large asset loss, 7 to 9 forced outage or major overhaul, 4 to 6 degraded output or accelerated wear, 1 to 3 minor effect
- Occurrence: 8 to 10 frequent in fleet history, 5 to 7 periodic, 2 to 4 rare with controls, 1 remote
- Detection: 8 to 10 low chance to detect before impact, 5 to 7 moderate chance, 2 to 4 high chance with current program, 1 certain detection

Current controls checklist

- Drawings and PIDs available
- OEM limits and chemistry specs available
- Sampling procedures defined
- CMMS history attached
- Online sensor tags mapped
- Oil and coolant labs aligned on methods and limits

TABLE 12.15
FMEA Transformers and OLTC

Item	Failure Mode	Cause	Effect	S	O	D	RPN	Current Controls	Detection Methods	Recom-mended Actions	Owner	Due Date	Residual S	Residual O	Residual D	Residual RPN
T1	Loss of dielectric strength	Moisture ingress, seal leaks, cold trapping	Trip, insulation breakdown	10	4	3	120	Sealed tank, desiccant breather, nitrogen blanket	Karl Fischer ASTM D6304, BDV IEC 60156, tan delta IEC 60247	Hot oil dry out, seal repair, online moisture alarm	Substation					
T2	Copper sulfur corrosion	Reactive sulfur in oil, passivator depletion	Winding shorts	9	3	5	135	Approved oil per IEC 60296, passivator	Copper strip ASTM D130, visual staining, DGA CO CO2 trend	Oil change to non-corrosive grade, passivator addition, copper coupon check	Eng.					
T3	OLTC contact wear and debris	Switching duty, inadequate filtration	Switching failure	8	4	4	128	OLTC filtration skid, periodic oil change	ISO 4406 particle code, metal debris on patch	Offline clean, finer filter, duty review	Maint.					

TABLE 12.16
FMEA for Steam and Gas Turbines

Item	Failure Mode	Cause	Effect	S	O	D	RPN	Current Controls	Detection Methods	Recommended Actions	Owner	Due Date	Residual S	Residual O	Residual D	Residual RPN
TG1	Varnish deposits	Oxidation, soft contaminants, high film temperature	Servo sticking, bearing heat	9	6	4	216	Depth filters, reservoir cooling	MPC ASTM D7843, FTIR oxidation, RPVOT ASTM D2272	Electrostatic polishing, AO top treat, reservoir cleanout	Reliability					
TG2	Air entrainment and foam	Return agitation, low air release oil	Pump cavitation, control instability	8	5	5	200	Baffles, antifoam control	Air release ASTM D3427, foam ASTM D892, ultra-sound	Rework returns, lower return velocity, review antifoam dose	Maint-enance					
TG3	Water contami-nation	Cooler leak, breather fault	Corrosion, film loss	8	4	4	128	Leak tests, desiccant breathers	Karl Fischer, rust on patch micro-scopy	Isolate cooler, dehydrate oil, breather replacement	Ops					

TABLE 12.17
FMEA for Hydraulic Systems and Compressors

Item	Failure Mode	Cause	Effect	S	O	D	RPN	Current Controls	Detection Methods	Recommended Actions	Owner	Due Date	Residual S	Residual O	Residual D	Residual RPN
HY1	Cavitation pitting	Low NPSH, suction leaks, hot oil	Pump failure, efficiency loss	8	6	5	240	Strainers, proper hose routing	Inlet pressure, ultrasound, spherical wear on ferrograms	Seal suction joints, raise supply head, larger inlet line	Maintenance					
HY2	Wear from solids	Poor filtration, ingress	Valve sticking, actuator wear	7	6	4	168	Beta rated filters, breathers	ISO 4406 counter, patch microscopy	Finer filtration, offline kidney loop, seal upgrades	Reliability					
HY3	Zinc soap deposits	AW depletion, ester interactions	Spool stiction	7	4	6	168	Approved additive chemistry	FTIR phosphate ester bands, valve delta P	Flush, change to compatible fluid	Engineering					

TABLE 12.18
FMEA for Wind Turbines

Item	Failure Mode	Cause	Effect	S	O	D	RPN	Current Controls	Detection Methods	Recommended Actions	Owner	Due Date	Residual S	Residual O	Residual D	Residual RPN
WT1	Grease incompatibility	Mixed thickener chemistries	Main bearing overheating	8	6	5	240	Lube IDs, color coded guns	ASTM D6185, D217 penetration shift, D6184 bleed	Purge, single approved grease, training	Site lead					
WT2	Micropitting gearbox	Low film due to viscosity loss at cold start	Gear noise, early overhaul	9	5	4	180	Correct ISO grade, heaters	Viscosity ASTM D445, PQ index, wear metals	Raise winter grade, heater control, filter upgrade	Reliability					
WT3	Water in generator bearings	Breather failure, condensation	Corrosion, spalling	7	5	4	140	Sealed housings	Karl Fischer on purge grease, visual rust	Replace breathers, set purge schedule	Maint.					

TABLE 12.19
FMEA for Diesel and Gas Engines

Item	Failure Mode	Cause	Effect	S	O	D	RPN	Current Controls	Detection Methods	Recommended Actions	Owner	Due Date	Residual S	Residual O	Residual D	Residual RPN
EN1	Fuel dilution	Injector leak, cold starts, low load	Low viscosity, wear	9	6	3	162	Injector PM	GC ASTM D3524, flash point, viscosity	Injector service, raise load profile, condition based drain	Ops					
EN2	Soot overload	Low load, EGR faults	Abrasive wear	9	7	3	189	API category match	TGA soot, ICP Fe Cr, FTIR nitration	Correct EGR, periodic high load run, higher dispersant oil	Reli					
EN3	Turbo coking	Hot shutdowns	Oil coke in bearings	8	4	5	160	Cooldown run	Borescope, housing delta T	Post lube pump, improved drain routing	Maint.					
EN4	Dry sump starvation	Scavenge fault, aeration	Seizure risk	10	4	4	160	Level alarms	Return flow check, pressure trend	Vent fix, screen clean, line reroute	Maint.					

TABLE 12.20
FMEA for Cooling Loops and Thermal Oils

Item	Failure Mode	Cause	Effect	S	O	D	RPN	Current Controls	Detection Methods	Recommended Actions	Owner	Due Date	Residual S	Residual O	Residual D	Residual RPN
CL1	Corrosion in water glycol	Low inhibitors, low pH	Leaks, fouling	7	6	4	168	OAT program, DI control	pH ASTM D1287, nitrite or molybdate residual, ion chromatography for anions	Retreatment of inhibitors, makeup water quality fix	Water chem					
CL2	Glycol misblend	Wrong percent	Freeze or cavitation	8	5	4	160	Refractometer checks	Refractometer ASTM D3321, D1120 boiling point	Drain and refill to spec, label tanks	Ops					
CL3	Thermal oil cracking	High film temperature	Varnish, viscosity rise	8	5	5	200	Vent condenser, nitrogen blanket	GC low and high boilers, TAN, MPC	Lower film temperature, polish filtration, bleed and feed	Reli.					

TABLE 12.21
FMEA for Nuclear Auxiliaries and Shielded Systems

Item	Failure Mode	Cause	Effect	S	O	D	RPN	Current Controls	Detection Methods	Recommended Actions	Owner	Due Date	Residual S	Residual O	Residual D	Residual RPN
NA1	Ester acid build in actuator circuits	Hydrolysis, radiation effects	Valve stiction, control loss	9	3	5	135	Filtration, dry control	TAN, resistivity, water ppm	Dehydrate, acid scavenger top treat, changeout plan	Engineering					
NA2	Fine particulate in auxiliaries	Wear in spools, corrosion	Spool jam	8	4	4	128	Fine filtration	ISO 4406, patch microscopy	Upgrade filter rating, bypass polishing	Maintenance					

Action Log

No.	Action Description	Type S or O or D	Owner	Cost Level	Due Date	Status	Evidence of Effectiveness

Verification Checklist

- Ratings reviewed and agreed by team
- RPN ranking validated against fleet history
- High RPN items added to CMMS as corrective work orders
- Residual risk documented
- Review date scheduled

13 Electrical Fluid Failures and Diagnostics

ELECTRICAL FLUID FAILURES AND DIAGNOSTICS

Electrical fluid systems in transformers, switchgear, HV cables, and other energized equipment rely on insulating and cooling fluids to maintain dielectric integrity, dissipate heat, and protect internal components. Failures in these fluids can originate from contamination, chemical degradation, thermal stress, and electrical discharge activity. Each mechanism produces specific, detectable changes in fluid chemistry and physical properties. The ability to identify these changes early is central to preventing dielectric breakdown, overheating, and component damage (see Table 13.1).

Failure Mechanisms in Electrical Fluids

Oxidation and Acid Formation
Insulating oils and synthetic dielectric fluids degrade under oxygen exposure, high operating temperatures, and catalytic metal surfaces. Oxidation produces acids, sludge, and varnish that reduce dielectric strength and increase dissipation factor. Acids attack cellulose insulation in transformers, accelerating aging.

Moisture Contamination
Water ingress from seals, gaskets, atmospheric exchange, or condensation reduces breakdown voltage and promotes partial discharge. In ester fluids, hydrolysis can occur, producing free fatty acids and alcohols. Moisture solubility differs across fluid types, making threshold limits application-specific.

Particle Contamination
Wear debris from mechanical movement, corrosion products, carbonized material from arcing, and fibers from insulation contribute to conductive particulate loading. High particle counts can bridge electrodes under high-voltage stress.

Dissolved Gas Formation
Electrical and thermal faults in transformers generate characteristic gases. Partial discharges produce hydrogen and methane, arcing generates acetylene, and

DOI: 10.1201/9781003775201-18

TABLE 13.1
Failure Mode–Diagnostic Matrix in Electrical Fluids

Failure Mode	Primary Indicators	Test Methods
Oxidation	Increased acidity, sludge, low IFT, darker color	ASTM D974, D971, D1500, RULER
Moisture ingress	Reduced breakdown voltage, ppm water increase	ASTM D1533, D1816
Particle contamination	High particle count, dielectric breakdown	ISO 4406, ASTM D1816
Thermal degradation	Gas profile (ethylene, ethane), color change	DGA, ASTM D1500
Arcing	Acetylene spikes, high particulate carbon	DGA, particle count
Cellulose degradation	Furan compound increase	ASTM D5837

overheating generates ethylene and ethane. Gases dissolve into the fluid and can be analyzed to identify fault type and severity.

Thermal Degradation

Overheating of fluids due to overloading, cooling system failure, or poor circulation causes molecular breakdown, producing low molecular weight hydrocarbons and darkening of the fluid. This reduces dielectric strength and increases viscosity in mineral oils.

Electrical Discharge and Arcing

Localized high-voltage stress, partial discharge activity, or full arcing degrade fluids through pyrolysis and ionization. This results in rapid generation of gases, conductive carbon, and permanent loss of insulating performance.

Diagnostic Methods in Electrical Fluids

Dissolved Gas Analysis (DGA)

Gas content in transformer oils is analyzed using ASTM D3612 methods (headspace, vacuum extraction, or stripper column techniques). The relative ratios of key gases are interpreted using methods such as Rogers Ratios, Duval Triangle, and Dornenburg Ratios.

- *Application*: Identifies electrical faults (partial discharge, arcing), thermal faults (hot spots), and combined mechanisms.
- *Limitation*: Requires historical data for trend accuracy.

Moisture Measurement – Karl Fischer Titration

Determines water content in insulating fluids in ppm. ASTM D1533 specifies methods for both coulometric and volumetric titration.

- *Application*: Detects seal failures, ambient ingress, and paper insulation moisture migration.

Interfacial Tension (IFT)
Measures the force between oil and water interface. Low IFT indicates dissolved polar contaminants from oxidation and degradation products. ASTM D971 is the standard method.

Furan Analysis
Detects 2-furfural and related compounds from cellulose insulation degradation. ASTM D5837 outlines test methods.

- *Application*: Estimates paper insulation life in transformers.

Breakdown Voltage Testing
Measures the AC or DC voltage at which the insulating fluid electrically fails. ASTM D1816 (small gap) and ASTM D877 (flat electrode) are common standards.

- *Application*: Immediate measure of dielectric condition.

Acidity (Neutralization Number)
Measured by ASTM D974. Rising acidity indicates oxidation in mineral oils and ester hydrolysis.

Sludge and Color Assessment
Sludge detection and color change provide visual evidence of oxidation and contamination. ASTM D1500 color scale is often applied for trending.

Online Monitoring Systems
Integrate dissolved gas, moisture, and temperature sensors for continuous evaluation of fluid condition. Online chromatographs provide near real-time gas generation trends.

Transformer Explosions from Insulation Breakdown

Insulation breakdown in transformers can initiate a chain of events leading to catastrophic mechanical rupture and explosion. The insulation system consists of dielectric fluid, cellulose-based solid insulation, and structural spacers. Its primary functions are to prevent electrical discharge between energized components and to transfer heat from windings and core to the cooling system. When insulation loses dielectric strength, high-voltage stresses can initiate partial discharge, arc propagation, and localized heating.

PRIMARY INITIATION MECHANISMS IN ELECTRICAL FLUIDS

Moisture Contamination

Water ingress from seal failures, gasket degradation, or condensation reduces breakdown voltage. Even ppm-level moisture in mineral oil can lower dielectric strength to the point where partial discharge occurs at normal operating voltages. Moisture also accelerates cellulose aging through hydrolysis, further reducing mechanical strength.

Dielectric Fluid Degradation

Oxidation, thermal stress, and contamination from wear metals or particles reduce the dielectric strength of insulating oils. As acid number increases and interfacial tension decreases, sludge deposits form in cooling ducts, leading to localized overheating.

Overheating and Hot Spots

Overload, blocked cooling ducts, or high ambient temperatures can cause localized heating above 140 °C in mineral oil systems. Elevated temperature accelerates paper insulation depolymerization and gas generation. Hot spots are common initiation sites for electrical breakdown.

Electrical Discharge

Partial discharge activity, when unchecked, carbonizes insulation and produces conductive paths. Once the breakdown channel is established, arc energy rapidly vaporizes surrounding dielectric fluid, producing large volumes of gas.

Contamination by Conductive Particles

Metallic debris from mechanical wear, loose hardware, or corrosion can migrate in oil and bridge conductors under electric stress. This provides a direct path for arcing and rapid fault propagation.

Explosion Sequence

1. **Initial Breakdown**
   A dielectric weak point fails, creating an arc between energized components or to ground.
2. **Gas Generation**
   The arc energy pyrolyzes oil and insulation, producing hydrogen, methane, acetylene, and other fault gases. Gas generation rates can reach hundreds of liters in milliseconds.
3. **Pressure Rise**
   The sudden vaporization of oil and expansion of gases increases internal tank pressure.

4. **Mechanical Rupture**
   If the pressure relief device cannot vent fast enough, the tank ruptures at seams or welds. In worst cases, structural failure results in complete tank disintegration.
5. **Fire and Explosion**
   Expelled oil can atomize and ignite from arc energy or hot surfaces, resulting in a fireball and secondary explosions from oil vapor ignition.

Preventive Diagnostics

- **Dissolved Gas Analysis (DGA)** for early detection of fault gases such as acetylene from arcing.
- **Moisture monitoring** with Karl Fischer titration or online probes to maintain safe water levels.
- **Dielectric breakdown voltage testing** to verify oil insulating capacity.
- **Thermal imaging and load profiling** to detect hot spots before insulation damage occurs.
- **Particle count and wear debris analysis** to identify conductive contamination sources.

SF_6 LEAKAGE AND DIELECTRIC LOSS

Sulfur hexafluoride (SF_6) is widely used in high-voltage switchgear, gas-insulated substations (GIS), and circuit breakers due to its high dielectric strength and arc-quenching capability. Its performance depends on maintaining proper gas pressure, purity, and moisture content. Leakage or contamination reduces dielectric margin and increases the risk of electrical breakdown.

Mechanisms of SF_6 Leakage

Seal and Gasket Degradation
Elastomeric seals and gaskets lose elasticity due to thermal cycling, UV exposure, or chemical incompatibility. Microcracking and hardening lead to gradual gas escape.

Flange and Joint Loosening
Vibration, thermal expansion, and improper torque on bolted joints allow small but continuous leakage at interfaces.

Porosity and Microcracks in Welds
Manufacturing defects or fatigue-induced microcracks in welds can provide leak paths.

Fitting and Valve Leaks
Maintenance activities involving gas filling or sampling introduce opportunities for slow leaks through valve stems, O-rings, or improper closures.

Effects on Dielectric Performance

Reduced Gas Pressure

SF_6 insulation effectiveness is directly related to gas density. Lower pressure reduces dielectric withstand capability, increasing the probability of partial discharge under operating voltage.

Contamination by Moisture

Moisture ingress forms corrosive byproducts such as HF when combined with arc energy. This degrades internal surfaces and increases dielectric losses.

Byproduct Accumulation

Arc decomposition produces SO_2, SOF_2, and SO_2F_2, which increase the loss factor (tan δ) and reduce breakdown strength.

Diagnostic and Monitoring Methods

Gas Density Monitoring

Continuous density monitors or pressure-temperature compensated gauges trigger alarms when density falls below safe limits.

Moisture Analysis

Portable dew point meters per IEC 60376 detect moisture content in ppmv, allowing early intervention before HF formation.

Gas Chromatography (GC)

Laboratory GC per IEC 60480 identifies and quantifies arc byproducts, indicating internal discharge activity.

Leak Detection

Handheld infrared SF_6 detectors or fixed monitoring systems identify and locate leaks by detecting IR absorption at 10.6 μm.

Dielectric Loss Measurement

Tan δ measurement in GIS systems quantifies insulation loss factor. Rising values indicate contamination or moisture ingress.

Preventive Actions

- Scheduled torque verification for flanges and joints.
- Periodic seal inspection and replacement with compatible materials.
- Gas handling procedures that maintain high purity per IEC 60376 specifications.
- Installation of fixed leak detection for critical GIS installations.

MOISTURE INGRESS AND INSULATION AGING IN SWITCHGEAR

Moisture contamination in switchgear insulation systems is a primary cause of dielectric deterioration, partial discharge initiation, and premature failure. Both air-insulated and gas-insulated switchgear (AIS and GIS) are susceptible, although the pathways and mechanisms differ. In AIS, moisture is often absorbed directly by insulating materials such as epoxy, phenolic laminates, or paper-based composites. In GIS, moisture typically enters via seal failures, gas handling errors, or from insufficiently dried components during assembly (see Table 13.2).

Pathways for Moisture Ingress

Seal and Gasket Degradation
Aged elastomer seals lose compression set and allow ambient humidity to enter. In GIS, degraded O-rings at flanges and inspection ports are frequent leak points.

Breathing and Ventilation Systems
In AIS, natural breathing of enclosures due to thermal cycling can draw in humid air, especially if desiccant breathers are missing, saturated, or bypassed.

Installation and Maintenance Practices
Improper storage of insulation components before assembly, or reassembly with insufficient drying after maintenance, leaves residual moisture.

Condensation from Temperature Cycling
Switchgear in unheated or outdoor environments can develop internal condensation when equipment cycles through dew point conditions.

Effects on Insulation Performance

Reduced Dielectric Strength
Moisture lowers the breakdown voltage of solid and gas mediums, increasing the likelihood of partial discharge at normal operating voltages.

Surface Tracking and Flashover
Moisture films on insulating surfaces promote leakage current and conductive path formation, leading to carbonized tracking.

Accelerated Aging of Organic Insulation
Moisture triggers hydrolytic degradation in paper and resin systems, breaking molecular bonds and reducing mechanical strength.

Corrosion of Conductors and Hardware
Moisture presence accelerates oxidation of copper and aluminum, increasing contact resistance and localized heating.

AGING MECHANISMS

Thermal Aging
Moisture accelerates thermal oxidative aging by providing a medium for ionic conduction, which increases localized heating under load.

Chemical Attack
In SF_6 or other gas-filled switchgear, moisture reacts under discharge conditions to form acids such as HF, which chemically attack epoxy and metal surfaces.

Mechanical Fatigue
As insulation swells and contracts with moisture cycling, microcracks form, allowing further ingress and weakening dielectric boundaries.

Diagnostic Methods

- **Dew Point Measurement** in GIS per IEC 60376 to assess water vapor content in the insulating gas.
- **Insulation Resistance (IR) Testing** to detect moisture-lowered resistance values.
- **Dielectric Loss (tan δ) Testing** to identify increased power factor from conductive moisture paths.
- **Visual Inspection** for condensation, corrosion, and surface tracking during outages.
- **Partial Discharge (PD) Measurement** to detect activity linked to localized moisture pockets.

Preventive and Mitigation Actions

- Maintain desiccant breathers in AIS and replace when saturated.
- Verify seal integrity and replace degraded elastomers with moisture-resistant compounds.
- Dry components to specified moisture content before assembly.
- Keep GIS compartments above minimum gas dew point margins to prevent condensation.
- Apply space heating or controlled ventilation in moisture-prone environments.

Partial Discharge and Corona Inception Due to Contamination
Contamination on insulation surfaces in high-voltage equipment initiates localized ionization at electric field stress points. In air-insulated systems, this can develop into corona activity. In solid or composite insulation systems, the result is partial discharge (PD) within voids or along surfaces. Both mechanisms erode insulation, generate heat, and produce chemical byproducts that accelerate further degradation.

Contamination Sources

Dust and Particulate Deposition
Metallic dust from switchgear operation, carbonized arc byproducts, and airborne industrial particulates deposit on insulation surfaces, altering surface resistivity and concentrating the electric field.

Salt and Conductive Pollution
Marine environments and chemical processing plants introduce hygroscopic salts that form conductive films when moisture is present, significantly lowering inception voltage.

Moisture Films
Humidity or condensation combines with surface contamination to create leakage paths that enable corona inception at lower field strengths.

Gas and Oil Byproducts
In GIS, SF_6 decomposition products (SOF_2, SO_2F_2) and polymerized particles from arcing events settle on insulators, altering dielectric properties. In oil-insulated systems, oxidation sludge and varnish deposits on solid insulation can trap air pockets and moisture, promoting PD.

Mechanisms of Discharge Initiation

Corona Inception
Occurs when the local electric field exceeds the ionization threshold of the surrounding gas, producing a visible glow or audible noise. Typically initiated at sharp edges, protrusions, or contamination ridges.

Surface Discharge
Develops along an insulating surface contaminated with conductive or semi-conductive films. The film distorts the electric field, allowing microarcs to propagate along the surface.

Internal Partial Discharge
Forms within voids, cracks, or delaminations in solid insulation where gas-filled cavities are exposed to high field stress. Contamination accelerates cavity formation by chemical or mechanical degradation.

Effects on Equipment

- **Chemical Degradation**: Discharges in SF_6 or air produce ozone, NOx, and acids that attack insulation and metal surfaces.
- **Thermal Damage**: Localized heating from repetitive discharges causes carbonization, embrittlement, and eventual flashover.

TABLE 13.2
Failure Pathway Table: Contamination, Chemical Degradation, Thermal Stress, Electrical Discharge

Pathway	Typical Initiators and Sources	Fluid or Insulation Response	Field Evidence and Primary Tests	Immediate Equipment Effects	Diagnostics and Thresholds	Corrective and Preventive Actions
Contamination	Particles from wear, corrosion fines, dust ingress, salt spray, maintenance debris, incompatible top-up fluids, water from coolers and breathers	Increased conductivity, reduced breakdown strength, surface tracking, filter loading, sludge carryover	ISO 4406 particle code, patch microscopy, Karl Fischer water ppm per ASTM D6304, BDV per IEC 60156 or ASTM D1816, chloride and sulfate by ion chromatography, color per ASTM D1500	Valve spool stiction, relay misoperation, pump wear, partial discharge onset in contaminated gaps	Particle codes above target, BDV trending down, moisture above OEM limit, anion spikes against site baseline	Upgrade filtration and breathers, seal repair, dehydration, fluid reclaim, controlled fill and sampling, contamination control procedures
Chemical degradation	Oxidation from air ingress and hot metal, hydrolysis in esters, additive depletion, SF6 byproduct formation after interruptions	TAN rise per ASTM D974, IFT drop per ASTM D971, antioxidant depletion, varnish precursors, furan increase in transformer oil, SF6 acids and SOxFy in GIS	TAN trend, IFT trend, MPC per ASTM D7843 for turbine oils, RPVOT per ASTM D2272, furan HPLC per ASTM D5837, SF6 purity and byproducts per IEC 60480	Varnish on bearings, cooling duct fouling, rising winding hot-spot, spacer embrittlement, spacer surface attack in GIS	TAN increase above site alarm, IFT below site alarm, MPC above 30, antioxidant remaining below site limit, SF6 byproducts above reuse criteria	Reclaim or change oil, antioxidant top treat where approved, nitrogen blanket on reservoirs, oxygen control, GIS gas cleanup and filter replacement, temperature control and loading review

Thermal stress	Overload, loss of cooling, maldistributed flow, film overheating in thermal oils, hot spots in windings, valve bypass causing local recirculation	Cracking of base stock, viscosity shift, low and high boiler formation by GC, color darkening, accelerated paper depolymerization, gas generation in oil	Viscosity per ASTM D445, GC low or high boiler split for thermal oils, infrared thermography hot spots, DGA ethylene and ethane rise per ASTM D3612 and IEC 60599, insulation resistance drift	Trip on temperature, rapid aging of cellulose, varnish plating, pump cavitation from viscosity change	Viscosity deviation beyond control band, DGA thermal fault signature T1 to T3, thermal map deviations from baseline	Restore cooling capacity, correct flow distribution, reduce load, set film temperature limits, bleed and feed for thermal oils, clean heat transfer surfaces
Electrical discharge activity	Geometry defects, protrusions, voids, contaminated spacers, moisture pockets, low gas density in GIS, conductive debris in oil	Ionization, carbonization tracks, rapid gas evolution, polymer chain scission on surfaces, higher tan delta	DGA acetylene and hydrogen for arcing and PD per ASTM D3612 and IEC 60599, PD tests per IEC 60270, TEV and ultrasonic surveys, dew point in GIS per IEC 60376, tan delta and resistivity per IEC 60247	Flashover risk, OLTC contact damage, spacer tracking, tank pressure rise, nuisance trips	Rising acetylene or rapid TCG rate, PD magnitude and phase-resolved patterns above acceptance, dew point above spec, tan delta above acceptance at temperature	Deburr and rework electrodes, clean and dry insulation, restore gas density and purity, repair seals, install online DGA or PD monitors, set action limits and outage criteria

- **Acoustic and EM Emissions**: PD generates high-frequency electromagnetic noise and ultrasonic signals, interfering with nearby control systems if not shielded.
- **Progressive Dielectric Weakening**: Over time, repeated PD erodes insulation thickness and creates conductive tracks.

Diagnostic Methods

- **Ultrasonic Detection**: Portable ultrasonic probes detect airborne noise from corona or PD events.
- **Electromagnetic (TEV) Measurement**: Transient earth voltage sensors detect high-frequency pulses associated with PD.
- **Phase-Resolved PD Analysis (PRPD)**: Provides patterns for distinguishing between corona, surface, and internal PD sources.
- **Optical Inspection**: In darkened enclosures, corona discharge can be visually observed with UV-sensitive cameras.
- **Gas Analysis**: In GIS, chemical analysis of SF_6 for byproducts indicates PD activity.
- **Oil Analysis**: In oil-filled systems, dissolved gas analysis detects hydrogen and hydrocarbons from discharges.

Preventive and Corrective Actions

- Maintain cleanliness of insulation surfaces, especially in coastal and industrial zones.
- Install and replace filters, purifiers, or dehumidifiers to reduce airborne contaminants.
- Verify creepage distances and smoothness of HV conductor surfaces during maintenance.
- Remove or polish sharp metallic edges that may concentrate the electric field.
- Periodically test for PD activity and trend results to detect worsening conditions before failure.

Notes for Application

- Use site limits with trend analysis rather than single-point judgments.
- Interpret multi-parameter changes together. Example: TAN rise with IFT drop and MPC rise indicates oxidation-driven varnish risk in turbine oils.
- Apply at-temperature dielectric tests for liquids to avoid false confidence at 25 °C.
- For GIS, treat dew point and gas purity as leading indicators.
- Dissolved gas analysis of transformer insulating fluids. CIGRÉ.

14 Thermal System Failures

Thermal systems in power generation and electrification such as heat exchangers, cooling loops, thermal oil systems, and phase-change units fail when heat transfer performance degrades to the point where operating temperatures exceed design limits. Failures arise from fluid property changes, fouling of heat transfer surfaces, mechanical defects, or improper control of flow and temperature.

FAILURE MECHANISMS IN THERMAL SYSTEMS

Fouling and Scaling

Deposition of minerals, corrosion products, biofilms, or degraded fluid residues on heat transfer surfaces create thermal resistance. This increases temperature differentials and raises surface metal temperatures, accelerating oxidation and mechanical stress. In water-based systems, scaling from calcium and magnesium salts is a primary contributor.

Corrosion and Erosion

Chemical attack from oxygen ingress, low pH, or incompatible additives corrodes tubes and channels, thinning walls and increasing leakage risk. High-velocity particulates in fluid streams erode heat exchanger surfaces, altering geometry and reducing efficiency.

Fluid Degradation

Thermal oils oxidize, forming acids and varnish that reduce heat transfer and increase viscosity. Glycols degrade thermally or oxidatively, producing acids that attack system metals. In molten salt systems, contamination with moisture leads to freezing point shifts and corrosion of carbon steel.

DOI: 10.1201/9781003775201-19

Flow Imbalance and Maldistribution

Improper valve settings, pump wear, or blockage can create uneven flow across heat exchanger surfaces, causing localized overheating and thermal fatigue.

Pump and Circulator Failures

Loss of fluid circulation from bearing failure, seal leaks, or impeller damage leads to rapid temperature rise in heat-loaded equipment.

Insufficient Expansion Control

In closed systems, failure of expansion tanks or relief devices can result in overpressure during heat-up cycles, stressing piping and vessel welds.

Diagnostic Methods

- **Temperature Profiling**: Use of infrared thermography to detect hot spots, bypass flow, or blocked passages.
- **Flow Measurement**: Ultrasonic or magnetic flow meters to detect reduced or uneven flow.
- **Fluid Analysis**: Periodic testing of viscosity, acid number, oxidation stability, particulate load, and water content.
- **Pressure Drop Monitoring**: Tracking pressure loss across heat exchangers to identify fouling accumulation.
- **Corrosion Monitoring**: Coupons or probes to measure corrosion rate in high-risk circuits.

Preventive Actions

- Maintain fluid chemistry within OEM and industry-specified limits.
- Implement filtration and side-stream conditioning for continuous contaminant removal.
- Use chemical inhibitors for corrosion control and anti-scaling agents in water-based loops.
- Schedule cleaning or chemical descaling based on monitored fouling factors.
- Replace degraded thermal oils before acid number or varnish potential reaches critical levels.
- Protect idle systems from moisture ingress by nitrogen blanketing or draining.

THERMAL SYSTEM FAILURES: GENERATOR HYDROGEN COOLING LEAKS

Hydrogen-cooled generators rely on a sealed gas circuit to remove heat from stator windings, core iron, and other internal components. Hydrogen is selected for its high thermal conductivity, low density, and low windage losses. Leakage in the hydrogen containment system reduces cooling efficiency and can create explosive mixtures with air if concentrations enter the flammable range (4–75% H_2 by volume) (see Table 14.1).

Failure Mechanisms

Seal Wear and Degradation
Shaft seals, often oil-film type, degrade from wear, contamination, or oil breakdown. Loss of seal oil pressure relative to hydrogen pressure allows gas to escape along the shaft (see Table 14.2).

Weld and Flange Defects
Thermal cycling, vibration, or fabrication flaws can produce cracks in welded joints or loosened flanged connections on the generator casing, coolers, or gas piping.

Cooler Tube Failures
Hydrogen-to-water heat exchangers can develop tube wall thinning from corrosion or erosion. Perforations allow hydrogen to enter the cooling water system or vice versa, leading to cross-contamination.

Instrumentation and Piping Leaks
Hydrogen manifold fittings, pressure relief devices, and sample ports may develop leaks from improper torque, gasket degradation, or mechanical damage during maintenance.

Pressure Control Failures
Automatic hydrogen makeup systems can malfunction, leading to excessive pressure fluctuations that strain seals and fittings, increasing leakage probability.

Field Evidence and Diagnostics

- **Gas Pressure Trends**: Continuous monitoring of generator casing pressure relative to ambient identifies abnormal loss rates.
- **Gas Purity Readings**: Declining hydrogen purity indicates air ingress through the leak path.
- **Acoustic Leak Detection**: Ultrasonic detectors identify pressurized gas leak noise signatures at flanges, seals, and tubing.
- **Soap Solution Testing**: Localized leak detection on accessible joints.
- **Mass Spectrometry (Helium Tracer)**: Precision leak location during outages using helium as a tracer gas.
- **Seal Oil Monitoring**: Drop in seal oil differential pressure or contamination with hydrogen indicates seal leakage.

Immediate Equipment Effects

- Reduced cooling efficiency, leading to higher winding temperatures.
- Increased stator and rotor insulation aging rates.

TABLE 14.1
Failure Pathway Table – Generator Hydrogen Cooling Leaks

Pathway ID	Initiators and Sources	System Response	Field Evidence and Tests	Immediate Equipment Risks	Diagnostic Thresholds	Corrective and Preventive Actions
H1	Shaft seal wear, oil film degradation, loss of seal oil pressure	Hydrogen escapes along rotor shaft path, air may ingress	Seal oil differential pressure drop, oil contaminated with gas, ultrasonic leak detection, hydrogen purity drop	Reduced cooling efficiency, insulation overheating, flammable mixture in casing	Seal oil differential < OEM spec, purity < 98%, O_2 content > 1%	Restore seal oil margin, replace seals, flush oil system, inspect seal faces and housings
H2	Weld fatigue cracks or flange gasket failure on casing or piping	Gas leaks from casing or manifold, possible air ingress	Pressure decay test during outage, soap solution on suspect joints, ultrasonic leak detection	Hot spots from reduced gas density, localized oxidation on conductor surfaces	Leak rate > OEM limit (Nm^3/hr), rapid purity loss > 1% per day	Re-weld or replace cracked sections, retorque flanges, replace gaskets, conduct NDT on welds
H3	Cooler tube wall thinning from corrosion or erosion	Hydrogen migrates into cooling water, water ingress to gas circuit	Eddy current inspection, hydrostatic pressure test, water chemistry anomalies, DGA moisture rise	Cross-contamination, hydrogen loss, corrosion of stator iron	Tube wall loss > 20% nominal, water content > OEM ppm limit	Plug or replace defective tubes, implement corrosion inhibitors, review water treatment program
H4	Faulty fittings, valves, or relief devices on gas manifold	Localized leak points with pressure loss	Acoustic detection, soap bubble test, gas sniffer survey	Increased make-up gas consumption, risk of fire in vent areas	Make-up rate above normal baseline, pressure drop > 5 kPa/hr	Replace or reseat fittings, calibrate PRVs, verify valve shutoff integrity
H5	Pressure control regulator malfunction	Overpressure or underpressure cycles stressing seals and joints	Pressure fluctuation log, regulator calibration records	Seal blow-out, accelerated leak formation, instability in cooling capacity	Pressure fluctuation beyond ±5% of setpoint	Repair/replace regulator, inspect for seal distortion, re-balance gas control loop

TABLE 14.2
FMEA Hydrogen Cooling Leak
Use a 1 to 10 scale for Severity, Occurrence, and Detection. RPN equals S multiplied by O multiplied by D.

Item	Failure Mode	Cause	Effect	S	O	D	RPN	Current Controls	Detection Methods	Recommended Actions	Owner	Review Interval
H1	Loss of hydrogen at shaft seals	Seal wear, low seal oil differential pressure, oil varnish on faces	Reduced cooling, rising winding temperatures, risk of flammable mixture	10	5	3	150	Seal oil console with DP alarms, clean oil program	Seal oil DP trend, hydrogen purity, ultrasonic survey, oil analysis for varnish	Restore DP margin, replace seal rings and face, flush and filter seal oil, verify seal housings and runout	Maint.	Quarterly
H2	Casing or manifold leak	Weld fatigue, flange gasket damage, improper torque	Hydrogen loss, air ingress, local hot spots	9	4	5	180	Bolting procedure, periodic torque checks	Pressure decay test, soap solution, ultrasonic leak mapping	Repair welds, replace gaskets, retorque to spec, NDT of suspect seams	Mech.	Semiannual
H3	Cooler tube breach	Corrosion or erosion of tube walls, vibration fretting	Cross contamination with water, purity drop, corrosion risk	9	3	4	108	Water chemistry control, vibration supports	Eddy current tube inspection, hydrotest, conductivity and dissolved gases in cooling water	Plug failed tubes or retube bundle, adjust water chemistry, add inlet strainers if needed	Reli.	Annual

(continued)

TABLE 14.2 (Continued)
FMEA Hydrogen Cooling Leak

Use a 1 to 10 scale for Severity, Occurrence, and Detection. RPN equals S multiplied by O multiplied by D.

Item	Failure Mode	Cause	Effect	S	O	D	RPN	Current Controls	Detection Methods	Recommended Actions	Owner	Review Interval
H4	Fitting or valve leakage	Stem packing wear, O ring hardening, improper closure after service	Increased makeup gas use, local ignition risk near vents	8	5	4	160	Tagged closure steps, valve service schedule	Handheld IR SF6 style H2 detector, soap test, acoustic tool	Repack or replace valves, renew O rings, set closure torques, leak retest before return to service	Ops	Quarterly
H5	Regulator or control fault	Faulty pressure regulator, transmitter drift, control loop instability	Pressure swings, seal damage, repeat leak formation	8	4	5	160	Redundant transmitters, periodic calibration	Pressure trend with alarm limits, regulator bench test	Repair or replace regulator, recalibrate instruments, tune control loop, verify relief device setpoints	I and C	Semiannual

- Potential hydrogen-air explosive mixture formation in casing or ventilation spaces.
- Cooling water contamination if leak path exists through hydrogen coolers.

Corrective and Preventive Actions

- Maintain seal oil pressure at OEM-specified margin above hydrogen pressure.
- Inspect and replace worn shaft seal components during scheduled outages.
- Conduct periodic ultrasonic surveys on casing, cooler headers, and associated piping.
- Hydrotest and eddy current inspect cooler tubes to detect thinning or perforations.
- Verify torque and gasket condition on all bolted gas joints.
- Maintain gas purity monitoring with alarms for low purity and high oxygen content.
- Train operators in rapid hydrogen purge and inerting procedures for leak response.

Application Notes

- Track **hydrogen purity** and **oxygen content** as the leading indicators of air ingress.
- Use **mass spectrometer helium tracer testing** during outages to pinpoint micro-leaks that acoustic or soap testing cannot detect.
- Maintain **continuous monitoring** for purity, pressure, and seal oil parameters, with alarm thresholds based on OEM specifications.
- Store **spare seal assemblies, gaskets, and cooler tubes** for rapid leak repair.

Rating Notes

- Severity 10 represents personnel hazard or major generator damage.
- Occurrence rating uses fleet history and recent events.
- Detection rating reflects the chance to detect before effect using current monitoring.

Implementation Checklist

- Set hydrogen purity low alarm at or above 98%.
- Set oxygen high alarm at or below 1%.
- Maintain seal oil differential pressure above OEM minimum at all loads.
- Trend leak rate in normal cubic meters per hour against a site baseline.
- Add helium tracer testing in outages for chronic small leaks.

THERMAL SYSTEM FAILURES: COOLANT DROPOUT OR INHIBITOR FAILURE IN GENSETS

Genset cooling systems rely on a stable coolant mixture to control metal temperatures, prevent boiling, and limit corrosion in the engine block, cylinder head, oil cooler, and ancillary heat exchangers. Failures involving coolant dropout or corrosion inhibitor depletion reduce the system's protective capability, leading to scaling, localized overheating, and corrosion pitting (see Tables 14.3 and 14.4).

Failure Mechanisms

Precipitation and Dropout

Improper glycol concentration, mixing incompatible coolant chemistries, or high hardness makeup water can cause silicate, phosphate, or carbonate dropout. Solid precipitates settle in low-flow areas, reducing heat transfer and promoting hot spots.

Inhibitor Depletion

Organic acid technology (OAT), nitrite, molybdate, or silicate inhibitors neutralize corrosive species and form protective films. Extended service intervals without replenishment or incorrect top-ups with water dilute inhibitors below effective concentrations.

pH Drift and Acid Formation

Oxidation of glycol and contamination from combustion gas leakage (blown head gasket or liner seal failure) lower pH and increase acidity. This accelerates metal attack, particularly on aluminum and copper alloys.

Biological Growth

Low inhibitor levels combined with warm standby conditions can permit microbial growth in coolant, producing organic acids and sludge that exacerbate corrosion and block passages.

Thermal Cycling Stress

Scale and deposit formation on internal surfaces increases differential expansion and stresses thin-walled heat exchanger tubes, increasing risk of cracking or leak formation.

Field Evidence and Diagnostics

- **Coolant Analysis**: pH, reserve alkalinity, nitrite/molybdate content, and glycol concentration.
- **Visual Inspection**: Rust, scale, or gel formation in reservoir or heat exchanger tubes.
- **Conductivity Testing**: High conductivity indicates dissolved solids buildup or contamination.
- **Filter Inspection**: Capture of particulate dropout in coolant filters or strainers.

TABLE 14.3
Failure Pathway Table – Coolant Dropout or Inhibitor Failure in Gensets

Pathway ID	Initiators and Sources	System Response	Field Evidence and Tests	Immediate Equipment Risks	Diagnostic Thresholds	Corrective and Preventive Actions
C1	Mixing incompatible coolant chemistries (e.g., OAT with silicate-based)	Precipitate formation, solids settle in low-flow areas	Visual inspection of reservoir, filter debris analysis, coolant turbidity	Blocked passages, localized overheating	Turbidity > 20 NTU, solids visible in sample	Drain, flush, and refill with OEM-approved coolant, use compatibility lab test before top-up
C2	Hardness from untreated makeup water	Calcium/magnesium salts precipitate under heat load	Coolant analysis for hardness, scale found in heat exchanger	Reduced heat transfer, liner hot spots	Hardness > 50 ppm $CaCO_3$ equivalent	Use deionized or distilled water, add chemical softeners where allowed
C3	Inhibitor depletion from extended service interval or dilution	Loss of nitrite/molybdate/silicate protective film	Coolant analysis: low inhibitor ppm, low reserve alkalinity	Accelerated corrosion of aluminum, copper, and ferrous metals	Nitrite < 500 ppm, RA < 3 mL	Add inhibitor package or replace coolant, reinstate scheduled analysis program
C4	pH drift from glycol oxidation or combustion gas ingress	Coolant becomes acidic, attacks metal surfaces	pH test, FTIR for glycol oxidation products, oil analysis for coolant markers	Corrosion pitting, elastomer degradation	pH < 7.0, TAN > 5 mg KOH/g	Flush and replace coolant, investigate head gasket/cooler leaks, correct source of contamination
C5	Microbial growth in warm standby system	Sludge formation, biofilm coating surfaces	Microscopy of coolant sample, ATP test	Acid production, deposit blockage	Microbial count > 104 CFU/mL	Add approved biocide, clean and flush circuit, eliminate stagnation zones
C6	Thermal cycling with existing scale deposits	Differential expansion cracks thin tubes, leaks develop	IR thermography for hot spots, ultrasonic thickness of tubes	Coolant-to-oil or coolant-to-combustion gas cross-contamination	Tube wall loss > 20%, ΔT > OEM limit	Descale chemically under controlled procedure, replace damaged tubes

TABLE 14.4
FMEA – Coolant Dropout or Inhibitor Failure in Gensets

Item	Failure Mode	Cause	Effect	S	O	D	RPN	Current Controls	Detection Methods	Recommended Actions	Owner	Review Interval
CD1	Precipitate formation from incompatible coolant mixing	OAT mixed with silicate or phosphate coolant. Improper top-up	Blocked passages. Localized overheating	8	4	4	128	OEM coolant spec. Operator training	Visual reservoir check. Filter debris inspection. Turbidity	Drain. Flush. Refill with approved coolant. Lab check compatibility before top-up	Maintenance	Semiannual
CD2	Scale from hard makeup water	High Ca or Mg in makeup water	Reduced heat transfer. Liner hot spots	8	3	5	120	DI or distilled water use	Hardness test. Scale inspection at outage	Switch to DI water. Add softener if allowed. Chemical treatment per OEM	Reliability	Annual
CD3	Inhibitor depletion	Overdue interval. Dilution by water top-ups	Corrosion of aluminum. Copper. Ferrous metals	9	5	3	135	Scheduled coolant analysis	Nitrite or molybdate ppm. Reserve alkalinity	Add inhibitor package or replace coolant. Reset PM frequency	Maintenance	Quarterly

CD4	pH drift from glycol oxidation or combustion gas ingress	Thermal aging. Head gasket leak. Cooler tube leak	Acidic coolant. Elastomer degradation	9	3	4	108	Oil analysis for glycol. pH trending	pH. TAN. Sodium or potassium in oil	Flush and replace coolant. Repair leak source	Mechanical	Quarterly
CD5	Microbial growth	Warm standby with stagnation. Low biocide	Biofilm sludge. Flow restriction. Acid formation	7	4	6	168	Biocide dosing plan	ATP test. Microbial culture. Visual sludge	Dose approved biocide. Flush and sanitize. Periodic circulation during standby	Maintenance	Semiannual
CD6	Thermal cycling on scaled surfaces	Deposits causing differential expansion	Tube cracking. Cross-contamination	8	3	5	120	Descaling during outages	IR thermography hot spots. Ultrasonic wall thickness	Chemical clean to spec. Replace damaged tubes. Restore flow balance	Reliability	Annual

- **Infrared Thermography**: Identifies localized hot zones indicating blocked flow.
- **Oil Analysis**: Presence of glycol or coolant-derived sodium/potassium in engine oil suggests head gasket or cooler leak linked to corrosive coolant conditions.

Immediate Equipment Effects

- Reduced cooling efficiency leading to elevated cylinder head and oil temperatures.
- Accelerated corrosion of liners, water pumps, and heat exchanger components.
- Risk of catastrophic failure from cracked cylinder heads or liner cavitation if deposits localize boiling.
- Shortened life of rubber hoses and elastomer seals from acid exposure.

Corrective and Preventive Actions

- Maintain OEM-specified glycol concentration and inhibitor levels through scheduled coolant analysis.
- Use deionized or distilled water for mixing to avoid mineral scaling.
- Avoid mixing coolant types unless verified compatible by laboratory testing.
- Replace coolant filters per manufacturer schedule and inspect for solids.
- Flush and refill cooling circuit after significant inhibitor loss or precipitation events.
- Incorporate coolant condition sensors in high-reliability installations to trigger alarms before failure thresholds are reached.

Application Notes

- Apply **coolant condition monitoring** at the same maintenance interval as oil analysis for critical gensets.
- Trend **nitrite, molybdate, and pH** to forecast inhibitor service life.
- Deploy **inline filter element inspections** during PMs to detect early precipitation before full blockage occurs.
- For units in standby, circulate coolant periodically to prevent stratification and microbial growth.

Rating Notes

Severity 10 means catastrophic engine damage. Occurrence based on fleet history. Detection reflects the likelihood to catch before failure with current controls. RPN equals S × O × D.

THERMAL SYSTEM FAILURES: FOULING IN SOLAR THERMAL COLLECTORS

Fouling in solar thermal collectors reduces heat transfer, raises absorber and heat transfer fluid temperatures, drives glycol oxidation, and shortens component life. Failures arise from particulate deposition, scaling, biofilm growth, corrosion products, and degradation residues. Impacts include lower optical efficiency, higher thermal resistance across heat exchangers, flow maldistribution, stagnation events, and leaks from differential thermal stress (see Tables 14.5 and 14.6).

Principal Fouling Mechanisms

- **Particulate deposition**
  Dust, sand, pollen, soot, and construction debris settle on glazing and absorber surfaces. In concentrating collectors, specular reflectance loss at mirrors reduces flux at the receiver. Internal particulate can accumulate where strainers are absent.
- **Scaling and crystallization**
  Hard makeup water drives calcium carbonate, calcium sulfate, and silica scaling in absorber tubes and balance-of-plant heat exchangers. High surface temperatures and flash evaporation accelerate precipitation at low velocity zones. Risk is predictable from Langelier or Ryznar indices and silica solubility limits.
- **Biofouling**
  In open loops and poorly treated closed loops, microorganisms form biofilms on heat exchange surfaces. Biofilms trap solids and create under-deposit corrosion. Stagnant headers and oversized piping increase risk.
- **Corrosion products**
  Oxygen ingress, low inhibitor reserve, and pH drift generate iron oxides, copper oxides, and aluminum corrosion fines that lodge in headers, manifolds, and plate exchangers.
- **Glycol degradation residues**
  Propylene or ethylene glycol thermally degrades at high film temperature and during stagnation. Acids and oligomers increase viscosity and form soft polymeric deposits. Inhibitor depletion accelerates metal attack and sludge formation.

Field Evidence and Diagnostics

- **Optical losses**
  Drop in optical efficiency from baseline per ISO 9806 performance test. Mirror reflectance declined with portable reflectometer. Visual dust and bird droppings on glazing.
- **Hydraulic symptoms**
  Rising pressure drop across collector field at fixed flow. Flow imbalance between parallel strings. Strainer loading. Pump differential pressure trending up.

TABLE 14.5
Failure Pathway in Solar Thermal Collectors

Pathway ID	Initiators and Sources	System Response	Evidence and Tests	Risk to Equipment	Thresholds to Act	Controls and Fixes
ST1 Scaling in absorber tubes	Hard makeup water, high wall temperature, low velocity	Mineral scale on hot surfaces, rising pressure drop	Hardness and alkalinity high, DP trend up, IR hot stripes, tube cuts show carbonate or sulfate	Hot spots, gasket hardening, tube creep, leaks	Hardness above 50 ppm CaCO3, DP rise more than 15% from baseline	Switch to DI makeup, chemical clean to vendor spec, rebalance flow, set velocity target
ST2 Biofouling in exchangers	Warm standby, nutrients, low biocide	Biofilm on plates and headers, under deposit corrosion	ATP high, culture counts high, pressure drop up, approach temperature up	Corrosion pitting, exchanger performance loss	ATP above program limit, approach temperature increase more than 5 K	Shock biocide, mechanical clean, dose control, eliminate dead legs
ST3 Dust and soiling on glazing or mirrors	Arid climates, construction dust, pollen	Optical losses, lower useful heat gain	Transmittance or reflectance drop, soiling index high, visual fouling	Reduced output, higher field temperature, stagnation events	Optical loss more than 5 percent vs baseline or ISO 9806 retest	Scheduled cleaning, verified water quality, antisoiling coatings where proven
ST4 Glycol degradation residue	Stagnation, high film temperature, inhibitor depletion	TAN rise, viscosity rise, polymeric sludge	TAN up, FTIR oxidation bands, filters load fast, fluid dark	Pump overload, valve sticking, seal damage	TAN increase above site alarm or OEM limit, viscosity shift more than 10%	Partial drain and feed or full change, reinstall inhibitor package, reduce stagnation frequency
ST5 Corrosion product fouling	Oxygen ingress, low inhibitor reserve, pH drift	Iron and copper oxides in headers and strainers	Particle count elevated, SEM EDS confirms oxides, pH off target	Abrasive wear, under deposit corrosion, leaks	Inhibitor below spec, pH outside 8 to 9.5 for glycol systems	Fix air ingress, restore inhibitor, side stream filtration, set pH to target

TABLE 14.6
FMEA for Fouling in Solar Thermal Collectors
Use 1–10 scale for Severity (S), Occurrence (O), Detection (D).
RPN = S × O × D. Adjust ratings to your site history.

Item	Failure Mode	Causes	Effects	S	O	D	RPN	Current Controls	Detection Methods	Recommended Actions	Owner	Review Interval
ST-F1	Scaling inside absorber tubes	Hard makeup water, high wall temperature, low velocity	Heat transfer loss, hot spots, gasket hardening, tube leaks	8	5	4	160	DI water policy, strainers, flow balancing	DP trend across field, IR hot stripes, water hardness, cut-open tube inspection	Chemical clean to vendor spec, switch to DI, raise velocity, rebalance strings	O&M	Quarterly
ST-F2	Biofouling in HX and headers	Warm standby, low biocide, nutrient ingress	Approach temperature rise, under-deposit corrosion, flow loss	7	5	6	210	Biocide program, periodic circulation	ATP test, culture count, HX ΔP and ΔT trend	Shock biocide, mechanical cleaning, eliminate dead legs, maintain circulation	Water chem	Quarterly
ST-F3	External soiling on glazing/mirrors	Dust, pollen, ash, bird droppings	Optical efficiency loss, stagnation events	7	7	3	147	Cleaning schedule, wind screens	Reflectance/transmittance checks, ISO 9806 performance test, visual survey	Adjust cleaning frequency from measured soiling rate, use approved wash water quality	O&M	Monthly (seasonal)

(*continued*)

TABLE 14.6 (Continued)
FMEA for Fouling in Solar Thermal Collectors
Use 1–10 scale for Severity (S), Occurrence (O), Detection (D).
RPN = S × O × D. Adjust ratings to your site history.

Item	Failure Mode	Causes	Effects	S	O	D	RPN	Current Controls	Detection Methods	Recommended Actions	Owner	Review Interval
ST-F4	Glycol degradation residues	Stagnation, high film temperature, inhibitor depletion	TAN rise, viscosity rise, soft deposits, valve sticking	8	4	5	160	Inhibitor monitoring, stagnation controls	TAN (ASTM D664), FTIR, viscosity, filter loading	Partial bleed-and-feed or changeout, restore inhibitors, reduce stagnation frequency	Reliability	Semiannual
ST-F5	Corrosion product fouling	O_2 ingress, low inhibitor reserve, pH drift	Oxide fines load strainers, abrasive wear, leaks	8	4	5	160	Air separators, nitrogen blanket, inhibitors	Particle count, SEM/EDS on debris, pH, inhibitor residual	Fix ingress, restore inhibitor chemistry, add side-stream filtration	Water chem	Quarterly
ST-F6	Flow maldistribution from internal deposits	Scale/sludge in headers and risers	Uneven panel temps, local overheating, tube creep	9	4	6	216	Commissioned balance, orifice plates	String-by-string flow test, IR mapping of arrays, ΔP per branch	Clean and flush, rebalance, add balancing valves/orifices, update commissioning sheets	Engineering	After outage

ST-F7	HX fouling on process side	Poor process water quality, silica scaling	Reduced heat transfer, higher pump head, trips	8	4	5	160	Process water specs, strainers	HX ΔP and approach temp, silica analysis, visual bundle check	Descale HX, improve process water treatment, install finer pre-filtration	O&M	Semiannual
ST-F8	Filter/strainer bypass due to clogging	High particulate load, overdue service	Unfiltered flow, downstream fouling spike	6	6	4	144	ΔP gauges, PM task	Filter ΔP, inspection of elements, debris mass logged	Increase filter area or fineness, add duplex housings, tighten PM frequency	O&M	Monthly

- **Thermal symptoms**
  Higher collector outlet temperature at normal irradiance for a given flow indicates internal resistance. Stagnation frequency increase. IR thermography shows hot strips along fouled risers.
- **Chemistry indicators**
  Glycol TAN rise, inhibitor residual low, and UV absorbance growth. Hardness and alkalinity in makeup water above control limits. Microbial ATP counts above program limits.
- **Inspection findings**
  Scale on tube cuts, brown or black sludge in headers, biofilm on plate heat exchanger plates, opaque coolant, particulate in strainers and filters.

Tests and Methods

- Collector performance test per ISO 9806 to quantify optical and thermal efficiency loss.
- Glazing and mirror cleanliness audit with reflectance or transmittance checks per ISO 9050 or manufacturer procedure.
- Differential pressure trending across collector field and exchangers.
- Coolant analysis: glycol percent by refractometry, TAN per ASTM D664, inhibitor reserve per OEM method, FTIR screening for glycol oxidation products, particle count per ISO 4406 for closed loops.
- Water quality: hardness, alkalinity, chloride, sulfate, silica per Standard Methods for the Examination of Water and Wastewater; Langelier and Ryznar indices calculation.
- Microbiology: ATP luminometry and culture counts for open loops and storage tanks.
- Materials inspection: SEM or EDS of scale and sludge samples to identify mineralogy and corrosion species.

Immediate Equipment Risks

- Overtemperature at absorber leading to gasket hardening and seal failure.
- Stagnation triggers safety valve lift and glycol venting.
- Tube or plate heat exchanger fouling raises approach temperature and forces pumps to higher head.
- Under-deposit corrosion and pinhole leaks in copper and stainless components.
- Tracking errors in CSP receivers due to mirror fouling cause hot spots and tube creep.

Corrective Actions

- Clean glazing and mirrors with approved methods and water quality controls. No abrasive media. Verify runoff management.
- Chemical descaling of exchanger plates and absorber circuits with vendor-approved acid or chelant programs. Neutralize and rinse to conductivity targets.

- Glycol reclaim or replacement when TAN exceeds site limit or inhibitors are depleted. Flush to clear polymeric residues.
- Shock biocide treatment and biofilm removal in affected loops. Eliminate stagnation zones with piping changes or purge routines.
- Strainer and filter maintenance with finer media where particulate load persists.
- Repair oxygen ingress. Install low-permeation hoses and verified air separators.

Preventive Controls

- Maintain glycol concentration to OEM specification to control boiling margin and corrosion protection.
- Enforce deionized water for makeup. Hardness less than 50 ppm as CaCO3, chloride and sulfate below OEM limits, silica within solubility limits at operating temperature.
- Continuous or periodic filtration on the collector loop. Target cleanliness codes appropriate for pump and valve clearances.
- Inhibitor monitoring and top up per ASTM D1384 corrosion glassware guidance and OEM program.
- Dust and soiling management plan based on local fouling rate, tilt angle, and rainfall data. Set cleaning frequency from measured transmittance or reflectance loss rather than calendar time.
- Stagnation control through control logic, night sky cooling use, and dump radiators where feasible.
- Flow balancing of collector strings with commissioning test sheets and periodic verification.

Program setup for O and M

- Establish commissioning baselines for optical efficiency, hydraulic pressure drops, collector outlet temperature at reference irradiance, fluid chemistry, and water quality.
- Set trend limits and action thresholds per site design and OEM.
- Use a test matrix that pairs each failure symptom with a test method and a decision rule.
- Record after clean and after chemical clean data to update baseline when hardware changes.
- Tie alarms for TAN, inhibitor residual, hardness, and DP to computerized maintenance management system work orders.

Rating Guidance

- Severity 9–10: tube failure, safety valve lift, prolonged outage
- Occurrence based on site fouling rate and climate

- Detection low (high D) where only periodic tests exist; improve with continuous ΔP or optical sensors

THERMAL SYSTEM FAILURES: PUMP SEAL FAILURES FROM FLUID INCOMPATIBILITY

Pump seal failures tied to fluid incompatibility are chemistry driven. The fluid extracts plasticizers, swells or shrinks elastomers, attacks polymer or metal faces, alters lubrication at the seal interface, and changes vapor pressure at the film. The result is leakage, heat checking, blistering, rapid wear, or brittle fracture. Failures concentrate at secondary seals, dynamic O-rings, faces, springs, and gland hardware (see Table 14.7).

Common Seal Designs and Vulnerable Materials

- **Mechanical face seals** with hard or soft face pairs. Typical pairs are carbon graphite against silicon carbide, silicon carbide against silicon carbide, tungsten carbide against silicon carbide. Secondary seals are NBR, HNBR, FKM, FFKM, EPDM, PTFE, or PTFE blends.
- **Lip seals** in booster and small circulating pumps. Elastomer only contact with shaft.
- **Gland packings** in legacy installations. Compatibility issues show as severe leakage and glazing.

Fluids That Frequently Drive Incompatibility

- **Water glycol and amine-inhibited coolants**. Attack certain fluoroelastomers and carbon faces in poor lubrication service. EPDM and PTFE preferred.
- **Phosphate ester fire resistant fluids**. Require EPDM. NBR and many FKM grades soften or crack.
- **Polyol ester and natural ester oils**. Swell NBR and HNBR. FKM and FFKM stable.
- **Aromatic solvents and high polarity cleaners**. Extract plasticizers from NBR and HNBR and craze some thermoplastics.
- **Hydrocarbon thermal oils**. Compatible with NBR, HNBR, FKM. Poor with EPDM.
- **Deionized water**. Poor lubricity. Requires hard face pairs, generous flush, and compatible elastomers.

Failure Mechanisms

- **Swelling, softening, or shrinkage of elastomers**. Plasticizer extraction or absorption alters squeeze and compression set. Leakage starts cold and worsens hot.
- **Hydrolysis of ester-based fluids**. Acid formation increases corrosion of springs and metallized carbon faces.

TABLE 14.7
Failure Pathway Table from Fluid Incompatibility

Pathway ID	Initiators and Sources	System Response	Evidence and Tests	Risks	Action Thresholds	Corrective and Preventive Actions
PS1 Elastomer swell or shrink	Wrong elastomer for fluid; solvent or ester exposure	Loss of squeeze, extrusion, leakage at startup	Volume change from coupon test, hardness drift, visual cracks	Progressive leakage, shaft scoring	Volume change beyond ±10 percent or hardness shift beyond 10 Shore A	Select compatible elastomer. Typical pairs: EPDM for phosphate ester and water glycol, FKM or FFKM for hydrocarbon and POE oils, avoid EPDM in hydrocarbons. Replace O-rings and gaskets.
PS2 Poor lubricity fluid at faces	Water or lean glycol, low viscosity	Heat checking, blistering, rapid face wear	Microscopy shows pits and heat checks; high face temperature; no solids	Sudden leak, face fracture	Visible pits or radial checks, rise in flush temperature	Use hard face pair SiC against SiC. Increase flush per API 682 Plan 11 or 21. Lower face load if allowed.
PS3 Hydrolysis and acidity	Ester fluid with water or high temperature	Spring and metal corrosion, carbon face oxidation	TAN rise, KF water rise, rust particulate on patch	Spring fracture, hang up, leakage	TAN increase above site limit, water above OEM limit	Dry fluid, install moisture control, select corrosion resistant springs, passivate metallics.
PS4 Inhibitor or additive incompatibility	Mixed coolant families or additive packs	Gel or varnish deposits in seal chamber	FTIR of deposits, turbidity, filter loading	Face hang up, leakage	Turbidity above control, deposit mass in filters	Flush, charge one approved chemistry, install finer sidestream filtration.
PS5 Chemical stress cracking	Amines or aromatics with sensitive polymers	Cracks in cages or secondary components	Fractography, solvent exposure history	Sudden failure, face misalignment	Any crack observed	Replace with chemically resistant grades. Use PTFE or PEEK with verified resistance.
PS6 Vaporization at faces	Inadequate NPSHA, chamber pressure too close to vapor pressure	Film collapse, intermittent lift-off	Cavitation noise, temperature spikes, sheen of sprayed fluid	Catastrophic leak	Margin to vapor pressure less than 1 bar at chamber	Raise seal chamber pressure, cool flush, re-route flush to reduce heat, verify Plan 11 or Plan 21 performance.

- **Chemical stress cracking** of thermoplastics in PTFE filled composites or PEEK cages when exposed to amines or aromatics.
- **Face blistering**. Rapid vaporization in microasperities with poor-lubricity fluids.
- **Corrosion and galvanic attack** of stainless hardware in low pH or chloride service.
- **Coking or polymer varnish** on faces in hot thermal oils after additive incompatibility.

Field Evidence and Diagnostics

- **Leak path signature**. Clean wet track at split line indicates secondary seal failure. Sprayed film around gland indicates face lift-off.
- **Elastomer condition**. Measure hardness per ASTM D2240. Inspect for volume change and cracks.
- **Face inspection**. Look for radial heat checks, blister pits, transfer films, corrosion, or grooving.
- **Solids and deposits**. Patch test and FTIR on deposits to identify varnish or phosphate residue.
- **Fluid analysis**. TAN by ASTM D664, water by ASTM D6304 for hydrolysis, glycol percentage by refractometer, anions by ion chromatography for aggressive species.
- **Compatibility tests**. Immersion per ASTM D471 or ISO 1817 on candidate elastomers in the actual fluid at operating temperature.
- **Seal chamber review**. Verify API 682 piping plan, flush temperature, pressure margin to vapor pressure, and circulation.

Selection and Application Rules

- **Match elastomer to chemistry**.
  EPDM for water glycol and phosphate ester.
  FKM or FFKM for hydrocarbons and polyol esters.
  HNBR for aliphatic hydrocarbons at moderate temperature.
  Avoid EPDM in hydrocarbon oils. Avoid many FKM grades with strong amines. Confirm with ISO 1817 or ASTM D471 immersion data.
- **Face pairs**.
  Use silicon carbide against silicon carbide for water and water glycol.
  Use carbon against silicon carbide for hydrocarbon oils with good lubricity.
  Use tungsten carbide where solids are present.
- **Piping plan verification** per API 682.
  Plan 11 for recirculation flush. Plan 21 with cooler for hot fluids. Plan 23 for controlled circulation through heat exchanger in thermal oil service. Plan 32 clean external flush for poor lubricity fluids. Dual seals with Plan 53 for hazardous service.

- **Pressure and temperature margins**.
  Maintain seal chamber pressure at least 1 bar above fluid vapor pressure at face temperature. Keep flush temperature under the limit set by elastomer and fluid stability.
- **Surface speed and runout**.
  Check shaft surface finish and total indicated runout to seal vendor limits. Excess speed amplifies heat generation in poor lubricity service.

Diagnostic Workflow

1. Capture fluid and debris samples before teardown.
2. Inspect faces and elastomers with photos and notes.
3. Run fluid TAN, water, glycol percent, and FTIR.
4. Review operating logs for suction pressure, seal chamber pressure, flush temperature, and pump load.
5. Compare elastomer compound and face materials to fluid Bill of Materials.
6. Select corrective materials and update the piping plan or flush conditions.
7. Recommission with leak test and temperature check at faces.

Section VI

Emerging Trends and Future Systems

15 Advanced Fluids and Smart Materials

SELF-HEALING LUBRICANTS

Self-healing lubricants integrate microencapsulated additive packages or polymeric networks that release repair agents upon detection of mechanical stress or chemical degradation. Microcapsules can contain antioxidants, antiwear agents, or viscosity modifiers. When a shear event, wear scar, or oxidative breakdown ruptures the capsule, the additives are locally released into the tribological interface. This targeted release maintains boundary film integrity and slows wear propagation.

Another approach uses reversible polymer cross-linking, such as Diels–Alder chemistry, where the molecular structure reforms after thermal or mechanical disruption. This technique sustains viscosity stability and film strength across repeated stress cycles. Some experimental formulations use ionic liquids combined with nano-additives that re-orient under electrochemical potential, re-establishing lubrication films without the need for capsule rupture.

Key operational benefits include reduction in lubricant changeout frequency, lower particle generation rates, and sustained component surface protection under variable load conditions. Testing requires modified four-ball wear tests (ASTM D4172) with post-test surface characterization using scanning electron microscopy (SEM) and energy-dispersive X-ray spectroscopy (EDS) to confirm additive redeposition. Long-term field validation requires integration with in-service oil analysis to measure retention of viscosity, antiwear element levels, and oxidation stability after controlled damage events.

Self-Healing Dielectric Materials

In dielectric applications, self-healing capability has been implemented for decades in metallized film capacitors, where local dielectric breakdown vaporizes conductive layers and isolates the defect. This principle is being extended to transformer insulating fluids and solid dielectrics. Polymer matrix composites with embedded microcapsules containing reactive monomers can polymerize upon electrical puncture, sealing voids that could otherwise grow into partial discharge sites.

For liquid dielectrics, encapsulated oxygen scavengers and antioxidant packages can be dispersed within the fluid. Electrical or thermal stress triggers capsule rupture,

DOI: 10.1201/9781003775201-21

releasing stabilizers that slow oxidation and polymerization of degradation products. In ester-based fluids, reversible esterification reactions are under study to reverse mild hydrolytic breakdown. In gas-insulated systems, dielectric coatings infused with microcapsules containing fluorocarbon repair agents can reseal microfissures caused by arcing or particle impact.

Field validation requires accelerated aging studies under IEC 60247 (dielectric dissipation factor and resistivity) and IEC 60156 (breakdown voltage) before and after induced electrical stress. Monitoring dissolved gas profiles (IEC 60567) provides additional data on whether repair events alter fault gas generation patterns.

Smart Material Integration

Both lubricant and dielectric self-healing systems are moving toward integration with embedded sensing. Nanoparticle tracers detectable by inductive or optical sensors can confirm additive release events in real time. Coupled with predictive maintenance algorithms, these materials can extend intervals between interventions while preserving reliability margins.

Integration into rotating machinery, switchgear, and transformers must consider chemical compatibility with elastomers, copper alloys, and coatings. Self-healing additives must not cause precipitation, varnish, or loss of dielectric strength under normal service conditions. ASTM D7155 (compatibility between fluids and materials) and ISO 15380 (biodegradability for lubricants) will be part of qualification protocols as environmental requirements tighten.

SELECTION AND QUALIFICATION PROTOCOL

Self-Healing Lubricants and Dielectric Materials

1. **Scope and objectives**
   - Select candidate fluids and solids for rotating machinery, transformers, GIS, switchgear, and thermal loops that require self-healing behavior.
   - Verify chemical compatibility with metals, elastomers, coatings, varnishes, composites, and legacy fluids.
   - Prove functional performance under electrical, thermal, and mechanical stress.
   - Define condition monitoring requirements and acceptance criteria for deployment.
2. **Inputs and constraints**
   - Equipment duty cycle, temperatures, pressures, electrical stress, and clearances.
   - Material bill of materials for wetted parts.
   - Legacy fluid specifications and cleanliness targets.
   - Site environmental limits and regulatory requirements.
   - Safety classification for test rigs and energized trials.

3. **Candidate down-selection**
   - Review vendor data for self-healing mechanism type: microcapsules, reversible cross-links, ionic liquid packages, nanoparticle triggers, metallized self-clearing dielectrics.
   - Screen for base chemistry: mineral, PAO, POE, ester, silicone, fluorinated, natural ester for dielectrics.
   - Reject if known incompatibility exists with critical elastomers or coil enamels.
4. **Chemical compatibility program**

   4.1 Elastomers and plastics
   - Immersion per ASTM D471 or ISO 1817 at low, nominal, and high temperature setpoints.
   - Measure mass change, volume change, durometer hardness per ASTM D2240, tensile per ASTM D412.
   - Acceptance: volume change within ±10%, hardness drift within 10 Shore A or as per equipment vendor limits.

   4.2 Metals and coatings
   - Coupon exposure in quiescent and flowing conditions.
   - Mass loss, pitting inspection, SEM and EDS on select coupons.
   - For copper and silver use passivation checks where applicable.

   4.3 Electrical insulation and varnish systems
   - Soak tests of magnet wire enamels, epoxy spacers, paper and press-board in liquid candidates.
   - Dielectric dissipation factor and resistivity per IEC 60247 before and after soak.
   - Mechanical integrity check for swelling or softening.

   4.4 Cross-compatibility with legacy fluids
   - Blend testing per ASTM D7155 for turbine and circulating oils or vendor method for dielectric fluids.
   - Evaluate haze, sediment, MPC patch color for oils, BDV for dielectrics.
   - Reject if precipitation or BDV loss exceeds action limits.
5. **Contamination and stability screening**
   - Oxidation stability: RPVOT per ASTM D2272 for oils. D943 for long-term oxidation tendency.
   - Hydrolysis screening for esters with water content steps using TAN per ASTM D664 and Karl Fischer per ASTM D6304.
   - Storage stability of microcapsules: microscopy for shell integrity, release curve verification after thermal cycling.
   - Particle shedding from capsules or fillers: ISO 4406 count in fresh and sheared samples.
6. **Functional bench tests**

   6.1 Lubricants
   - Boundary and mixed regime wear: ASTM D4172 four-ball with pre-scratch protocol to trigger self-healing release.

- SRV or Mini Traction Machine for film durability with cyclic temperature profile.
- Post-test SEM and EDS on scars to confirm repair film deposition.

6.2 Dielectric liquids and solids

- Breakdown voltage per IEC 60156 for liquids before and after induced micro-faults.
- Partial discharge inception and extinction per IEC 60270 for solids and model spacers.
- Track tan delta at temperature per IEC 60247 during repeated stress cycles.
- For metallized films or coatings validate self-clearing behavior on a puncture matrix.

7. Trigger and response verification

- Define trigger events: shear burst, temperature threshold, voltage spike.
- Instrument sample cells with temperature, conductivity, and optical probes to confirm release timing.
- Quantify additive release via ICP for Zn, P, B, Mo if applicable or GC for capsule payload.
- Acceptance: repeatable activation within target band and no uncontrolled release under normal conditions.

8. Cleanliness and filtration impact

- Evaluate interaction with site filtration media and electrostatic separators.
- Measure removal rate of capsules or nano-packages across beta-rated filters.
- Adjust filter grade or add bypass polishing if functional agents are being removed.

9. System-level rig tests

- Closed-loop rig for oils: circulation, heat input, shear device, by-pass filtration, contaminant injection for credible faults.
- Energized cell for dielectrics: controllable field stress, moisture steps, thermal profile.
- Run for a minimum of 500 hours or as set by equipment class with periodic sampling.

10. Condition monitoring requirements

10.1 For lubricants

- Viscosity per ASTM D445 at two temperatures.
- TAN and RPVOT trend for oxidation control.
- FTIR for oxidation and unique self-healing spectral markers if defined.
- Particle count ISO 4406 and patch microscopy.
- Elemental analysis by ICP to track additive retention.
- Optional online sensors for viscosity, dielectric constant, moisture.

10.2 For dielectric liquids

- BDV per IEC 60156 routine.
- Moisture ppm by ASTM D6304 or IEC 60814.

- Dissipation factor and resistivity per IEC 60247 at operating temperature.
- Dissolved Gas Analysis per ASTM D3612 and interpretation per IEC 60599 if used in transformers.
- Online moisture and multi-gas sensors where critical.

10.3 Event confirmation

- Define signatures that indicate a heal event, for example reduction in wear particle rate after pre-defined stress, recovery of BDV after a controlled micro-fault, decrease in tan delta after repair cycle.

11. Test sequencing

1. Paper review of materials and hazard data.
2. Compatibility immersion on elastomers, plastics, metals, and insulation systems.
3. Cross-compatibility with legacy fluids.
4. Stability tests for oxidation, hydrolysis, storage, and capsule integrity.
5. Bench tribology or dielectric tests with induced triggers.
6. Filtration interaction study.
7. System-level rig run with sampling schedule.
8. Data review against acceptance metrics.
9. Pilot installation on one low-risk unit with online monitoring.
10. Fleet rollout with documented controls.

12. Acceptance criteria

- No unacceptable swelling or softening of critical materials.
- No precipitation or haze on blend testing at operating temperature.
- Wear scar reduction relative to control in trigger tests with proof of repair film.
- BDV and tan delta recover to within defined margin after microfault series for dielectrics.
- Stable viscosity and TAN trends within site limits.
- No significant removal of functional agents by standard filtration.
- Safe handling classification and compliance with site environmental constraints.

13. Pilot installation protocol

- Install on one asset with full bypass sampling port and return filter.
- Baseline sampling before charge.
- Weekly sampling for the first month, then monthly for six months.
- Online alarms for moisture, conductivity, and gas where applicable.
- Post-outage inspection of bearings, spacers, and contact surfaces.
- Go or no-go decision based on trend analysis and inspection findings.

14. Documentation and control

- Record all test conditions, materials, lot numbers, and lab methods.
- Store baseline spectra and micrographs.
- Issue a controlled specification sheet with fluid ID, compatible materials list, filtration class, and monitoring plan.
- Link to CMMS for sampling tasks and alarm setpoints.

15. **Safety and environmental checks**
    - Review SDS for capsule payloads and ionic liquids.
    - Verify seal and hose material compatibility against any fluorinated or amine components.
    - Confirm disposal and spill procedures for test residues and changeouts.
    - For dielectric liquids confirm conformance with IEC reuse limits if gas by-products are generated.

Nanoparticle-Enhanced Oils and Coolants

Function and Mechanisms

Nanoparticle-modified fluids target specific gaps in base oil and coolant performance. Dispersed solids act as rolling or mending agents at contacts, phonon or electron heat carriers in the bulk, and nucleation promoters at heat exchange surfaces. Typical size ranges fall between 5 nm and 100 nm. Surface functionalization keeps particles suspended and prevents agglomeration. Tribological benefit comes from transfer film formation, micropolishing, third-body load sharing, and chemical tribofilm growth. Heat transfer benefit comes from increased effective thermal conductivity, higher critical heat flux, altered interfacial wettability, and stabilized microconvection.

Candidate Materials and Use Cases

- Oxides for coolants. Al_2O_3, SiO_2, ZnO, TiO_2 for water and glycols. Use when fouling risk is controlled and filtration is sized correctly.
- Nitrides and carbides for oils. BN, Si_3N_4, WC, Mo_2C where sliding contacts need higher scuff resistance.
- Lamellar solids. h-BN, MoS_2, graphene derivatives for boundary lubrication and low-friction films.
- Metallics. Cu, Ag for boiling and two-phase cooling when corrosion control is in place.
- Diamond and DLC fragments. Low wear and high thermal conductivity in critical bearings with tight filtration control.

Formulation Constraints

Base fluid chemistry sets stability and compatibility boundaries. Oils include mineral, PAO, POE, and esters. Coolants include water, water-glycol, and dielectric fluorinated fluids. Dispersants must match particle surface chemistry and base fluid polarity. Treat rates typically range from 0.01 wt% to 2 wt% depending on target property and filtration limits. Excess treat rate increases viscosity, raises differential pressure across filters, and can seed deposits on hot surfaces.

Property Impacts to Validate

- Thermal conductivity. Verify using transient hot-wire or guarded hot plate at service temperatures. Expect modest gains at low shear rates in single-phase flow.

- Viscosity. Measure per ASTM D445 at 40 °C and 100 °C for oils and at the operating range for coolants. Trend shear stability under ASTM D6278 or equivalent.
- Friction and wear. Use ASTM D4172 four-ball, SRV tests, and mini traction machine. Inspect scars by SEM and EDS for particle-derived films.
- Heat transfer. For coolants test convective coefficient, critical heat flux, and boiling curve on a lab rig with controlled surface roughness and water chemistry.
- Electrical properties for dielectric coolants. Test tan delta and resistivity per IEC 60247 and BDV per IEC 60156 if applied near energized parts.
- Stability. Quantify sedimentation rate, zeta potential, and accelerated aging with thermal cycles and vibration. Verify re-dispersion after storage.

Integration and Compatibility

Check chemical compatibility with elastomers, seals, varnishes, and coatings. Run immersion per ASTM D471 or ISO 1817 on NBR, HNBR, FKM, EPDM, FFKM, and PTFE. Confirm no softening, swelling, or cracking. Verify corrosion on copper, aluminum, and steel coupons in the actual fluid with and without nanoparticles. Add copper passivators only if they do not strip or poison particle surfaces. For glycol coolants, validate inhibitor package function since some nanoparticles adsorb inhibitors and shift pH or reserve alkalinity.

Filtration and Cleanliness

Filtration must retain wear debris and let the functional fraction pass. Set beta ratios using a staged approach. Place coarse protection upstream and fine polishing downstream. Establish a target ISO 4406 code that accounts for the particle population. Confirm that filters do not strip the treat rate. Run breakthrough tests with particle size distribution before and after filters.

Failure Modes and Controls

- Agglomeration and sedimentation. Cause by incorrect dispersant, water ingress, or thermal cycling. Control by surfactant selection, moisture control, and recirculation.
- Abrasive wear. Appears when hard particles exceed film thickness or when filter cut is too coarse. Control by lowering treat rate, changing particle size, and upgrading filters.
- Deposit formation on hot surfaces. Occurs with metallic particles or when inhibitor chemistry collapses. Control by water chemistry management and heat flux limits.
- Electrical losses in dielectric zones. Caused by conductive particles or moisture. Control by dielectric testing and segregation from energized volumes.

Test Sequencing for Qualification

1. Composition screening. Confirm particle chemistry, mean size, size distribution, and surface treatment by TEM or DLS.

2. Compatibility immersion with elastomers, plastics, copper alloys, aluminum, and coatings.
3. Base fluid stability and oxidation tests. RPVOT for oils and hydrolysis or TAN for esters and glycols.
4. Bench tribology or heat transfer tests with and without nanoparticles. Compare to neat fluid and additive-only controls.
5. Filtration interaction testing. Determine retention and pressure drop at expected flow and temperature.
6. Loop-rig trials. Closed circuit with controlled heat input, duty cycling, and representative materials. Minimum 500 hours with periodic sampling.
7. Pilot asset trial. One machine or subloop with online sensors and short sampling interval.
8. Review against acceptance criteria. Proceed to controlled rollout if trends meet limits.

Condition Monitoring in Service

- **Oils:** Viscosity at two temperatures, particle counts, ICP for metals, FTIR for oxidation and any nanoparticle markers, patch microscopy for deposits, MPC for varnish potential when thermal loads are high.
- **Coolants:** pH, reserve alkalinity, inhibitor residuals, glycol percent, ion chromatography for chloride and sulfate, hardness for makeup water, turbidity for dropout detection.
- **Thermal duty:** Differential pressure across heat exchangers, approach temperature, and infrared mapping for hot spots.
- **Electrical duty for dielectric coolants:** BDV, tan delta, and moisture ppm at operating temperature.
- **Treat rate tracking:** Use elemental tracers or UV-visible tags when permissible to confirm nanoparticle concentration over time.

Application Guidance by Asset Class

- **Bearing and gearboxes:** Target boundary-to-mixed regime gains using lamellar solids at low treat rates with fine filtration and short trial intervals.
- **Hydraulic systems:** Use only if filter architecture is validated and servo valve clearances can tolerate the particle distribution.
- **Water-glycol coolants:** Favor oxide nanoparticles with pH and inhibitor control. Confirm pump seal compatibility with EPDM or FKM as specified.
- **Dielectric cooling:** Use insulating particles only. Verify tan delta and BDV after aging. Keep conductive or semiconductive particles out of energized volumes.

EHS and Lifecycle

Document SDS for particle powders and dispersions. Control airborne exposure during make-down and maintenance. Plan recovery and disposal to prevent sewer release. Verify waste handling for filters charged with nanoparticles. For esters and glycols prefer chemistries that meet local biodegradability goals.

ADVANCED FLUIDS AND SMART MATERIALS FOR IONIC LIQUIDS (ILS) AND DEEP EUTECTIC SOLVENTS (DES)

Functional Basis and Mechanisms

Ionic liquids (ILs) are molten salts with melting points typically below 100 °C, composed of bulky asymmetric organic cations and organic or inorganic anions. Their negligible vapor pressure, high thermal stability, and tunable polarity make them candidates for lubrication, cooling, and electrical insulation in power systems. Deep eutectic solvents (DES) are formed by mixing a hydrogen bond donor and a hydrogen bond acceptor to produce a eutectic mixture with a melting point far below that of either component. DES share many IL properties but are often cheaper and more biodegradable due to common raw materials like choline chloride and urea. Both systems can be tailored chemically to match dielectric, lubricating, or heat transfer requirements.

Candidate Chemistries and Use Cases

- **Lubricants and tribology**: Imidazolium- and phosphonium-based ILs with halide or phosphate anions used in gearboxes and bearings for high-temperature stability and low volatility. DES from choline chloride and polyols or acids applied in light-load mechanisms and biodegradable contexts.
- **Dielectric fluids**: Hydrophobic ILs with bis(trifluoromethylsulfonyl) imide ($[NTf_2]^-$) or dicyanamide ($[DCA]^-$) anions used in transformers and capacitors where oxidative stability and low flammability are critical.
- **Thermal fluids**: Low-viscosity ILs with high specific heat capacity for concentrated solar power loops; DES with glycols or glycerol for corrosion-controlled water replacement in cooling circuits.
- **Hybrid formulations**: IL or DES doped with nanoparticles for improved thermal conductivity or tribological performance.

Property Impacts to Validate

- **Viscosity and shear stability**: Measured per ASTM D445 at the operating temperature range; ILs often have higher viscosity than mineral oils and may require viscosity tuning.
- **Thermal stability**: TGA to determine onset of mass loss; ILs can exceed 300 °C stability, DES generally lower depending on donor/acceptor chemistry.
- **Lubricity**: Four-ball or SRV testing to assess wear scar diameter and friction coefficient; IL cation–anion combinations directly influence boundary film chemistry.
- **Electrical insulation**: IEC 60156 breakdown voltage, IEC 60247 dielectric dissipation factor; select ILs can match or exceed mineral oil BDV.

- **Corrosion tendency**: Copper strip test (ASTM D130) and long-term metal coupon immersion; halide-containing ILs require passivation strategies.
- **Compatibility**: Elastomer immersion tests (ASTM D471) for NBR, FKM, EPDM, PTFE; IL polarity can cause swelling in certain elastomers.
- **Water sensitivity**: Karl Fischer moisture analysis; some ILs and DES absorb water readily, altering conductivity and viscosity.

Integration and Compatibility

Material compatibility depends on polarity, acidity, and specific ion chemistry. Phosphonium ILs show better corrosion resistance toward steel and copper than imidazolium ILs with halide anions. DES with acidic donors can attack aluminum unless corrosion inhibitors are included. Both ILs and DES can soften natural rubber and certain polyurethanes; fluoropolymers and high-grade EPDM are more resistant. Coating compatibility testing is required in painted transformer tanks or turbine housings.

Filtration and Cleanliness

Particulate contamination can affect dielectric performance and lubrication film quality. Use filters rated for the fluid's viscosity at operating temperature. ILs with high viscosity may require positive displacement pumps for filtration loops. For DES and hygroscopic ILs, drying columns with molecular sieves can be integrated into filtration circuits to control water content.

Failure Modes and Controls

- **Hydrolysis and anion breakdown**: Moisture ingress can decompose sensitive anions like BF_4^-, producing HF and corroding metals; control by desiccant breathers and tight sealing.
- **Corrosion**: Aggressive halide species or acidic DES donors can initiate pitting; controlled by inhibitor packages.
- **Oxidation**: High-temperature exposure can oxidize organic cations; monitor TAN and color change.
- **Loss of dielectric integrity**: Water uptake or ionic contamination can reduce breakdown voltage; monitor with periodic dielectric testing.

Test Sequencing for Qualification

1. **Composition screening**: Confirm ion identity by NMR, water content by Karl Fischer.
2. **Material compatibility**: Elastomer, metal, and coating immersion tests.
3. **Thermal stability**: TGA and DSC to map stability and phase transitions.
4. **Tribology and wear**: ASTM D4172, D2783 for extreme pressure if relevant.
5. **Dielectric testing**: IEC 60156, IEC 60247 for electrical applications.
6. **Corrosion testing**: ASTM D130, ISO 2160, and long-term coupon tests.
7. **System simulation**: Loop-rig or bearing test rig with representative duty cycles.

8. **Pilot deployment**: Monitor with oil analysis (viscosity, TAN, particle count, water content) and electrical tests.

Condition Monitoring in Service

- **Viscosity**: ASTM D445 at reference temperature.
- **Moisture**: Karl Fischer in ppm.
- **TAN/TBN**: Acid–base balance tracking.
- **Wear metals**: ICP or RDE spectroscopy for mechanical systems.
- **Dielectric properties**: BDV, dissipation factor for insulating applications.
- **Particle count**: ISO 4406 for mechanical or hydraulic service.

Application Guidance by Asset Class

- **Transformers and switchgear**: Use hydrophobic ILs with high BDV, low water affinity, and corrosion inhibitors.
- **Wind turbine gearboxes**: ILs with antiwear anions and nanoparticle doping; monitor seals for swelling.
- **Solar thermal loops**: ILs with high heat capacity; verify pump compatibility and thermal oxidation resistance.
- **Closed-loop liquid cooling in electronics**: DES with low electrical conductivity and biocide additives.

EHS and Lifecycle
Some ILs have toxicity concerns due to persistent cations; select biodegradable DES where feasible. Disposal must meet local hazardous waste regulations. Monitor occupational exposure, particularly with fine mist generation. IL synthesis by green routes and DES from food-grade inputs can reduce lifecycle impact.

Molten Salts with Extended Temperature Ranges

Functional Basis and Mechanisms
Molten salts used in thermal and electrical applications are ionic compounds that exist in a liquid state at high temperatures, providing high volumetric heat capacity, chemical stability, and in some cases electrical insulation. In solar thermal, nuclear, and advanced heat storage systems, nitrate- and carbonate-based salts dominate due to availability, cost, and well-understood behavior. Emerging chemistries focus on extending operational temperature limits beyond the traditional 290–565 °C range of binary nitrate systems, targeting both lower melting points for cold-start capability and higher decomposition thresholds for ultra-high-temperature service in next-generation concentrating solar power (CSP) and high-temperature nuclear reactors.

Candidate Chemistries and Extended Range Formulations

- **Ternary nitrate/nitrite blends**: Mixtures of $NaNO_3$, KNO_3, and $NaNO_2$ that lower melting point below 140 °C while maintaining stability to ~600 °C.

- **Chloride-based systems**: $NaCl$-KCl-$MgCl_2$ eutectics and $LiCl$-KCl mixtures with melting points ~350 °C and upper stability approaching 800 °C; suitable for high-temperature storage and some molten salt reactors (MSRs).
- **Carbonate-based systems**: Li_2CO_3-Na_2CO_3-K_2CO_3 mixtures for intermediate to high temperature (~400–750 °C) with good thermal stability but higher corrosivity toward carbon steel.
- **Fluoride-based systems**: LiF-NaF-KF eutectic (FLiNaK) with melting point ~454 °C and stability beyond 900 °C; high thermal conductivity and neutron moderation potential in advanced nuclear designs.
- **Hybrid salt systems**: Incorporation of nitrates with carbonates or chlorides to reduce melting point while retaining higher stability; often under development for CSP with integrated thermal-to-electric cycles exceeding 650 °C turbine inlet.

Key Property Considerations

- **Melting point and freeze protection**: Cold-start risk is critical in CSP and storage systems; low melting eutectics reduce auxiliary heating needs.
- **Thermal stability**: Defined by decomposition temperature; nitrate systems limited by nitrite/nitrogen oxide evolution, chlorides/carbonates limited by vaporization and reaction with moisture or CO_2.
- **Viscosity**: Lower viscosity improves pumpability and heat transfer; chloride and fluoride salts tend to have lower viscosity at temperature than nitrates.
- **Thermal conductivity and heat capacity**: Directly affect heat transfer coefficient; fluoride salts have highest conductivity, nitrates have moderate conductivity but high specific heat.
- **Corrosivity**: Strongly influenced by salt chemistry, impurities, and operating atmosphere; chlorides and fluorides are more aggressive, requiring nickel-based alloys or ceramics.

Material Compatibility and Containment

- **Nitrate/nitrite salts**: Compatible with stainless steels and low-alloy steels at <600 °C with oxygen control; aluminum alloys avoided due to pitting risk.
- **Chloride/carbonate salts**: Require Inconel, Hastelloy, or ceramic-lined piping; oxygen/moisture exclusion is critical to limit active corrosion.
- **Fluoride salts**: Often paired with Hastelloy-N in MSR applications; rigorous moisture exclusion required to prevent HF formation.

Filtration and Maintenance

Molten salts accumulate insoluble corrosion products and particulates from system wear. Continuous filtration through high-temperature ceramic or metal filters is recommended. Salt chemistry should be monitored via sampling and analysis for composition shifts, oxygen content, and corrosion product levels (see Table 15.1).

TABLE 15.1
Failure Modes and Degradation Pathways for Molten Salts

Failure Mode	Mechanism	Mitigation
Solidification blockages	Salt cooling below melting point	Heat tracing, freeze protection procedures
Thermal decomposition	Oxidation/reduction of salt ions	Atmosphere control, chemistry adjustment
Corrosion of containment	Dissolution of metal in molten salt	Alloy selection, inhibitors, oxygen control
Contamination	Ingress of water, CO_2, or reactive gases	System sealing, inert gas cover
Fouling of heat transfer surfaces	Deposition of solids from corrosion or additives	Filtration, salt chemistry control

Condition Monitoring

- **Chemical composition**: Ion chromatography or ICP-OES to confirm salt ratio and detect contaminants.
- **Oxidation state control**: Potentiometric sensors or redox potential monitoring to maintain salt in stable range.
- **Particulate content**: High-temperature filtration with differential pressure tracking.
- **Corrosion product analysis**: Metal ion concentration in salt indicates containment wear.
- **Thermal performance**: Heat exchanger ΔT and flow rate tracking for fouling detection.

Test Sequencing for Qualification in Extended Temperature Systems

1. **Composition analysis** to verify melting point and eutectic range.
2. **Thermal cycling tests** to simulate start–stop operations.
3. **High-temperature corrosion tests** with candidate containment alloys.
4. **Long-term stability trials** under inert cover gas at design temperature.
5. **System integration trials** in loop rigs with representative pumps and heat exchangers.

Application Contexts

- **Next-generation CSP**: Chloride-based salts for >700 °C operation to match supercritical CO_2 turbine cycles.
- **Molten salt reactors**: Fluoride-based salts for fuel and coolant; stability under neutron flux.
- **Thermal storage retrofits**: Hybrid nitrate–carbonate salts for lower freeze risk in cold climates.

Selection and Qualification Protocol

Advanced Fluids and Smart Materials: Molten Salts with Extended Temperature Ranges

1. **Scope and objectives**
    1. Select molten salt chemistries for high temperature thermal storage and cooling in CSP, high temperature industrial loops, and advanced reactor auxiliaries.
    2. Verify compatibility with candidate alloys, ceramics, elastomers, and coatings.
    3. Prove thermal, chemical, and hydraulic performance across cold starts, steady state duty, and upset conditions.
    4. Define condition monitoring, operating limits, and acceptance criteria for fleet deployment.
2. **Candidate chemistry screen**
    - Binary and ternary nitrates or nitrite-modified nitrate blends for 140 to 600 °C service.
    - Carbonate eutectics for 400 to 750 °C with attention to CO_2 control.
    - Chloride eutectics based on NaCl, KCl, $MgCl_2$, or LiCl-KCl for 350 to 800 °C.
    - Fluoride eutectics such as FLiNaK for 450 to 900+ °C in specialized systems.
    - Hybrid systems that target reduced freeze point or higher decomposition temperature.

    Reject any option that cannot meet site freeze protection, materials, or environmental constraints.
3. **Property targets and verification tests**
    - Melting point and freeze margin, DSC with defined supercooling tolerance.
    - Thermal stability limit, isothermal TGA and long hold tests at design temperature in controlled atmosphere.
    - Viscosity and density versus temperature, capillary or oscillating U-tube methods for pump sizing.
    - Heat capacity and thermal conductivity, drop calorimetry or laser flash analysis.
    - Impurity profile, halides, moisture, oxides, carbonates by ion chromatography or ICP-OES.
    - Redox potential range, potentiometric probes with reference salts.

    Set internal specification sheets with test methods, temperature points, and acceptance bands.
4. **Materials compatibility program**

    4.1 Alloy Selection Matrix

Use numeric ratings: 1 acceptable with standard controls, 2 conditional with qualification testing and special controls, 3 generally unsuitable (see Table 15.2).

TABLE 15.2
Alloy Selection Matrix

Alloy class	Nitrate ≤600 °C	Carbonate 400–750 °C	Chloride 350–800 °C	Fluoride 450–900 °C
Carbon steel	1 at ≤565 °C with oxygen control	2 at lower range, corrosion rises with temp	3	3
304/316 SS	1 at ≤600 °C with low chloride and controlled oxygen	2, carburization risk at upper range	3, pitting and hot corrosion	3
347H/321 SS	1	2	3	3
Ni-base (Inconel 625, 617)	2 economics driven	1 to 2 with CO_2 control	1 with low O_2 and low moisture	1 with low HF risk
Hastelloy-N class	3	2	1 to 2 dependent on impurities	1 reference alloy for fluoride systems
Ceramics, SiC	2 thermal shock risk	1	1	1

Notes: rating reflects salt purity, oxygen and moisture control, and velocity. Validate with corrosion coupons.

4.2 Coupon and loop tests
- Static exposure tests at three temperatures for 1000 hours minimum. Measure mass loss, pit depth distribution, and scale composition by SEM-EDS and XRD.
- Flowing loop tests at design velocity and heat flux, measure uniform corrosion rate, erosion-corrosion, and deposition tendency.
- Oxygen and moisture control, ppm-level control using getter additions or cover gas management.

5. **Redox and impurity control**
   - Establish redox setpoints for each chemistry. For nitrates manage nitrite fraction. For chlorides and fluorides manage metal oxide and moisture content.
   - Validate getters or sparging agents in bench reactors.
   - Approve analytical methods for O_2, H_2O, and acid gas ingress.
   - Define corrective actions for off-spec conditions, for example chemical polishing additions or salt reprocessing.
6. **Freeze protection and cold start strategy**
   - Determine freeze point with DSC and confirm in a scaled freeze tank.
   - Define heat tracing density, ramp rates, and soak times for lines, valves, and exchangers.

- Write a freeze recovery procedure with localized re-melt steps, pressure relief checks, and thermal gradient limits to protect welds.

7. **Filtration and cleanliness**
    - Specify high temperature metal or ceramic filter elements, pore size selection from debris spectrum.
    - Prove filterability of corrosion products and insoluble fines in a circulating rig at temperature.
    - Track differential pressure and filter loading rate.
    - Qualify a salt polishing procedure for off-line cleanup.
8. **System hydraulics and heat transfer validation**
    - Map pump head curve using high temperature test fluid or validated surrogate.
    - Verify NPSH margin with salt vapor pressure data at operating temperature.
    - Measure heat transfer coefficients and approach temperature in a representative heater-to-storage-to-user loop.
    - Confirm no maldistribution in manifolds by temperature uniformity checks and tracer tests.
9. **Qualification sequence and gates**
    1. Paper screen of chemistry against duty, environment, and materials list.
    2. Bench property testing of candidate salts from two qualified suppliers.
    3. Static coupon compatibility at three temperatures.
    4. Flowing corrosion loop with on-line sampling and filtration tests.
    5. Redox control demonstration with upset and recovery cycles.
    6. Freeze test and recovery demonstration.
    7. Integrated pilot loop, minimum 1000 hours or 100 thermal cycles, with performance and corrosion trending.
    8. Site pilot on one subsystem with full instrumentation.
    9. Final specification with operating limits, monitoring plan, and spares.
10. **Condition monitoring and limits**
    - Salt composition by ICP-OES and ion chromatography, weekly during commissioning then monthly.
    - Moisture in salt by coulometric methods on hot samples, limit in low ppm range per chemistry.
    - Redox potential, on-line probe with calibration checks.
    - Particulate load, high temperature sampling with gravimetric or laser methods.
    - Corrosion rate probes or witness coupons in high risk zones.
    - Thermal performance, approach temperature across exchangers and ΔP across filters and critical lines.

    Alarm limits and actions must be equipment specific. Tie alarms to work orders.
11. **Acceptance criteria**
    - Mass loss on primary alloy less than program limit, for example ≤50 μm per year equivalent in coupon and loop tests.
    - No localized pitting above acceptance depth on pressure boundary alloys.

- Stable thermal performance within design approach temperature and pressure drop budgets.
- Freeze recovery within procedure time without deformation or leak.
- Salt analysis within impurity and redox bands for three consecutive cycles.
- Successful 1000-hour pilot loop with no corrective maintenance outside planned tasks.

12. **Operations and maintenance rules**
    - Maintain cover gas quality, nitrogen or argon with dew point and oxygen limits.
    - Keep make-up salt in sealed containers with pre-heat and bakeout steps.
    - Exercise bypasses and drain-fill lines during outages to avoid dead legs.
    - Inspect insulation, heat tracing, and supports each outage.
    - Replace filters based on differential pressure, not calendar time.
    - Calibrate redox and temperature instrumentation on schedule.
13. **Failure mode controls**
    - Solidification blockage, control with heat tracing density checks, freeze point margin, and recovery drills.
    - Decomposition and gas evolution, control with redox management and temperature caps.
    - Hot corrosion, control with alloy upgrade, impurity control, and flow velocity limits.
    - Fouling, control with polishing filtration and chemistry adjustments.
    - Leak during heat-up, control with conservative ramp rates and verified expansion allowances.
14. **Documentation and handover**
    - Controlled salt specification, operating limits, sampling plan, and analytical methods.
    - Materials dossier with alloy heat numbers, weld procedures, and corrosion test results.
    - Freeze recovery and upset response procedures.
    - Spares list for filters, instruments, heaters, and valves.
    - Training package for operators and maintenance teams.

MATERIALS TEST PLAN AND DATA SHEET TEMPLATES

Molten Salts with Extended Temperature Ranges

1. Test Plan Overview

Project ____________________ Loop ID ____________________ Salt chemistry ____________________

Objective ____

Operating window Tmin ______ °C Tmax ______ °C Nominal ______ °C

Cover gas ☐ N_2 ☐ Ar Dew point limit ______ °C O_2 limit ______ ppm

Duration ______ h Thermal cycles ______ Velocity ______ m/s Heat flux ______ kW/m²

Gate schedule

1. **Property verification complete ☐**
2. **Static coupon exposure complete ☐**
3. **Flowing loop exposure complete ☐**
4. **Freeze test completed ☐**
5. **Redox control demonstration completed ☐**
6. **Pilot loop completed ☐**

2. Static Coupon Exposure Data Sheet

Alloy ____________________ Form ☐ Plate ☐ Tube ☐ Weldment Heat no. __________

Surface prep ____________________ Initial roughness Ra ______ µm

Exposure temp ______ °C Hold time ______ h Atmosphere ☐ Inert ☐ Salt cover only

Coupon ID	Initial mass g	Final mass g	Mass change mg/cm²	Avg pit depth µm	Max pit depth µm	Scale type (XRD)	Notes

Post test tasks

- **Cross-section microscopy at hot face and cold face ☐**
- **SEM EDS on scale and base metal ☐**
- **Photolog saved to repository ☐**

Acceptance check

- **Uniform corrosion rate ≤ ______ μm/y ☐**
- **No pits deeper than ______ μm on pressure boundary alloys ☐**

3. Flowing Corrosion Loop Data Sheet

Loop fluid ____________________ Pump type ____________________ Filter media ____________________

Setpoints T_in ______ °C T_out ______ °C Flow ______ kg/s Pressure ______ bar

Daily log

Date	T_in °C	T_out °C	Flow kg/s	dP filter kPa	Redox mV	Moisture ppm	O_2 ppm	Notes

Witness coupons in flow

Location	Alloy	Start mass g	End mass g	Corrosion rate μm/y	Remarks

Events

- **Thermal cycle count updated ☐ New total ______**
- **Filter changeout ☐ Element ID __________ Debris mass ______ g**
- **Off-spec alarm ☐ Parameter __________ Peak value ______ Time ______**

4. Redox Control and Impurity Log

Target redox window ______ to ______ mV Reference ____________________

Date	Temp °C	Redox mV	Adjuster added	Quantity	Salt sample ID	ICP metals mg/kg	Halides mg/kg	Action

Acceptance

- **Redox within window for 3 consecutive days ☐**
- **Impurities within spec sheet limits ☐**

5. Freeze Protection and Recovery Drill Record

Salt freeze point ______ °C Lowest circuit point ______ °C

Heat tracing density ______ W/m Verified operable ☐

Drill steps

1. **Induce controlled cool-down to ______ °C.**
2. **Hold for ______ min.**
3. **Execute recovery procedure.**
4. **Record time to first flow ______ min.**
5. **Record full recovery time ______ min.**
6. **Inspect for deformation or leakage.**

Results

- **Differential expansion within limits ☐**
- **No leaks ☐**
- **Insulation cold spots corrected ☐**

6. Filtration Performance Log

Element type ____________________ Micron rating ______ Material ☐ Metal ☐ Ceramic

Date	dP clean kPa	dP change kPa	Flow kg/s	Element hours	Debris mass g	Dominant species (SEM EDS)	Notes

Acceptance

- **Element life ≥ ______ h at design dP ☐**
- **Debris composition consistent with expected corrosion products ☐**

7. Thermal Performance Sheet

Heat exchanger ID __________ Surface area ______ m²

Date	T_hot in °C	T_ hot out °C	T_ cold in °C	T_ cold out °C	Flow hot kg/s	Flow cold kg/s	Approach K	UA kW/ K	Notes

Action threshold

- **Approach increase > ______ K from baseline triggers inspection ☐**

8. Salt Composition and Moisture Report

Sampling method at temperature recorded ☐

Sample ID	Date	Na wt%	K wt%	Li wt%	Cl wt%	F wt%	NO_3 wt%	NO_2 wt%	H_2O ppm	O_2 ppm	Notes

Acceptance

- **Composition within spec bands ☐**
- **Moisture below ______ ppm ☐**

9. Nonconformance and Corrective Action

Event ____ **Date** __________

Parameter off-spec ____________________ **Measured** ______ **Limit** ______

Immediate action ____

Root cause ____

Corrective action ____

Verification result ____

10. Acceptance Matrix

Ratings: 1 acceptable, 2 conditional with controls, 3 reject

Category	Metric	Target	Result	Rating	Notes
Thermal stability	No mass loss at ______ °C for ______ h	Pass			
Corrosion uniform	≤ ______ µm/y on primary alloy	Pass			
Localized attack	Max pit depth ≤ ______ µm	Pass			
Freeze recovery	Full flow ≤ ______ min	Pass			
Redox control	Within window for ______ days	Pass			
Moisture	≤ ______ ppm	Pass			
Filtration	Element life ≥ ______ h	Pass			
Thermal duty	Approach within ______ K	Pass			

Approval

Engineer ____________________ Date __________ Signature __________

11. Sampling Plan

- **During commissioning: daily for 7 days.**
- **Pilot loop steady state: twice per week.**
- **After thermal cycle events: within 2 hours.**
- **Post maintenance: within 4 hours of restart.**
- **Routine operation: monthly or per site risk class.**

Sample set each event

- **Composition by ICP OES ☐**
- **Redox mV ☐**
- **Moisture ppm ☐**
- **Particulate gravimetric ☐**
- **Corrosion probe reading ☐**

12. Data Management

- **Store raw data, photos, SEM images, and lab certificates in controlled repository.**
- **File naming convention: Site-LoopID-TestType-Date-SampleID.**
- **Link acceptance matrix to gate review minutes.**

13. Quick Reference Limits
Populate from your internal spec sheet before testing.

- **Redox window: ______ to ______ mV**
- **Moisture limit: ______ ppm**
- **O_2 limit: ______ ppm**
- **Uniform corrosion limit: ______ µm/y**
- **Pit depth limit: ______ µm**
- **Filter dP change limit: ______ kPa**
- **Heat exchanger approach change: ______ K**
- **Freeze recovery time: ≤ ______ min**

14. Checklists
Pre-test
☐ **Calibrations current**
☐ **Leak check complete**
☐ **Cover gas purged to limits**
☐ **Sampling hardware hot and clean**
☐ **Emergency cool-down verified**

Post-test

- ☐ **Salt samples archived**
- ☐ **Coupons labeled and stored**
- ☐ **Filters weighed and logged**
- ☐ **Instruments post-checked**
- ☐ **Gate review scheduled**

16 Digital Twins, AI, and Fluid Monitoring

Digital twins in electrification and power generation depend on accurate, time-aligned datasets to represent the actual condition of assets. Fluid monitoring is a primary data stream for these systems. Lubricants, insulating fluids, and coolants carry chemical, thermal, and contamination signatures that directly reflect operational stresses, component wear, and incipient failures. Predictive analytics uses this data to forecast degradation trends, detect anomalies, and simulate future asset performance under varying conditions.

DATA ACQUISITION

Sensors and laboratory measurements provide viscosity, dielectric strength, oxidation levels, water content, particulate contamination, and additive depletion rates. For lubricants, wear metal concentration from ICP or RDE spectroscopy, soot loading, and base/acidity balance (TBN/TAN) are high-value indicators. For insulating fluids, dissolved gas analysis, moisture content, and interfacial tension link directly to insulation aging and fault detection. For coolants, pH, glycol content, inhibitor levels, and conductivity changes indicate chemical stability and corrosion protection status. Data must be tagged with operating context such as load, temperature, and run hours to support valid correlation analysis.

DATA INTEGRATION INTO DIGITAL TWINS

Condition data is imported into the digital twin through OPC-UA or equivalent protocols. Time-series alignment is essential to ensure operational states match fluid measurements. The twin models the equipment and fluid as coupled systems, incorporating thermal models, chemical degradation kinetics, and contamination ingress pathways. Fluid health indices are calculated from parameter weightings, specific to the fluid type and the asset's criticality. The twin tracks these indices against predefined failure thresholds and predicts remaining service life.

DOI: 10.1201/9781003775201-22

AI MODEL DEVELOPMENT

Machine learning algorithms such as gradient boosting, random forests, and recurrent neural networks process historical and live data to detect patterns that precede failures. For example, in transformer oil, a combination of rising ethylene content, decreasing breakdown voltage, and increased moisture may indicate winding insulation breakdown before catastrophic failure. In turbines, viscosity shear loss correlated with rising ferrous debris signals bearing wear progression. AI models are trained with labeled failure events to strengthen fault classification accuracy and false positive rejection.

Predictive Analytics Process

1. Acquire and validate raw sensor and lab data.
2. Normalize units and remove outliers that result from sampling or sensor errors.
3. Calculate derived parameters such as oxidation stability index, contamination severity codes (per ISO 4406), and additive depletion percentages.
4. Weight parameters by asset type to form a composite health score.
5. Feed scores and individual parameters into AI models for trend forecasting.
6. Compare forecast curves with maintenance planning windows and operational risk tolerances.
7. Issue recommendations for fluid changeout, filtration, additive replenishment, or hardware inspection.

Condition-Based Maintenance Triggers

Digital twin-driven predictive analytics replaces fixed-hour fluid changes with condition-based intervals. The system triggers maintenance only when projected degradation crosses intervention thresholds before the next inspection cycle. This prevents premature disposal of serviceable fluids while reducing the probability of in-service failures. Critical parameters such as TAN rise rates, dissolved gas growth rates, or particle count acceleration are weighted higher in critical assets like nuclear reactor coolant pumps or HVDC converter transformers.

Failure Mode Linkages

By correlating fluid condition changes with known mechanical and electrical fault modes, predictive analytics can determine the most probable root cause before teardown. For example, sudden coolant conductivity increases in inverter systems can be linked to heat exchanger leak-through from process water. SF_6 density loss trends in GIS switchgear can indicate seal deterioration long before electrical faults occur.

Continuous Learning and Feedback

Once maintenance actions are performed, post-service fluid samples and equipment inspection results are re-integrated into the AI dataset. This closes the feedback loop, refining the predictive model and improving accuracy for future events. For large fleets, cross-unit analysis identifies systemic weaknesses in fluid management or equipment design.

INTEGRATION WITH SCADA AND CMMS FOR FLUID CONDITION MONITORING IN ELECTRIFICATION AND POWER GENERATION SYSTEMS

SCADA Integration

Supervisory control and data acquisition (SCADA) systems provide the live operational data stream necessary to interpret fluid condition trends within their operating context. Real-time measurements such as load, temperature, vibration, and electrical output are paired with fluid monitoring parameters from online sensors and laboratory uploads. Typical integration uses OPC-UA, Modbus TCP/IP, or IEC 61850 protocols. Fluid sensors, viscosity, moisture (ppm), particle count (ISO 4406), dielectric strength, and dissolved gas, are polled at scheduled intervals. Data is time-synchronized within the SCADA historian to maintain accurate correlation between process events and fluid property changes. Alarm logic is configured within SCADA to generate real-time notifications when condition thresholds are exceeded. Trend plotting within SCADA allows field engineers to confirm whether excursions are transient or part of a developing failure mode.

CMMS Integration

Computerized maintenance management systems (CMMS) receive processed condition data and predictive analytics outputs from SCADA or digital twin platforms. This integration can be direct, using API calls or data exchange files (CSV, XML, JSON), or indirect through an enterprise data historian. CMMS uses this information to trigger automated work orders, schedule inspections, or initiate fluid changeouts based on condition codes rather than fixed calendar intervals. Historical maintenance records stored in CMMS are linked to prior fluid analysis results, allowing correlation between fluid degradation patterns and actual component failures. This linkage improves the predictive accuracy of both the CMMS maintenance rules and the AI models running in parallel.

Data Flow and Verification Process

1. Online or lab analysis data is ingested into the SCADA historian.
2. Condition indices and trend analysis results are calculated.
3. SCADA triggers alarms or notifications for out-of-spec readings.
4. Validated condition data is exported to the CMMS.

5. CMMS generates condition-based work orders with equipment, location, and parameter details.
6. After maintenance, the CMMS records service actions, which are fed back into the analytics system for model retraining.

Cybersecurity and Reliability Considerations

Fluid monitoring integration into SCADA and CMMS must meet NERC CIP requirements in North America and equivalent cybersecurity frameworks in other regions (see Figure 16.1). Segmentation of data networks and secure authentication protocols are mandatory to prevent operational disruptions. Redundant communication paths and buffered data logging are used to avoid data loss during network outages, ensuring continuity in predictive maintenance calculations.

Standards and Guidance

- IEEE Std 1854-2016: Guide for Smart Distribution Applications.
- ISO 55000:2014: Asset management, Overview, principles, and terminology.
- ISA-95: Enterprise-Control System Integration.
- NERC CIP Standards for Critical Infrastructure Protection in power systems.

ML-DRIVEN DGA AND OIL ANALYSIS INTERPRETATION

Use dissolved gas analysis and oil condition data to predict fault type, fault energy, and time to intervention. Combine lab and online sensors in an engineered pipeline

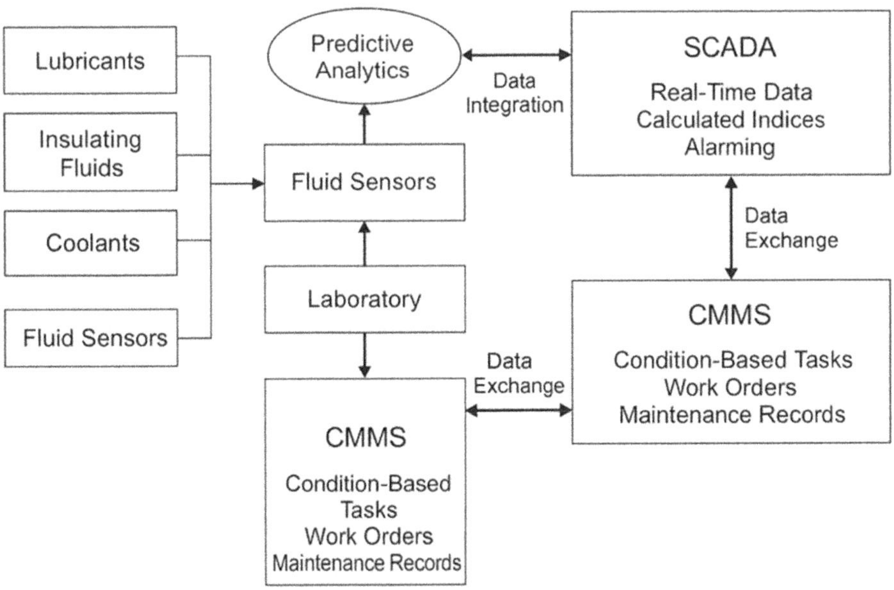

FIGURE 16.1 SCADA–CMMS integration architecture diagram.

TABLE 16.1
Example Feature Map

Feature Block	Examples	Purpose
Ratio suite	C_2H_2/C_2H_4, CH_4/H_2, C_2H_6/CH_4, Duval barycentric coords	Fault type separation
Trend suite	7-day and 30-day slopes, burst detectors, CUSUM on TCG	Early change detection
Health indices	Oil Health Index, Insulation Health Index	Single score for CMMS
Dielectric set	BDV at temperature, tan δ at temperature, moisture % saturation	Flashover risk
Aging set	2-FAL, 5-HMF, CO/CO_2, DP estimate	Cellulose life tracking
Context	Load factor, ambient, tap duty	Normalize gas growth

that supports anomaly detection, fault classification, and remaining life estimation. Integrate outputs with SCADA and CMMS for automated work orders and evidence trails (see Table 16.1).

Data Sources and Tags

- **DGA**: H_2, CH_4, C_2H_6, C_2H_4, C_2H_2, CO, CO_2, O_2, N_2 from ASTM D3612.
- **Transformer oil condition**: BDV, tan δ and resistivity per IEC 60247, moisture ppm, IFT, acidity, furan compounds for paper aging.
- **Turbine and industrial oils**: viscosity, TAN, RPVOT, MPC, FTIR oxidation and nitration, ISO 4406 particle code, water by Karl Fischer.
- **Context tags**: load, hot-spot temperature, ambient, tap position, switching count, maintenance events, oil batch, hours since last intervention.

Feature Engineering

Build both instantaneous and trend features. Normalize by temperature and sampling lag.

- **Gas ratios**: Rogers, Dornenburg, IEC ratio sets, Duval triangle coordinates and distance to zone boundaries.
- **Growth rates**: d[TCG]/dt, per-gas exponential slope, half-life since last purge, burst magnitude over trailing window.
- **Oxidation and aging**: TAN slope, IFT slope, furan indices and DP estimate, antioxidant reserve where available.

- **Moisture and dielectric**: moisture ppm vs saturation curve at oil temperature, BDV drift, tan δ at service temperature.
- **Contamination and wear**: ISO code deltas, ferrous PQ index, FTIR peak areas, MPC ΔE.
- **Operational covariates**: per-unit load factor, thermal gradients, on-load tap changer duty markers.
- **Quality flags**: sampling method, transport time, lab method version, instrument serial to de-bias lab effects.

Model Classes and Where to Apply

- **Unsupervised anomaly detection** for fleets with sparse failures
  Isolation Forest or autoencoder on standardized feature vectors. Output an anomaly index with confidence bands.
- **Supervised fault classifiers** when labeled events exist
  Gradient boosted trees or random forests trained on expert-verified labels: PD, low energy thermal, high energy thermal, arcing, stray gassing, paper aging.
- **Time-series forecasters** for gas growth
  State-space models or LSTM/Temporal Convolution for TCG and key gas trajectories.
- **Survival and renewal models** for remaining useful life
  Cox models or Random Survival Forests using hazard features such as acetylene growth, moisture supersaturation time, TAN rate, MPC trend.

Ground Truth and Labeling

- Use outage reports, borescope photos, DLRO and power factor tests, OLTC inspections, relay operations, and teardown findings.
- Match each label to the closest sample by time, with a lockout window to prevent leakage of post-failure data into training.
- Add a "non-fault operating change" label for load or temperature driven excursions.

Validation Protocol

- Split by asset, not by sample, to avoid leakage.
- Stratify by manufacturer, voltage class, and climate zone where fleet is diverse.
- Score with precision-recall, class-wise F1, and **false alarm rate per 1000 asset-days**.
- Backtest on rolling windows. Require stability across vintages and lab changes.
- Calibrate alarms to hit **miss rate ≤10%** for high-energy arcing at a **false alarm rate ≤1 per 100 asset-months**, then tune for site tolerance.

Decision Logic and Actions

Create a rules-plus-model layer. Do not discard physics rules that guard safety.

- **Immediate trip risk**: high C_2H_2 spikes with concurrent BDV drop and rapid TCG growth. Action: forced inspection plan, load reduction, or outage.
- **Thermal hotspot**: rising C_2H_4 with load-correlated slope, normal moisture, stable BDV. Action: thermal survey, cooling checks, load management.
- **Paper aging**: furan rise with CO/CO_2 ratio change, moisture trending high at constant load. Action: dry-out plan, moisture control, oil reclaim schedule.
- **Stray gassing**: early life H_2-CH_4 growth without fault ratios, stable BDV. Action: monitor only, extend sampling interval after stabilization.
- **Turbine oil varnish**: MPC and FTIR oxidation growth with rising tan δ. Action: electrostatic polish, bleed-and-feed, AO top-up if permitted.

Handling Imbalanced Data

- Weighted loss or focal loss for rare arcing.
- Synthetic minority oversampling at the asset level only for training.
- Cost matrix in thresholding to reflect outage cost vs false truck roll.

Explainability and Sign-Off

- Use SHAP or permutation importance to show which features drove a prediction.
- Attach plots to the CMMS work order: gas trend, model index, and top features.
- Keep a human-readable rule trace. Example: "Alarm fired because C_2H_2 slope is in the 95th percentile while BDV dropped 12% and Duval point moved into zone 3."

Drift Control and Model Governance

- Track input drift for each feature and lab method.
- Shadow new models for at least one heating season before cutover.
- Quarterly governance review with confusion matrices, alarm audit, and economic outcomes.
- Retrain when drift exceeds set limits or when a new lab method or sensor firmware is introduced.

Deployment Pattern

- **Edge**: simple ratio rules, moving averages, and anomaly index calculation near the asset for fast alarms.
- **Cloud or data center**: full models, survival analysis, fleet benchmarking, retraining.
- **SCADA link**: OPC-UA tags for live moisture, load, and temperatures.

- **CMMS link**: API to open work orders with asset ID, fault class, evidence links, and SLA.

Oil Analysis Model Extensions for Rotating Equipment

- Predict viscosity shear loss using temperature-corrected viscosity deltas and shear severity factors.
- Classify soot-dominated diesel wear using soot %, ICP Fe/Cr/Ni, and nitration index.
- Forecast varnish risk from MPC slope, RPVOT decline, and FTIR oxidation bands.
- Detect aeration issues from air release time, foam tendency, and acoustic cavitation signatures.

Pilot Sequence

1. Six-month data harvest on a representative set of assets.
2. Offline model build and cross-site backtest.
3. Parallel run with existing rules.
4. Alarm setpoint tuning with operations.
5. CMMS workflow integration and training.
6. Fleet rollout with quarterly reviews.

DIGITAL TWIN MODELING OF HEAT TRANSFER AND DEGRADATION FOR POWER EQUIPMENT FLUIDS

The model must predict heat distribution, temperature-dependent viscosity changes, oxidation rates, and insulation aging while integrating real-time sensor data for calibration. The model is a physics-informed, data-driven model that replicates the thermal and chemical behavior of lubricants, insulating fluids, and coolants under operational conditions (see Tables 16.2, 16.3, and 16.4).

Core Model Components

1. **Geometry and Material Properties**
   - Equipment CAD models with fluid pathways (transformers, turbines, compressors, generators).
   - Thermal conductivity, specific heat capacity, density, and viscosity curves for the working fluid from ASTM D341 (viscosity–temperature) and ASTM E1952 (thermal conductivity).
   - Solid component properties for metals, ceramics, elastomers.
2. **Heat Generation Sources**
   - Electrical losses (I^2R in windings, eddy currents, dielectric losses).
   - Mechanical friction losses (bearings, seals, gear meshes).
   - Chemical exothermic reactions from oxidation or polymerization.

TABLE 16.2
Degradation Prediction Metrics

Parameter	Model Output	Threshold Reference	Standard
Oil oxidation life	% antioxidant depletion, TAN projection	End-of-life at TAN > 2.0 mg KOH/g	ASTM D664
Viscosity shift	% deviation from baseline at 40 °C	±10% warning	ASTM D445
Hot-spot temperature	°C above nameplate	Insulation class limit	IEEE C57.91
Moisture in oil	ppm and % saturation	60% saturation caution	IEC 60422
MPC index (varnish)	ΔE color scale	$\Delta E > 20$ warning	ASTM D7843

3. **Heat Transfer Modes**
   - **Conduction**: Fourier's law for metallic paths and thermal interfaces.
   - **Convection**: Forced or natural convection coefficients from Nusselt correlations, adjusted for viscosity and flow regime.
   - **Radiation**: Stefan–Boltzmann law with emissivity factors for external surfaces.
4. **Degradation Kinetics**
   - Oxidation modeled via Arrhenius equation with activation energy from ASTM D943 (oxidation stability) or RPVOT tests.
   - Polymerization and varnish formation modeled as second-order reactions with temperature acceleration factors.
   - Moisture-induced dielectric breakdown risk modeled via IEC 60296 moisture saturation curves.
5. **Boundary Conditions**
   - Ambient temperature, wind speed (outdoor equipment), and load cycles from SCADA.
   - Coolant inlet temperature, flow rates, and pump curve characteristics.

Data Integration

- **Sensors**: RTDs, thermocouples, infrared thermography for surface temperature mapping, flow meters, dissolved gas monitors, online viscometers.
- **Data ingestion**: OPC-UA or Modbus TCP into the digital twin platform.
- **Feedback loop**: Bayesian updating or Kalman filtering to correct simulation drift from measured reality.

Calibration and Validation

- **Initial calibration** using factory acceptance test thermal runs and baseline oil analysis.

TABLE 16.3
Digital Twin Heat Transfer and Degradation Failure Pathway Table

Model Layer	Failure Pathway	Leading physical indicators	Data features for Analytics	Diagnostics to Confirm	Twin Outputs	Action Criteria	Typical Actions
Thermal network and CFD	Hot spot due to fouling in exchanger or jacket	Rising approach temperature at constant load. Local surface hot bands on IR. ΔP across exchanger increasing	Rolling slope of approach temperature. CUSUM on ΔP. Seasonal baseline correction. UA decline per run-hour	IR thermography map. Chemical analysis for deposits. Clean-in-place delta performance test	Predicted hot spot location. Heat transfer coefficient map. Time to limit	Approach temperature increase above baseline by 5 K. UA drop more than 15 percent	Schedule cleaning. Side stream filtration. Restore flow distribution
Thermal network and hydraulics	Flow maldistribution in parallel branches	String flow imbalance. Outlet temperature spread across branches	Coefficient of variation on branch flows. K-means cluster of outlet temperatures	Flow measurement by ultrasonic meter. Header inspection	Branch-level Nusselt estimates. Predicted re-balance setpoints	Flow imbalance more than 10 percent across strings	Rebalance valves. Remove blockages. Update commissioning sheet
Fluid property model	Viscosity growth from oxidation or contamination	Pump power up at fixed flow. Start-up torque increase in cold conditions	Viscosity delta at 40 C and 100 C. TAN slope. Water ppm trend	ASTM D445 viscosity. TAN per D664. Karl Fischer water	Viscosity map vs temperature. Predicted pump margin	Viscosity deviation more than 10 percent from baseline	Polishing filtration. Partial bleed and feed. Leak root cause fix

(continued)

TABLE 16.3 (Continued)
Digital Twin Heat Transfer and Degradation Failure Pathway Table

Model Layer	Failure Pathway	Leading physical indicators	Data features for Analytics	Diagnostics to Confirm	Twin Outputs	Action Criteria	Typical Actions
Chemical kinetics	Oxidation and varnish formation in turbine or thermal oil	MPC rise. Tan delta rise. Filter loading increase	MPC ΔE slope. RPVOT decline rate. FTIR oxidation band growth	ASTM D7843 MPC. ASTM D2272 RPVOT. FTIR trend review	Remaining oxidation life. Varnish risk index	MPC ΔE above 20 or site limit. RPVOT below alarm band	Electrostatic polishing. AO top treat where approved. Lower film temperature
Electric and thermal coupling	Dielectric fluid hot spot and moisture rise	BDV downtrend under stable load. Moisture percent saturation rising	Moisture at temperature. BDV drift. DGA ethylene slope	IEC 60156 BDV. IEC 60247 tan delta. IEC 60599 DGA	Predicted insulation margin. Time to BDV limit	Moisture saturation above 60 percent. BDV drop more than 10 percent	Dry-out plan. Leak repair. Oil reclaim or change
Boiling and two phase	Local nucleate to film boiling transition in power electronics cold plates	Sudden rise in junction temperature. Acoustic cavitation signatures	Critical heat flux proximity index. Plate pressure pulsation RMS	High speed thermal probe. Acoustic emission. Visual endoscope if accessible	Margin to CHF by zone. Safe operating envelope	CHF margin under 20 percent	Reduce load. Increase flow. Clean microchannels. Change surface finish spec

Hydrogen or gas cooling	Gas purity drop with cooling loss	Generator winding temperature rise at steady load. Makeup gas rate increase	Purity and O_2 trend. Leak rate estimate. Seal DP trend	Ultrasonic leak survey. Purity meter. Seal oil console checks	Predicted winding temperature at purity setpoints	Purity under 98%. O_2 above 1 percent	Restore seal DP. Repair leaks. Verify regulators
Corrosion and erosion	Wall thinning in hot salt or glycol loops	ΔP drift at constant flow. Metal ion rise in fluid	ICP metals slope. Erosion risk index from velocity and solids	Coupon mass loss. UT thickness grid. SEM EDS on debris	Remaining wall estimate. Time to minimum thickness	Corrosion rate above limit. Forecasted thickness under minimum before next outage	Lower velocity. Chemistry correction. Alloy upgrade. Install strainers
Contamination ingress	Air entrainment and foam in lube circuits	Air release time increase. Pump cavitation noise	Air release time slope. Foam tendency score. Acoustic energy band	ASTM D3427 air release. ASTM D892 foam. Suction leak test	Gas volume fraction map. Cavitation risk index	Air release time above spec. Foam tendency rise across sequences	Fix suction leaks. Rework returns. Control antifoam dose
Thermal structure of solids	Insulation hot spot due to blocked ducts	Winding gradient rise with load constant. DGA thermal gas growth	Thermal gradient index. C_2H_4 slope. Load normalized temperature	IR scan. Endoscopic duct inspection	Predicted winding hot spot. Time to insulation class limit	Gradient above design by 10 K. DGA indicates T2 or T3 class	Clean ducts. Correct cooling path. Adjust load profile

TABLE 16.4
Alarm and Work Order Mapping

Alert Class	Trigger Set	Work Order Content	Evidence Pack
Thermal efficiency loss	UA drop and approach temperature rise on two consecutive windows	Heat exchanger inspection and clean	Trend plots, before and after UA, IR map
Insulation risk	Moisture percent saturation and BDV downtrend with load stable	Dry-out plan and leak check	BDV trend, moisture at temperature, DGA summary
Varnish risk	MPC slope and RPVOT decline beyond control band	Polish filtration and reservoir check	MPC images, RPVOT history, FTIR overlays
Flow maldistribution	Branch flow CV and outlet temperature spread above limit	Rebalance and obstruction removal	Flow table, ΔT map, CFD branch prediction

- **Periodic recalibration** using monthly oil and coolant analysis data and trending from online sensors.
- **Validation** by comparing predicted hot-spot temperatures and viscosity loss to field measurements over a defined load and ambient cycle.

Use Cases

- **Predictive maintenance**: Schedule oil change or filtration before oxidation products accelerate sludge formation.
- **Design evaluation**: Test the impact of alternate cooling fluid with higher thermal conductivity or different viscosity index.
- **Failure investigation**: Reconstruct overheating events and match fluid degradation profile to operational logs.
- **Efficiency optimization**: Evaluate heat exchanger fouling progression and predict when cleaning yields the best cost-to-performance ratio.

Example Workflow

1. Import equipment 3D model and material data.
2. Input fluid property curves and degradation constants.
3. Define operational load cycles and environmental conditions.
4. Run CFD and FEA simulations for steady-state and transient conditions.
5. Link to real-time SCADA data for continuous calibration.
6. Output: predicted fluid condition index, time to next intervention, and recommended load adjustments.

Integration Rules for the Twin

- Align sensor timestamps with process states. Use historian tags for load, ambient, valve position, pump speed, and duty cycle.
- Calibrate convection and fouling factors with step tests after maintenance to reset baselines.
- Use Arrhenius factors for oxidation and hydrolysis. Update activation energy with lab data from current oil batch.
- Combine model residuals with anomaly indices. Trigger alerts only when both exceed thresholds to reduce false calls.

Data Quality Gates

- Reject samples without a chain of custody or with transit time beyond shelf life.
- Flag sensors with calibration drift detected by reference checks.
- Apply temperature normalization for BDV, tan delta, viscosity, and moisture before analytics.

KPIs for Program Control

- False alarm rate per 100 asset months.
- Miss rate for high energy faults.
- Lead time to intervention in days.
- Post action delta in UA, BDV, viscosity, or MPC relative to model forecast.

17 Circular Economy and Sustainable Fluid Management

CIRCULAR ECONOMY AND SUSTAINABLE FLUID MANAGEMENT: REREFINED BASE STOCKS AND FLUID RECLAIMING IN ELECTRIFICATION AND POWER GENERATION

Circular economy practices for lubricants, insulating fluids, and coolants in electrification and power generation rely on rerefining, reclaiming, and closed-loop fluid management to reduce virgin resource consumption, lower disposal volumes, and maintain operational reliability. Rerefining converts used oils into base stocks meeting American Petroleum Institute (API) Group I, II, or III specifications. Fluid reclaiming restores in-service oils or coolants to near-original performance through targeted removal of degradation products and contaminants. Both methods require process control, condition monitoring, and adherence to regulatory standards to ensure equipment protection and compliance.

Rerefined Base Stocks

Rerefined base oils originate from used lubricants collected under controlled handling protocols to avoid cross-contamination with hazardous materials. Processing includes vacuum distillation, hydrotreating, hydrocracking, and clay polishing to remove polar compounds, oxidized species, metals, and particulates. The output must meet the same physical and chemical specifications as virgin base stocks for viscosity index, volatility, oxidation stability, and purity.

API 1509 licensing allows rerefined base oils to be formulated into lubricants carrying the same performance claims as those made with virgin base stocks. In power generation, these products are suitable for turbine oils, compressor lubricants, gear oils, and certain insulating fluids provide additive packages are correctly reintroduced and compatibility is confirmed. Long-term monitoring of oxidation stability, MPC values, and viscosity stability is necessary to validate performance parity.

Fluid Reclaiming

On-site or off-site reclaiming removes moisture, acids, varnish precursors, particulates, and dissolved gases from in-service fluids without disposal. Techniques

DOI: 10.1201/9781003775201-23

include Fuller's earth adsorption, vacuum dehydration, electrostatic separation, centrifugation, and chemical reconditioning. This process extends the operational life of insulating oils in transformers, turbine oils in combined-cycle plants, and synthetic coolants in wind and solar power applications.

Reclaiming requires baseline testing of acidity, interfacial tension, dielectric strength, and color for insulating fluids, or viscosity, membrane patch colorimetry (MPC), and total acid number (TAN) for lubricants. Process endpoints are defined by achieving original equipment manufacturer (OEM) or American Society for Testing and Materials (ASTM)-recommended property limits. For critical systems such as nuclear generator lubricants or high-voltage transformers, reclaimed fluids must be qualified through dielectric breakdown voltage (International Electrotechnical Commission (IEC) 60156), dissipation factor (IEC 60247), and oxidation stability (ASTM D2272).

Compatibility and Contamination Control

Fluid recovery programs must address material compatibility between rerefined or reclaimed fluids and system components. Elastomer swelling, additive precipitation, and seal degradation are risks if compatibility is not verified. Recovered fluids should be filtered to ISO 4406 cleanliness codes matching OEM specifications. When reclaiming insulating fluids, removal of corrosive sulfur compounds is essential to prevent copper conductor damage in transformers.

Cross-contamination with incompatible fluids, residual cleaning agents, or incompatible additive chemistries can lead to foam, deposit formation, or poor film strength. Dedicated transfer lines, filtration systems, and storage tanks should be used to maintain purity.

Monitoring Recovered Fluids in Service

Condition monitoring protocols for rerefined and reclaimed fluids mirror those for virgin fluids but require closer trending early in deployment. Parameters include the following:

- **Lubricants:** Viscosity at 40 °C and 100 °C (ASTM D445), TAN (ASTM D664), oxidation/nitration (Fourier transform infrared spectroscopy (FTIR)), wear metals (ICP), MPC (ASTM D7843).
- **Insulating Fluids:** Dielectric breakdown voltage (IEC 60156), moisture (IEC 60814), acidity (IEC 62021-1), DGA (IEC 60599), interfacial tension (ASTM D971).
- **Coolants:** pH, conductivity, inhibitor concentration, glycol ratio, and corrosion product metals.

Online sensors for moisture, particle count, and dielectric loss factor can provide real-time tracking, while lab testing confirms property restoration.

Environmental and Regulatory Considerations

Rerefined and reclaimed fluids must comply with environmental regulations including Registration, Evaluation, Authorisation and Restriction of Chemicals (REACH) (EC 1907/2006) for chemical safety, Restriction of Hazardous Substances (RoHS) for hazardous substances, and local waste management laws. In North America, the Environmental Protection Agency (EPA) defines specifications for rerefined oil under 40 Code of Federal Regulations (CFR) Part 279. End-of-life fluids not meeting recovery criteria must be disposed of according to hazardous waste classification or processed via energy recovery.

Rerefined oil production typically results in 50–85% lower lifecycle greenhouse gas emissions compared to virgin oil production, supporting sustainability targets for power generation operators. Fluid reclaiming further reduces waste transport and processing emissions by eliminating off-site disposal (see Table 17.1).

Implementation Protocol for Sustainable Fluid Programs

1. **Assessment:** Inventory all fluid systems and identify those suitable for rerefined or reclaimed fluids.
2. **Collection:** Establish segregated collection processes for used oils and coolants, preventing contamination.
3. **Qualification:** Define property acceptance limits for rerefined and reclaimed products based on OEM and industry standards.
4. **Processing:** Select processing technology and qualified service providers with documented performance data.
5. **Compatibility Testing:** Verify elastomer, coating, and metal compatibility before deployment.
6. **Baseline Testing:** Establish baseline fluid property data prior to system refill.
7. **Condition Monitoring:** Implement laboratory and online testing to track in-service performance.
8. **Documentation:** Maintain traceability records for all recovered fluids, including batch testing and service history.

PROCUREMENT AND QUALIFICATION CHECKLIST – VIRGIN, REREFINED, AND RECLAIMED FLUIDS FOR POWER GENERATION SYSTEMS

This checklist is structured for field reliability engineers and maintenance managers to evaluate fluid sourcing options for lubricants, insulating fluids, and coolants in thermal, nuclear, renewable, diesel, gas, and electrified infrastructure (see Table 17.2).

1. **Supplier Qualification**
   - Verify compliance with **API Base Oil Interchange Guidelines** for lubricants or **IEC/ASTM insulating oil standards**.
   - Confirm rerefining or reclamation facility holds **ISO 9001** quality certification.

TABLE 17.1
Technical Performance Comparison – Virgin vs Rerefined vs Reclaimed Fluids

Property/Performance Metric	Virgin Base Stocks	Rerefined Base Stocks	Reclaimed In-Service Fluids
Viscosity Stability (ASTM D445, % change from baseline over service life)	0–2%	0–3%	0–5% after reconditioning
Viscosity Index (VI) (ASTM D2270)	95–150 (Group I–III)	95–150 (same as virgin)	Matches original formulation if additives replenished
Oxidation Stability (ASTM D2272, minutes)	900–1500	900–1500	700–1200 depending on prior degradation
Dielectric Breakdown Voltage (IEC 60156, kV)	65–80	65–80	60–75 after dehydration and filtration
Total Acid Number (TAN) (ASTM D664, mg KOH/g)	≤0.05	≤0.05	≤0.08 after treatment
Moisture Content (IEC 60814, ppm)	≤10 ppm	≤10 ppm	≤20 ppm after dehydration
Cleanliness Level (ISO 4406)	≤16/14/11	≤16/14/11	≤18/16/13 unless upgraded filtration is applied
Additive System Integrity	100% OEM formulation	100% OEM formulation	Restored to ≥90% original concentration via readditization
Thermal Conductivity (Coolants, W/m·K)	Glycol-water: 0.38–0.45	Same as virgin	Same as virgin if composition restored
Pour Point (°C)	As formulated	As formulated	May shift ±1–2 °C if minor base oil degradation remains
Flash Point (°C)	200–240 (lubricants)	200–240	190–240 depending on prior contamination
Deposit Tendency (MPC Value) (ASTM D7843, ΔE units)	≤10	≤10	≤15 after varnish removal processes
Environmental Impact (Lifecycle GHG Reduction)	Baseline	50–85% reduction vs virgin production	60–90% reduction vs virgin production by avoiding disposal and refining
Typical Applications	All power generation systems	All power generation systems	Best suited for stationary systems with accessible maintenance intervals

TABLE 17.2
Decision Outcomes by Asset Class

Asset Class	Minimum Gate Outcome to Allow Rerefined	Minimum Gate Outcome to Allow Reclaimed	Default Choice if Gates Not Met
Generator and steam turbine bearings	Gates 1 to 5 passed. OEM letter for rerefined turbine oil. RPVOT and MPC margin proven	Gates 1 to 5 passed. Prior reclaim success on same oil family. Online varnish control present	Virgin
Large power transformers and shunt reactors	Gates 1 to 5 passed. IEC dielectric set met with margin. DGA baseline clean	Gates 1 to 5 passed. Reclaim performed by qualified vendor. Post-treatment BDV and tan delta meet limits	Virgin
GIS and GIL fluids where applicable	Not typical. If used, same as transformer rules	Not typical	Virgin
Reciprocating and gas engine oils	Gates 1 to 4 passed. Soot and fuel dilution history acceptable	Generally not advised except for low duty backup sets with close monitoring	Virgin
Hydraulic systems and servo controls	Gates 1 to 5 passed. ISO 4406 code held in service with fine filtration	Rare. Only with strong filtration and stable additive chemistry	Virgin
Closed loop coolants in power electronics	Gates 1 to 4 passed. Inhibitor package verified and compatible	Allowed with lab verified inhibitor restoration and DI makeup controls	Virgin if inhibitor control is weak
Cooling water glycol for gensets and solar thermal loops	Gates 1 to 4 passed. IC anions and hardness under control	Allowed with biocide program and hard limits on turbidity	Virgin if biofouling risk is high

- Audit environmental compliance records (REACH, RoHS, GHS, local waste handling).
- Review chain-of-custody documentation for incoming used oil feedstock (rerefined) or reclaimed fluid.

2. **Chemical and Physical Property Verification**
   - **Viscosity** (ASTM D445) within OEM specification for service application.
   - **Viscosity Index** (ASTM D2270) matches or exceeds baseline fluid.
   - **Oxidation Stability** (ASTM D2272 or IEC 61125) meets OEM minimum life requirement.
   - **Dielectric Breakdown Voltage** (IEC 60156) ≥ 65 kV for transformer fluids after treatment.

- **Total Acid Number (TAN)** (ASTM D664) ≤ OEM limit; **TBN** verified for lubricants where applicable.
- **Moisture Content** (IEC 60814) ≤ 20 ppm for insulating fluids.
- **ISO 4406 Cleanliness Code** verified for all circulating systems.
- **Thermal Conductivity** measured for coolant applications.

3. **Additive Package Integrity**
   - Laboratory confirmation of additive retention ≥ 90% for reclaimed fluids.
   - Readditization procedure documented and validated by supplier.
   - Compatibility testing performed between base oil and additive chemistry per **ASTM D7155**.
4. **Compatibility Testing**
   - **Elastomer compatibility** testing (ASTM D471 for lubricants, IEC 60836 for insulating oils).
   - **Metallurgical compatibility** with copper, aluminum, and steel components.
   - **Paint and coating interaction** checks for fluid contact surfaces.
5. **Condition Monitoring Integration**
   - Baseline oil analysis report stored in computerized maintenance management system (CMMS).
   - Inclusion in **fluid health index (FHI)** tracking program.
   - Online monitoring sensors calibrated for new or reclaimed fluid properties.
6. **Field Acceptance Testing**
   - Pre-fill fluid sample pulled for lab confirmation before commissioning.
   - System flushed or filtered to meet cleanliness targets prior to fill.
   - First scheduled analysis within 500 operating hours or 1 month of service.
7. **Environmental and Regulatory Compliance**
   - Documentation of lifecycle greenhouse gas savings for sustainability reporting.
   - End-of-life handling plan for subsequent reclaim/rerefine cycles.
   - Disposal pathways in compliance with local hazardous waste regulations.
8. **Risk Assessment and Contingency Planning**
   - Failure mode and effects analysis (FMEA) updated for rerefined/reclaimed fluid usage.
   - Contingency stock plan in case of batch rejection.
   - Supplier liability agreement covering off-spec delivery.

Flowchart Decision Matrix
Selection among Virgin, Rerefined, and Reclaimed Fluids

Stepwise logic with pass–fail gates
Gate 1. Equipment Criticality and Risk Tolerance

- If consequence of failure is personnel hazard or long forced outage, set Critical.
- If consequence is production loss with fast repair, set Noncritical. **Branch**

- Critical requires virgin or fully qualified rerefined with OEM approval and tight monitoring.
- Noncritical can consider reclaimed with proof of property recovery.

Gate 2. OEM and Standards Constraints

- Check OEM approvals, warranty terms, and fluid category standards.
- If any clause prohibits rerefined or reclaimed, select virgin or seek written variance.
- If allowed with conditions, proceed to Gate 3.

Gate 3. Property Compliance at Delivery

Mandatory test set before fill.

- Lubricants: viscosity at 40 C and 100 C, VI, TAN, RPVOT, ISO 4406, FTIR, MPC.
- Insulating fluids: breakdown voltage (BDV) per IEC 60156, tan delta and resistivity per IEC 60247, moisture ppm, interfacial tension (IFT), acidity, DGA baseline.
- Coolants: glycol percent, pH, inhibitor residuals, conductivity, hardness, anions by IC.
  Branch
- If any key property is out of spec, reject batch.
- If all properties meet spec, proceed to Gate 4.

Gate 4. Compatibility and Contamination Control

- Elastomers and coatings verified by immersion data or prior field proof.
- Cross-contamination risk controlled by dedicated tanks and lines.
  Branch
- If compatibility or segregation is not proven, select virgin or defer until controls are in place.
- If proven, proceed to Gate 5.

Gate 5. Monitoring Capability

- Online or frequent lab monitoring available.
- CMMS and supervisory control and data acquisition (SCADA) integration in place for alarms and sampling cadence.
  Branch
- If monitoring is limited, prefer virgin or OEM-approved rerefined on critical assets.
- If monitoring is strong, reclaimed or rerefined is acceptable per asset class.

Gate 6. Environmental and Regulatory Objectives

- Site targets for circularity and GHG reduction documented.
- Waste permits and used-oil segregation established.
  Branch
- If objectives require recycled content and risk is acceptable, favor rerefined.
- If disposal risk is high or segregation is weak, postpone reclaimed use.

Rapid Selection Tree

1. Is the asset critical
   - Yes. Go to 2
   - No. Go to 3
2. Are OEM approvals and standards satisfied for rerefined or reclaimed
   - Yes. Go to 4
   - No. Select virgin
3. Can monitoring detect degradation before failure
   - Yes. Candidate may be reclaimed or rerefined. Go to 4
   - No. Select virgin
4. Do delivered properties meet spec with test certificates
   - Yes. Go to 5
   - No. Reject batch
5. Are compatibility and contamination controls in place
   - Yes. Choose rerefined for recycled content target or reclaimed when life extension is the goal
   - No. Select virgin or delay until controls are ready

Action Limits for Qualification Decisions

- **Turbine oils**
  RPVOT within site alarm band. MPC ΔE less than 10 after treatment. ISO code at or better than 16/14/11 before fill.
- **Transformer oils**
  BDV not less than 70 kV. Tan delta at operating temperature within IEC limit. Moisture less than 20 ppm before energization. IFT above site limit.
- **Coolants**
  pH in OEM band. Nitrite or OAT residual within program limits. Chloride and sulfate near detection. Turbidity low and stable.
- **Hydraulic oils**
  Air release time within spec. Foam sequence I to III within limits. No haze or additive dropout.

Procurement Notes Tied to the Matrix

- Always pull a pre-fill retain sample. Store at controlled conditions for dispute resolution.

- Require supplier batch traceability and chain of custody for rerefined and reclaimed lots.
- Define a short early-life sampling interval. Thirty days is typical for stationary assets.
- Tie acceptance to CMMS gates. No work order closeout without lab pass on first sample.

Escalation Rules

- Any critical asset that fails a key property after fill triggers an immediate work order.
- Any reclaimed lot that shows rapid drift in TAN, BDV, or ISO code triggers root cause analysis on contamination control.
- Two consecutive batches with corrective actions trigger supplier audit.

Lifecycle Scoring Aid

Score each option from 1 to 5 for each category. Higher is better. Select the highest total that also passes all gates (see Table 17.3).

Use the score only after the pass–fail gates. Do not substitute the score for property compliance.

Implementation Checklist Tied to Selection

- Baseline test panel stored in CMMS before fill.
- Alarm thresholds in SCADA adjusted for the selected fluid source.
- Sampling plan issued. Early sample after 100 to 500 hours.
- Filter and breather specifications confirmed against expected cleanliness and moisture targets.
- Operator training completed on segregation and transfer procedures.

Rerefined base stocks and reclaimed fluids can meet the performance requirements of lubricants, insulating fluids, and coolants in electrification and power generation systems when applied through a structured qualification and monitoring protocol. Decision-making must be driven by asset criticality, OEM and standards compliance, delivered fluid properties, chemical compatibility, contamination control, monitoring capability, and site environmental objectives. Critical assets require tighter acceptance

TABLE 17.3
Lifecycle Scoring Aid

Category	Virgin	Rerefined	Reclaimed
Standards and OEM fit	5	4 to 5	3 to 4
Risk under limited monitoring	5	4	2 to 3
Circularity and GHG reduction	2	5	5
Cost per run hour	3 to 4	4 to 5	4 to 5
Lead time and availability	5	3 to 4	3 to 4

criteria, written OEM approval, and continuous monitoring, while noncritical systems with proven monitoring can accommodate reclaimed fluids with validated property restoration. All fluids must pass pre-fill laboratory testing, and acceptance is contingent on compatibility verification and contamination risk management. Post-fill sampling and ongoing condition monitoring are essential to verify stability in service and detect early degradation. Selection must balance operational risk with environmental and cost objectives, ensuring that any use of rerefined or reclaimed fluids does not compromise reliability or safety.

COOLANT RECYCLING, FILTERABLE ESTERS, ZERO-WASTE PRACTICES

Coolant recycling, filterable esters, and zero-waste practices are part of closed-loop fluid management strategies aimed at reducing disposal volumes and raw material demand without compromising thermal or dielectric performance.

Coolant recycling involves on-site or vendor-operated processes to remove particulates, degraded inhibitors, and dissolved contaminants from glycols, water-glycol blends, or dielectric coolants. Typical steps include coarse and fine filtration, ion exchange or reverse osmosis for ionic contamination, inhibitor replenishment, and biocide redosing. Performance acceptance requires that recycled coolants meet OEM freeze/boil point limits, pH specifications, conductivity thresholds, and inhibitor residual targets before reintroduction.

Filterable esters are synthetic or natural-based esters formulated for both lubrication and dielectric applications that can be regenerated through depth filtration, vacuum dehydration, and acid scavenging. Unlike some ester systems that degrade irreversibly via hydrolysis, filterable esters are engineered for repeat purification cycles. This makes them suitable for environments where access for full fluid changeouts is limited, such as wind turbine nacelles or offshore substations. Qualification requires hydrolytic stability testing, compatibility checks with elastomers and coatings, and monitoring of acid number, moisture, and dielectric strength after each filtration cycle.

Zero-waste practices integrate recycling and reclaiming into operational maintenance schedules so that no spent fluid is sent to landfill or incineration. This requires dedicated storage and segregation systems, fluid property trending in CMMS or SCADA platforms, vendor return programs for off-site reprocessing, and a documented chain of custody for all recovered fluids. For mixed fluid streams or applications where direct recycling is not possible, rerefining or chemical conversion into secondary products (e.g., industrial cleaners, process oils) can complete the zero-waste loop (see Table 17.4).

Procedural Flow – Coolant Recycling, Filterable Esters, and Zero-Waste Practices (see Figure 17.1)

1. **Fluid Identification and Application Review**
   - Record fluid type, OEM specification, service history, and operating environment.

TABLE 17.4
Decision and Qualification Matrix – Coolant Recycling, Filterable Esters, and Zero-Waste Practices

Step	Decision Criteria	Test/Verification	Acceptable Range/Requirement	Action if Out of Spec
1. Application Classification	Identify system: thermal, dielectric, mixed service	System review and P&ID check	N/A	Reclassify and assign fluid category
2. Criticality Assessment	Determine asset consequence of failure	Reliability impact scoring (RPN/FMEA)	≤ Moderate risk for recycled fluids unless OEM-approved	For high-risk assets, default to virgin or OEM-certified recycled fluid
3. Contamination Baseline	Particulate, water, ionic contamination, inhibitor level	ISO 4406, Karl Fischer, ion chromatography, inhibitor titration	Meet OEM baseline	Filter, dry, or treat before reuse
4. Recycling Feasibility	Compatibility of fluid with regeneration process	Bench test with proposed filtration/reclamation method	No property deviation >5%	Use alternate recycling method or fluid replacement
5. Performance Recovery Verification	Post-treatment property validation	ASTM/IEC tests for thermal capacity, dielectric strength, inhibitor level	≥ 95% recovery of key properties	Retreat or reject batch
6. Filterable Ester Qualification	Hydrolytic stability, compatibility with seals/metals	ASTM D2619, elastomer immersion, metal strip corrosion	No degradation >5% mass change or TAN >0.1 mg KOH/g in 168 hr	Reformulate or use alternate ester chemistry
7. Zero-Waste Pathway Confirmation	Verify downstream recycling, reclaim, or rerefining vendor	Vendor certification, disposal pathway audit	100% recovery / reuse documented	Source alternate certified vendor
8. Integration into Maintenance Plan	Define sampling frequency, treatment cycles	CMMS/SCADA scheduling	As per risk level: critical assets 3–6 months, noncritical up to 12 months	Increase monitoring frequency
9. Environmental Compliance	Check REACH, RoHS, GHS, PFAS status	Regulatory review and SDS audit	Fully compliant	Substitute compliant chemistry
10. Final Approval for Reuse	Reliability, safety, and compliance sign-off	Engineering review and documented approval	Signed authorization	Hold batch until approval complete

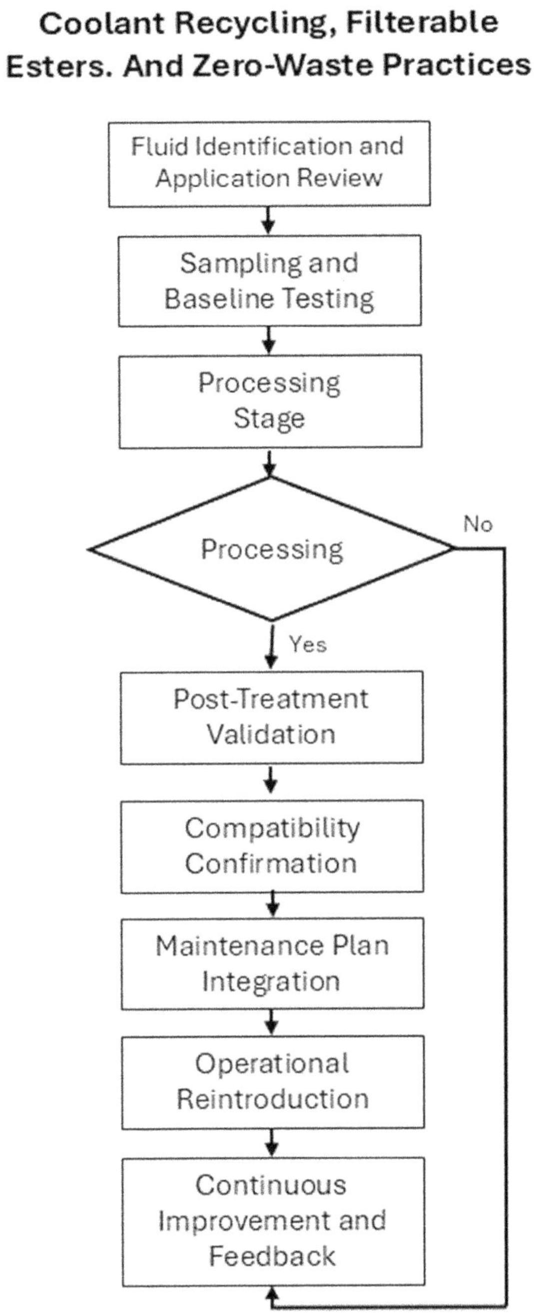

FIGURE 17.1 Coolant recycling, filterable esters, and zero-waste practices.

- Confirm whether the system is thermal (heat transfer), dielectric (insulation), or mixed service.
- Assign asset criticality rating from reliability database or FMEA.

2. **Sampling and Baseline Testing**
    - Pull representative samples from live system circulation points.
    - Conduct property tests: ISO 4406 particle count, Karl Fischer moisture, pH, conductivity, inhibitor concentration, TAN, glycol percentage, ion chromatography for ionic contamination.
    - For dielectric coolants, add breakdown voltage and dielectric dissipation factor.
3. **Recycling Feasibility Assessment**
    - Match fluid chemistry to available regeneration processes: filtration, ion exchange, vacuum dehydration, acid scavenging, inhibitor replenishment.
    - Bench-test fluid sample with proposed recycling method.
    - Verify no property drift beyond ±5% from OEM limits after trial treatment.
4. **Processing Stage**
    - Segregate by fluid chemistry to avoid cross-contamination.
    - Execute chosen treatment method under controlled parameters.
    - Record process variables: flow rate, filter media type and micron rating, temperature, chemical dosing, dwell times.
5. **Post-Treatment Validation**
    - Resample treated batch and repeat full property testing.
    - Compare against OEM specification, IEC/ASTM test standards, and baseline results.
    - For filterable esters, run hydrolytic stability (ASTM D2619), elastomer compatibility, and metal corrosion strip tests.
6. **Compatibility Confirmation**
    - Check treated fluid against all wetted materials in the system: elastomers, gaskets, coatings, metals.
    - Review chemical compatibility charts and conduct immersion testing if data is missing.
7. **Zero-Waste Compliance Review**
    - Verify full recovery or reuse pathway with vendor certification.
    - Document waste diversion metrics and chain of custody for fluid batches.
    - Ensure no portion is landfilled or incinerated unless non-recoverable by regulation.
8. **Maintenance Plan Integration**
    - Update CMMS or SCADA with fluid batch ID, treatment date, property data, and next sampling interval.
    - Set monitoring frequency based on asset criticality:
        - Critical systems: 3–6 month intervals
        - Noncritical: up to 12 months
    - Link sampling tasks to preventive maintenance work orders.
9. **Operational Reintroduction**
    - Secure engineering approval before refilling system.
    - Record volume and concentration adjustments.
    - Maintain 30-day follow-up sampling to confirm stability in service.

10. **Continuous Improvement and Feedback**
    - Review historical data for repeat failures or accelerated degradation.
    - Adjust treatment intervals, methods, or vendors based on trend analysis.
    - Feed failure or deviation data back into FMEA for asset class.

TOTAL FLUID MANAGEMENT (TFM) PROGRAMS IN ELECTRIFICATION AND POWER GENERATION SYSTEMS

A total fluid management (TFM) program is a structured, cradle-to-grave approach to selecting, storing, applying, monitoring, reclaiming, and disposing of all lubricants, insulating fluids, and coolants in a facility or fleet (see Figure 17.2). In electrification

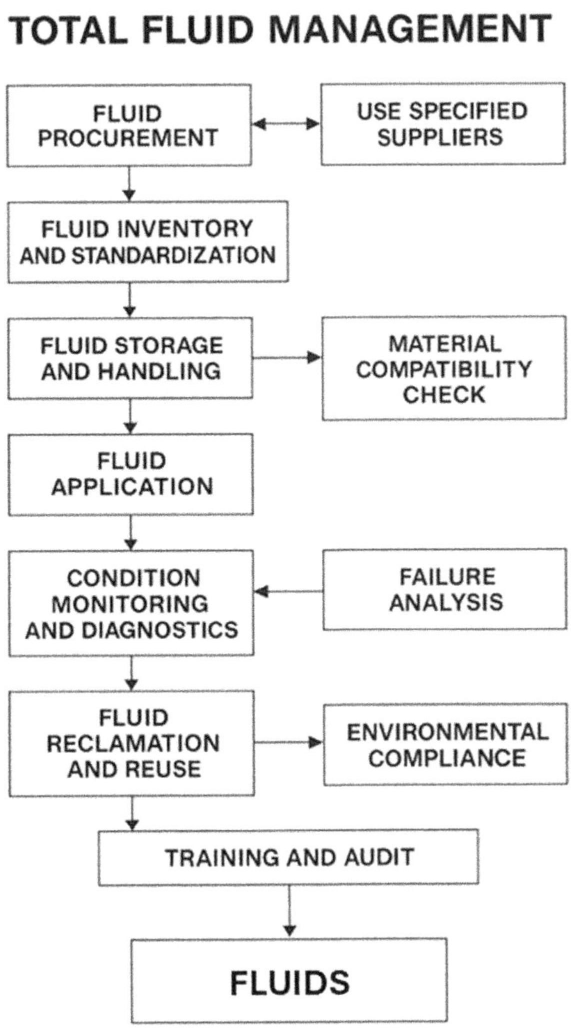

FIGURE 17.2 TFM program flow and control diagram.

and power generation, TFM integrates reliability engineering, condition-based maintenance, environmental compliance, and procurement control into a single management framework (sec Table 17.5).

TFT Core Elements

1. **Fluid Inventory and Standardization**
   - Maintain a master list of approved fluids by OEM specification and application.
   - Limit the number of fluid types to reduce cross-contamination risk and simplify storage.
   - Assign unique batch identifiers for traceability.
2. **Procurement Control**
   - Source fluids from qualified suppliers with ISO 9001 quality systems.
   - Require certificates of analysis for each delivery.
   - Include rerefined or reclaimed fluid options where reliability risk is acceptable.
3. **Storage and Handling**
   - Segregate fluids by category (lubricants, insulating fluids, coolants) and chemistry.
   - Use sealed, labeled containers and closed-transfer systems to avoid ingress of contaminants.
   - Monitor storage conditions: temperature, humidity, and light exposure.
4. **Application Protocols**
   - Match fluid to equipment using an application-compatibility matrix.
   - Verify material compatibility (metals, elastomers, coatings) before introduction.
   - Document fill volumes, methods, and flush procedures.
5. **Condition Monitoring and Diagnostics**
   - Implement scheduled sampling by asset criticality.
   - Use a full suite of tests: viscosity, TAN/TBN, particle count, water content, oxidation, wear metals, dielectric strength, inhibitor concentration, and specialized coolant or grease parameters.
   - Apply statistical trending and alarm limits within CMMS or SCADA systems.
6. **Reclamation and Reuse**
   - Apply on-site or vendor-based filtration, dehydration, readditization, and polishing for qualifying fluids.
   - Maintain property recovery validation through ASTM/IEC methods.
   - Ensure zero-waste or documented recovery pathways for spent fluids.
7. **Failure Analysis and Feedback**
   - Perform root cause analysis for fluid-related equipment failures.
   - Use findings to adjust fluid selection, maintenance intervals, and handling procedures.

TABLE 17.5
Comparative GHG Impact Table and Low-GWP Substitutions for Thermal and Insulation Fluids

Fluid Class	Common Fluids	Typical GWP100*	Notes Affecting Selection	Low-GWP Substitution Options
Dielectric gas	SF_6	**~25,200** (AR6 value; AR4 often 22,800–23,500 in legacy specs)	Highest warming impact among grid dielectrics; tight leak detection and recovery required	Fluoroketone blends such as **Novec 4710** or **Novec 5110** in N_2/CO_2 mixes, each **<1 GWP** when neat. Verify OEM performance maps and mixture ratios.
Dielectric liquid, immersion cooling	Perfluorocarbons such as $\mathbf{C_6F_{14}}$ (perfluorohexane, FC-72)	**~9,300** (AR4/US EPA inventory)	Very high GWP and long atmospheric lifetime	**FK-5-1-12** fluoroketone (**Novec 649**) with **GWP ≈1**; check materials and water-reactivity controls.
Electronic cooling solvent, test media	**HFE-7100**	**~320** (100-yr)	Mid-range GWP; reclaiming programs exist	For boiling or single-phase cooling, favor **FK-5-1-12** (GWP≈1) or qualified HFO diluents where dielectric spec allows.
Electronic cleaning or heat transfer	**HFE-7200**	**~55** (datasheet MSDS)	Lower than HFE-7100 but still nonzero	**FK-5-1-12** when heat flux and materials allow; evaluate ester or water-glycol where electrical clearance permits.
Refrigerant or secondary coolant examples in power electronics	**HFC-134a**; **HFC-245fa**	HFC-134a ~1,300–1,470; HFC-245fa ~858–950 depending on source and method	Use is shifting away due to policy limits	**HFO-1234yf** and **HFO-1234ze(E)** with **GWP ≤ 1–6** depending on method; confirm flammability class and TFA considerations.
Transformer insulating oil	Mineral oil	Not assigned a single GWP; use **LCA** for cradle-to-grave accounting	Emissions are indirect from losses, leaks, and end-of-life	**Natural or synthetic esters** with reclaim programs; target lower lifecycle emissions and improved fire point. *(Use site LCA.)*
Water-based coolants	Water, water-glycol	**1** for CO_2 reference; mixture impact is indirect	GWP is dominated by electricity usage and inhibitor production	Keep water-based where dielectric constraints are absent; push energy efficiency and inhibitor recycling.
Thermal storage and loops	Molten nitrates, carbonates, chlorides	No direct GWP; evaluate LCA for salt production and heaters	Emissions driven by auxiliary heating power and leaks of any co-used organics	Prefer salts when duty cycle fits; insulate, heat-trace efficiently, and avoid organic diluents.

*GWP100 values are 100-year time horizon. Use the same assessment basis when comparing options.

8. **Environmental and Regulatory Compliance**
   - Maintain documentation for REACH, RoHS, GHS, PFAS, and other applicable regulations.
   - Track disposal and recycling volumes for sustainability reporting.
9. **Training and Competency Development**
   - Provide technical training for operators and maintenance staff on handling, testing, and interpreting results.
   - Include updates on emerging fluid chemistries and regulatory changes.
10. **Program Auditing and Continuous Improvement**
    - Conduct annual audits against internal standards and external regulations.
    - Review key performance indicators: fluid consumption per operating hour, number of fluid-related failures, and recycling rate.

Reducing Greenhouse Gas (GHG) Impact from Thermal and Insulation Fluids

1. **Fluid Selection and Low-GWP Alternatives**
   - Replace high-GWP fluorinated fluids with low-GWP fluoroketones (e.g., Novec 649, 4710, 5110) or hydrofluoroolefins (HFOs).
   - For dielectric service, consider natural esters, synthetic esters, or high-purity mineral oils with documented low lifecycle GHG emissions.
   - In thermal systems, substitute synthetic aromatics and high-GWP siloxanes with eutectic molten salts, glycol-water blends, or nanoparticle-enhanced water-based coolants where performance allows.
2. **Lifecycle Assessment (LCA) Integration**
   - Perform LCA for each fluid option covering production, transport, operational emissions, and disposal/recycling impacts.
   - Prioritize fluids with low embodied energy and high recyclability rates.
3. **Leak Prevention and Containment**
   - For SF_6 and other fluorinated dielectric gases, implement leak detection with infrared sensors and automated alarm triggers.
   - Use hermetically sealed equipment designs where possible.
   - Maintain detailed leak history logs to quantify emissions for regulatory reporting.
4. **Recycling and Reuse Pathways**
   - Establish closed-loop reclamation for esters and mineral oils using filtration, dehydration, and readditization.
   - For fluorinated liquids, use certified reclaiming services to prevent venting.
   - Recover and reprocess glycol-based coolants with ion exchange and inhibitor replenishment.
5. **Operational Efficiency Gains**
   - Maintain optimal thermal conductivity and dielectric performance through proactive condition monitoring (viscosity, TAN/TBN, particle count, moisture, dielectric breakdown voltage).

- Avoid overcooling or overheating, which increases energy consumption and indirectly increases GHG emissions.

6. **End-of-Life Disposal and Carbon Accounting**
   - Document fluid disposal in line with GHS, REACH, and local environmental regulations.
   - Quantify avoided emissions through reclamation programs and integrate results into corporate carbon reporting.
7. **Regulatory Alignment**
   - Comply with EU F-Gas Regulation, U.S. EPA SNAP program, and Kyoto Protocol GWP reduction targets.
   - For PFAS-containing fluids, implement transition plans ahead of regulatory bans.

Field Guidance

For gas-insulated switchgear and lines, the largest GHG reduction comes from retiring SF_6 designs in favor of fluoroketone-based mixtures that meet dielectric duty with **<1 GWP** neat component values. Verify gas handling carts, leak detection thresholds, and IEC type tests for the exact mixture.

For immersion or direct-contact cooling where perfluorocarbons were used, qualify **FK-5-1-12** as the working fluid. Require moisture control, materials compatibility, and heat-transfer validation. The 3M datasheet lists **GWP = 1** and short atmospheric lifetime, which cuts Scope 1 risk materially compared to FC-72.

Where HFEs are still needed for cleaning or specialty heat transfer, pick **HFE-7200** over **HFE-7100** on a like-for-like basis due to the lower published GWP, and implement reclaim loops so solvent is not vented.

For secondary cooling and refrigeration around power electronics, move from HFCs to **HFO-1234yf** or **HFO-1234ze(E)** where flammability class A2L is acceptable and local code allows. Document risk controls and capture any TFA policy requirements in the environmental plan.

For transformer liquids, the climate lever is condition-based maintenance, leak prevention, and reclaiming programs. Where justified, migrate to natural or synthetic esters to reduce fire risk and improve LCA. Pair changes with online moisture and DGA controls.

Always cite the same GWP basis in specifications. If a vendor quotes AR4 values and your site policy uses AR6, restate the value set and require requoting to avoid false comparisons. The **GHG Protocol AR6 table** and **IPCC AR6** are the current references.

Circular Economy and Sustainable Fluid Management

Sustainable fluid management in electrification and power generation systems depends on closed-loop practices that minimize virgin fluid demand, extend service life, and reduce environmental liabilities. Circular economy strategies for lubricants, insulating fluids, and coolants include rerefining of mineral oils to API Group II/III quality, reclaiming of dielectric fluids through dehydration, degassing, and filtration, and chemical recovery of thermal fluids. Condition-based maintenance, guided by laboratory and online monitoring, enables fluids to remain in service until functional degradation reachcs defined thresholds, reducing unnecessary disposal. Coolant recycling programs incorporate inhibitor replenishment and contaminant removal to maintain thermal performance without full replacement. For ester-based systems, periodic filtration and acid number control allow multiple service cycles before end-of-life treatment.

End-of-life management prioritizes recycling over disposal, with waste fluid streams directed to licensed rerefiners or chemical processors that meet REACH, GHS, and local environmental compliance requirements. PFAS-containing fluids require documented containment, high-efficiency reclamation, and avoidance of open release. In systems with high GHG-impact fluids, substitution with low-GWP alternatives and rigorous leak management reduces Scope 1 emissions while maintaining performance. TFM programs integrate procurement, monitoring, reclamation, and disposal into a single operational framework with traceable lifecycle records, enabling compliance reporting and continuous improvement. The result is reduced material throughput, lower environmental footprint, and extended asset reliability through disciplined fluid stewardship supported by data-driven decision-making.

18 Lubricants, Insulating Fluids, and Coolants in Electrification and Power Generation Systems

Fluid management in electrification and power generation is a critical engineering discipline that ties mechanical reliability, electrical safety, and thermal control to long-term asset performance. Across lubricants, insulating fluids, and coolants, selection, application, and maintenance must be based on quantifiable property data, equipment-specific compatibility, and operational duty cycles. The interdependence of fluid function with mechanical, thermal, and electrical systems means that a weakness in one domain can propagate to others, producing compounded reliability risks.

Lubricants remain central to friction control, wear prevention, and powertrain efficiency in turbines, diesel engines, compressors, gearboxes, and electromechanical assemblies. Insulating fluids serve both dielectric and thermal management roles in transformers, switchgear, and high-voltage power electronics. Cooling fluids ensure temperature stability in thermal loops, renewable systems, generators, and battery packs, with properties tuned to specific heat loads and material compatibility.

Condition monitoring is the operational safeguard. Oil analysis, dielectric fluid testing, and coolant property verification provide actionable data to correct deviations before they cause damage. Laboratory methods such as viscosity, total acid number (TAN)/total base number (TBN), Fourier transform infrared spectroscopy (FTIR), dissolved gas analysis (DGA), Karl Fischer moisture, particle counts, and inhibitor concentration measurement must be coupled with online sensors and integrated into supervisory control and data acquisition (SCADA) or computerized maintenance management system (CMMS) for predictive intervention.

Failure forensics have shown that fluid-related breakdowns often originate from contamination, chemical degradation, or incompatibility with system materials. Case studies in varnish formation, grease phase collapse, moisture-induced dielectric breakdown, cavitation, and inhibitor depletion highlight the necessity for structured forensic analysis and corrective actions.

Emerging trends point toward self-healing lubricants, nanoparticle-enhanced fluids, ionic liquids, and molten salts with extended service ranges, along with AI-driven monitoring, digital twin simulations, and circular economy recovery systems. Sustainability is now a primary specification driver, with low-global warming

DOI: 10.1201/9781003775201-24

potential (GWP), recyclable, and biodegradable fluids replacing high-emission, persistent chemistry.

Long-term reliability in power systems depends on an engineered lifecycle approach to fluids: correct initial selection, verified material compatibility, standardized monitoring protocols, controlled replenishment, and environmentally compliant end-of-life processing. The integration of advanced fluids, predictive diagnostics, and closed-loop recovery supports both operational stability and compliance with tightening environmental and safety regulations.

Bibliography

Ambrosetti, G., Fedele, L., & Valenzano, L. (2020). Thermophysical properties of chloride salts for next-generation CSP. *Solar Energy*, 211, 316–329.

American Petroleum Institute. (2014). *API Standard 610: Centrifugal pumps for petroleum, petrochemical and natural gas industries* (11th edn). Washington, DC: API Publishing Services. ISBN 9781630810039.

American Petroleum Institute. (2014). *API Standard 682: Pumps shaft sealing systems for centrifugal and rotary pumps* (4th edn). Washington, DC: API Publishing Services. ISBN 9781630810121.

American Petroleum Institute. (2021). *API 1509: Engine oil licensing and certification system.* Washington, DC: API Publishing Services. ISBN 9781636710642.

API Standard 610. (2021). *Centrifugal pumps for the petroleum, petrochemical, and natural gas industries.* Washington, DC: API Publishing Services.

API Standard 682. (2014). *Pumps shaft sealing systems for centrifugal and rotary pumps.* Washington, DC: API Publishing Services.

ASME Section II and Section VIII. (2025). *Materials and pressure vessel rules.* New York, NY: American Society of Mechanical Engineers.

ASTM D2240. (2015). *Standard test method for rubber property Durometer hardness.* West Conshohocken, PA: ASTM International.

ASTM D4378-20. (2020). *Standard practice for in-service monitoring of mineral oil used in electrical equipment.* West Conshohocken, PA: ASTM International.

ASTM D471. (2021). *Rubber property effect of liquids.* West Conshohocken, PA: ASTM International.

ASTM D6210-21. (2021). *Standard specification for fully-formulated glycol base engine coolant for heavy-duty engines.* West Conshohocken, PA: ASTM International.

ASTM International. (2016). *ASTM D471-16: Standard test method for rubber property, effect of liquids.* West Conshohocken, PA: ASTM International. ISBN 9780803168377.

ASTM International. (2019). *ASTM D877-19: Standard test method for dielectric breakdown voltage of insulating liquids using disk electrodes.* West Conshohocken, PA: ASTM International.

ASTM International. (2021). *ASTM D974-21: Standard test method for acid and base number by color-indicator titration.* West Conshohocken, PA: ASTM International.

ASTM International. (2021). *ASTM D1500-21. Standard test method for ASTM color of petroleum products (ASTM Color Scale).* West Conshohocken, PA: ASTM International.

ASTM International. (2021). *ASTM D1816-21: Standard test method for dielectric breakdown voltage of insulating oils of petroleum origin using VDE electrodes.* West Conshohocken, PA. ASTM International.

ASTM International. (2016). *ASTM D7155: Evaluating compatibility of mixtures of turbine lubricating oils.* West Conshohocken, PA: ASTM International. ISBN 9780803171568.

ASTM International. (2016). *ASTM D7155-16: Standard practice for evaluating compatibility of mixtures of turbine lubricating oils.* West Conshohocken, PA: ASTM International. ISBN 9780803171568.

ASTM International. (2018). *ASTM D4172: Wear preventive characteristics of lubricating fluid four-ball method.* West Conshohocken, PA: ASTM International. ISBN 9780803176679.

ASTM International. (2018). *ASTM D4172-18: Standard test method for wear preventive characteristics of lubricating fluid (four-ball method).* West Conshohocken, PA: ASTM International. ISBN 9780803176679.

ASTM International. (2018). *ASTM D892: Foaming characteristics of lubricating oils*. West Conshohocken, PA: ASTM International.

ASTM International. (2018). *ASTM D943: Oxidation characteristics of inhibited mineral oils*. West Conshohocken, PA: ASTM International. ISBN 9780803175405.

ASTM International. (2018). *ASTM D943-18: Standard test method for oxidation characteristics of inhibited mineral oils*. West Conshohocken, PA: ASTM International. ISBN 9780803175405.

ASTM International. (2020). *ASTM D1533-20: Standard test method for water in insulating liquids by Coulometric Karl Fischer Titration*. West Conshohocken, PA: ASTM International.

ASTM International. (2019). *ASTM D2240-15e1: Standard test method for rubber property, Durometer hardness*. West Conshohocken, PA: ASTM International. ISBN 9780803186630.

ASTM International. (2020). *ASTM D2272: Standard test method for oxidation stability of steam turbine oils by rotating pressure vessel*. West Conshohocken, PA: ASTM International. ISBN 9780803180327.

ASTM International. (1999). *ASTM D3427: Air release of petroleum oils*. West Conshohocken, PA: ASTM International.

ASTM International. (2022). *ASTM D3612-22: Standard test methods for analysis of gases dissolved in electrical insulating oil by gas chromatography*. West Conshohocken, PA: ASTM International.

ASTM International. (2021). *ASTM D5837-21: Standard test method for furanic compounds in electrical insulating liquids by HPLC*. ASTM International.

ASTM International. (2020). *ASTM D6304: Water determination by Karl Fischer titration*. West Conshohocken, PA: ASTM International. ISBN 9780803177812.

ASTM International. (2020). *ASTM D6304-20: Standard test method for determination of water in petroleum products by Karl Fischer titration*. West Conshohocken, PA: ASTM International. ISBN 9780803177812.

ASTM International. (2021). *ASTM D664: Acid number of petroleum products by potentiometric titration*. West Conshohocken, PA: ASTM International. ISBN 9780803177140.

ASTM International. (2021). *ASTM D664: Standard test method for acid number of petroleum products by potentiometric titration*. West Conshohocken, PA: ASTM International. ISBN 9780803180914.

ASTM International. (2021). *ASTM D664-21: Standard test method for acid number of petroleum products by potentiometric titration*. West Conshohocken, PA: ASTM International. ISBN 9780803177140.

ASTM International. (2021). *ASTM D971: Standard test method for interfacial tension of oil against water by ring method*. West Conshohocken, PA: ASTM International. ISBN 9780803180143.

ASTM International. (2022). *ASTM D445: Standard test method for kinematic viscosity of transparent and opaque liquids*. West Conshohocken, PA: ASTM International. ISBN 9780803181058.

ASTM International. (2021). *ASTM D7843: Membrane Patch Colorimetry for turbine oils*. West Conshohocken, PA: ASTM International.

ASTM International. (2022). *ASTM D7843: Standard test method for measurement of lubricant generated insoluble color bodies*. West Conshohocken, PA: ASTM International. ISBN 9780803185681.

ASTM International. (2022). *ASTM D92-22: Standard test method for flash and fire points by Cleveland open cup tester*. West Conshohocken, PA: ASTM International. ISSN: 1520-7483.

ASTM International. (2023). *ASTM D5864-23: Standard test method for determining aerobic aquatic biodegradation of lubricants or their components*. West Conshohocken, PA: ASTM International. https://doi.org/10.1520/D5864-23 ISSN: 1520-7483.

Bloch, H. P., & Budris, A. R. (2010). *Pump user's handbook: Life extension* (3rd edn). Lilburn, GA: Fairmont Press. ISBN 9780881736135.

Boyce, M. P. (2021). *Gas turbine engineering handbook* (5th edn). Oxford: Elsevier. ISBN: 9780128236833.

Boyde, S. (2002). *Transformers: Selection, operation, and maintenance*. New York, NY: CRC Press. ISBN: 9780824707669.

Bradshaw, R. W., & Meeker, D. E. (1990). High-temperature stability of ternary nitrate molten salts for solar thermal energy systems. *Solar Energy Materials*, 21(1), 51–60.

Bruce, R. W., & Stambaugh, K. (eds.). (2020). *Lubrication and maintenance of industrial machinery: Best practices and reliability*. Boca Raton, FL: CRC Press. ISBN: 9780367331694.

Castro, H. F., & Ramos, L. P. (2019). *Industrial lubricants: Fundamentals and applications*. Cham: Springer. ISBN: 9783030199532.

CIGRÉ Technical Brochure 771. (2019). *Condition assessment of insulating liquids in transformers*. Paris: International Council on Large Electric Systems. ISBN: 9782858735388.

Ebbens, S., & Vincent, B. (2012). *Polymer microcapsules for self-healing materials*. Cambridge: RSC Publishing. ISBN 9781849733568.

Electric Power Research Institute. (2020). *Mechanical seal reliability in plant water and glycol services* (Report No. 3002019432). Palo Alto, CA: EPRI. ISBN 9781646177097.

European Chemicals Agency (ECHA). (2023). *Annex XV restriction report: Proposal for a restriction on per- and polyfluoroalkyl substances (PFASs)*. Helsinki: ECHA.

European Sealing Association & Fluid Sealing Association. (2021). *Mechanical seal handbook* (8th edn). Wayne, PA: ESA/FSA Joint Publication. ISBN 9781737492702.

Fernandez, A. G., Galleguillos, H. R., Perez, F. J., Fuentealba, E., & Ushak, S. (2014). Corrosion of materials in molten nitrate salts for thermal energy storage in CSP plants. *Renewable and Sustainable Energy Reviews*, 30, 788–799.

Ghaffari Mosanenzadeh, S., Naguib, H. E., & Park, C. B. (2016). Self-healing polymeric systems: A review of material design, properties, and applications. *Polymer Reviews*, 56(3), 131–175. ISBN 9781498744621.

Hemmerich, K. J. (2017). *Dielectric fluids for electrical engineering*. Berlin: De Gruyter. ISBN: 9783110463980.

IAEA-TECDOC series. (2023). *Materials and chemistry for molten salt reactor systems*. Vienna: International Atomic Energy Agency.

IEA SolarPACES. *Recommended practices for molten salt CSP operation and maintenance*. Paris: International Energy Agency.

IEC 60079-10-1. (2020). *Explosive atmospheres classification of areas*. Geneva: IEC.

IEC 60422:2013. (2013). *Mineral insulating oils in electrical equipment – Supervision and maintenance guidance*. Geneva: IEC. ISBN: 9782832216018.

IEC. (2025). *IEC 60156: Insulating liquids AC breakdown voltage test*. Geneva.

IEC. (2012). *IEC 60247: Measurement of relative permittivity, dielectric dissipation factor, and DC resistivity of insulating liquids*. Geneva: IEC. ISBN 9782832215197.

IEC. (2015). *IEC 60599: Mineral oil-impregnated electrical equipment in service, guide to the interpretation of dissolved and free gases analysis*. Geneva: IEC. ISBN 9782832237120.

IEC. (2018). *IEC 60156: Insulating liquids, Determination of the breakdown voltage*. Geneva: IEC. ISBN 9782832240335.

IEEE Power & Energy Society. (2021). *IEEE C57.147-2021: Guide for acceptance and maintenance of natural ester fluids in transformers*. Piscataway, NJ: IEEE. ISBN: 978-1-5044-7700-5.

IEEE Std C57.106-2015. *IEEE guide for acceptance and maintenance of insulating oil in equipment*. Piscataway, NJ: IEEE Standards Association.

International Electrotechnical Commission. (2017). *IEC 60567: Oil-filled electrical equipment, Sampling of gases and analysis of free and dissolved gases*. Geneva: IEC. ISBN 9782832250672.

International Electrotechnical Commission. (2018). *IEC 60156: Insulating liquids Determination of the breakdown voltage at power frequency*. Geneva: IEC. ISBN 9782832250696.

International Electrotechnical Commission. (2018). *IEC 60247: Insulating liquids Measurement of relative permittivity, dielectric dissipation factor, and DC resistivity*. Geneva: IEC. ISBN 9782832250702.

International Electrotechnical Commission (IEC). (2018). *IEC 60296: Fluids for electrotechnical applications – Unused mineral insulating oils for transformers and switchgear*. Geneva: IEC. ISBN: 9782832252825.

International Electrotechnical Commission. (2022). *IEC 60599: Mineral oil filled electrical equipment in service Guidance on interpretation of dissolved and free gases analysis*. Geneva: IEC. ISBN 9782832260602.

International Electrotechnical Commission. (2022). *IEC 61099: Insulating liquids – Specifications for unused synthetic organic esters for electrical purposes*. Geneva: IEC. ISBN: 978-2-8322-6232-5.

International Electrotechnical Commission. (2023). *IEC 60296: Fluids for electrotechnical applications – Unused mineral insulating oils for transformers and switchgear*. Geneva: IEC. ISBN: 978-2-8322-6245-5.

International Organization for Standardization. (2015). *ISO 1817: Rubber, Determination of the effect of liquids*. Geneva: ISO. ISBN 9789267103928 ISBN: 9780849324383 ISBN: 978-92-9481-846-2.

ISO 1817. *Rubber: Determination of the effect of liquids*. Geneva: ISO. ISSN: 0301-4797.

ISO 4406. *Hydraulic fluid power contamination code*. Geneva.

ISO 5167. *Flow measurement by differential pressure devices*. Geneva.

Jones, D. A. (2014). *Principles and prevention of corrosion* (2nd edn). Upper Saddle River, NJ: Pearson. ISBN: 9780133599930.

Jones, F. R., & Childs, P. R. N. (2017). *Mechanical design of pumps: Fundamentals with applications*. Amsterdam: Elsevier. ISBN 9780081020881.

Karassik, I. J., Messina, J. P., Cooper, P., & Heald, C. C. (2012). *Pump handbook* (4th edn). New York, NY: McGraw-Hill Education. ISBN 9780071832159.

Khanna, S. K., & Arora, R. K. (2018). *Power plant engineering* (4th edn). New Delhi: Dhanpat Rai & Sons. ISBN: 9789352167489.

Kuffel, E., Zaengl, W. S., & Kuffel, J. (2000). *High voltage engineering fundamentals* (2nd edn). Oxford: Newnes.

Kuravi, S., Trahan, J., Goswami, D. Y., Rahman, M. M., & Stefanakos, E. K. (2013). Thermal energy storage technologies and systems for concentrating solar power plants. *Progress in Energy and Combustion Science*, 39(4), 285–319.

Liscinsky, D. S., & Chorpening, B. T. (2017). *Combined heat and power systems: advanced applications*. Oxford: Elsevier. ISBN: 9780081013658.

Liu, X., & Wu, D. (2018). Nanoparticle-based smart lubricants for improving tribological performance. *Tribology International*, 119, 272–279. ISBN 9780128147794.

McLyman, W. T. (2011). *Transformer and inductor design handbook* (3rd edn). Boca Raton, FL: CRC Press. ISBN: 9781439810308.

Mortier, R. M., Fox, M. F., & Orszulik, S. T. (2011). *Chemistry and technology of lubricants* (3rd edn). Springer. ISBN: 9789400715998.

NACE SP0408. *Methods for corrosion control in high temperature process streams*. Houston, TX: AMPP.

National Fire Protection Association. (2023). *NFPA 850: Recommended practice for fire protection for electric generating plants and high voltage direct current converter stations.* Quincy, MA: NFPA. ISBN: 9781455929117.

Nelson, W. E. (2019). *Practical guide to industrial water treatment*. Tulsa, OK: PennWell Corporation. ISBN 9781593704666.

Otero, J., & Fernández, J. (2019). *Heat transfer fluids and systems for process and energy applications*. Boca Raton, FL: CRC Press. ISBN: 9781138554458.

Pollaczek-Geir, M. (ed.). (2000). *High voltage engineering: Fundamentals*. Boca Raton, FL: CRC Press.

Rakov, V. A., Kezunovic, M., & Zhang, X. (2018). *High-voltage insulation technology: Fundamentals, condition monitoring, testing, diagnosis*. Boca Raton, FL: CRC Press. ISBN: 9781138613315.

Rudnick, L. R. (ed.). (2020). *Synthetics, mineral oils, and bio-based lubricants: Chemistry and technology* (3rd edn). Boca Raton, FL: CRC Press. ISBN: 978-0-367-18714-4.

Samiee, S., & Ahmadi, G. (2017). Development and application of self-healing dielectric fluids. *IEEE Transactions on Dielectrics and Electrical Insulation*, 24(6), 3729–3738. ISBN 9781509040021.

Serp, J., Allibert, M., Beneš, O., Delpech, S., Feynberg, O., Ghetta, V., et al. (2014). Review of materials and chemistry issues for molten salt reactors. *Progress in Nuclear Energy*, 77, 308–319.

Shreve, R. N., & Brink, J. A. (2018). *Chemical process industries* (5th edn). New York, NY: McGraw-Hill Education. ISBN 9780070342079.

Speight, J. G. (2020). *Handbook of industrial hydrocarbon processes* (2nd edn). Oxford: Gulf Professional Publishing. ISBN: 9780128099230.

Stutz, B., & Colas, H. (2018). *Thermal energy storage with molten salts: Technologies and applications*. Hoboken, NJ: Wiley. ISBN: 9781118565985.

Thapar, B. L., & Prakash, R. (2006). *Power cable technology: Engineering fundamentals*. New Delhi: Springer. ISBN: 9788185330640.

Totten, G. E. (2017). *Handbook of lubrication and tribology: Volume I* (2nd edn). Boca Raton, FL: CRC Press. ISBN: 9781498708746.

Totten, G. E. (ed.). (2011). *Handbook of lubrication and tribology* (2nd edn, Vol. I–III). Boca Raton, FL: CRC Press. ISBN: 9781420089357.

Totten, G. E. (ed.). (2018). *Handbook of hydraulic fluid technology* (2nd edn.). Boca Raton, FL: CRC Press. ISBN: 978-1-4987-7076-1.

United Nations Environment Programme (UNEP). (2022). *Kigali Amendment to the Montreal Protocol: HFC phase-down schedules and obligations*. Nairobi: UNEP Ozone Secretariat. https://ozone.unep.org/treaties/montreal-protocol ISBN: 978-92-807-3899-8.

U.S. Department of Energy. (2021). *Transformer oil and fluid analysis for reliability-centered maintenance*. Washington, DC: DOE Office of Electricity. ISBN: 978-1-7329-1760-8.

U.S. Environmental Protection Agency (EPA). (2020). *SF_6 emission reduction strategies in electric power systems*. Washington, DC: US EPA 430-R-20-001.

United States Environmental Protection Agency. (2020). *Used oil management standards: 40 CFR Part 279*. Washington, DC: Government Printing Office. ISBN 9780160952749.

United States Environmental Protection Agency (EPA). (2022). *PFAS strategic roadmap: EPA's commitments to action 2021–2024*. Washington, DC: US EPA. www.epa.gov/pfas/pfas-strategic-roadmap-epas-commitments-action-2021-2024. ISBN: 978-1-7345-6825-3.

Venkanna, B. K. (2010). *Fundamentals of heat and mass transfer*. New Delhi: PHI Learning. ISBN: 9788120331963.

Viswanathan, R. (2018). *Power plant materials: Selection and behavior*. London: ASM International. ISBN: 9781627081661.

Wakil, M. M. E. (2017). *Power plant technology* (2nd edn). New York, NY: McGraw-Hill Education. ISBN: 9780070616950.

Wiley, R. S. (2015). *Maintenance engineering handbook* (8th edn). New York, NY: McGraw-Hill Education. ISBN 9780071826615.

Williams, D. F., Clarno, K. T., & Middlehurst, M. (2008). Technology and corrosion challenges in molten salts. *Journal of Nuclear Materials*, 376(3), 229–232.

World Bank. (2021). *Global program on sustainability: Guidance for measuring and reporting greenhouse gas emissions*. Washington, DC: The World Bank. ISBN: 978-1-4648-1690-1.

Zavoico, A. B. (2001). *Solar power tower design basis document*. SAND2001-2100. Albuquerque, NM: Sandia National Laboratories.

Zhang, Z., & Suresh, S. (2022). Environmental and thermal performance of dielectric coolants for electrified power systems. *Applied Thermal Engineering*, 201, 117718. https://doi.org/10.1016/j.applthermaleng.2021.117718 ISSN: 1359-4311.

Zolkos, S., Young, C., & McDonough, C. (2021). PFAS in industrial applications: Environmental persistence and substitution challenges. *Journal of Environmental Management*, 298, 113506. https://doi.org/10.1016/j.jenvman.2021.113506

Zubaidy, E. A. H., & Abouelnasr, D. M. (2020). Used lubricating oil management: Options for environmental sustainability. *Renewable and Sustainable Energy Reviews*, 131, 110023. https://doi.org/10.1016/j.rser.2020.110023 ISSN: 1364-0321.

APPENDICES

INDUSTRY STANDARDS CROSS-REFERENCE (ASTM, IEEE, ISO, SAE)

TABLE A.1
Lubricants in Electrified and Power Generation Equipment

Function/ Product Type	ASTM Standards	IEEE Standards	ISO Standards	SAE Standards
Turbine Oils	ASTM D4304 – Standard Specification for Turbine Oils for Steam, Gas, and Combined-Cycle Turbines	IEEE 1206 – Guide for Maintenance of Turbine Oils for Generators	ISO 8068 – Lubricating Oils for Turbines	SAE AMS 3150 – Lubricating Oil, Turbine Engine
Gear Oils	ASTM D7038 – Oxidation Stability of Lubricants	N/A	ISO 12925-1 – Industrial Gear Oils	SAE J2360 – Automotive Gear Lubricants (Non-Synchronized)
Greases	ASTM D4950 – Classification for Automotive Service Greases	N/A	ISO 6743-9 – Classification of Lubricants, Greases	SAE J310–Automotive Lubricating Greases
Transformer Tap-Changer Lubricants	ASTM D3487 (overlap with dielectric fluids for mineral-based lubricants)	IEEE C57.131 – Requirements for Tap-Changers	ISO 8068 (for lubricating fluids)	N/A

TABLE A.2
Insulating Fluids (Dielectric Fluids)

Application	ASTM Standards	IEEE Standards	ISO Standards	SAE Standards
Mineral Oil-Based Transformer Fluids	ASTM D3487 – Mineral Insulating Oils	IEEE C57.106 – Acceptance and Maintenance of Insulating Oil	ISO 20696-Insulating Liquids – Specifications for Unused Mineral Insulating Oils	N/A
Natural/Synthetic Ester Fluids	ASTM D6871 – Natural Ester Insulating Fluids; ASTM D2668 – Synthetic Esters	IEEE C57.147 – Acceptance and Maintenance of Natural Ester Fluids	ISO 15380 (synthetic esters, biodegradable fluids)	N/A
Silicone-Based Insulating Fluids	ASTM D2225 – Silicone Fluids for Electrical Applications	IEEE C57.111–Guide for Silicone Liquids	ISO 6427 (fluid testing methods)	N/A
Gas-Immersed (SF_6 Alternatives)	ASTM D2472 – SF_6 Purity	IEEE C37.122.3 – Application of SF_6 Alternatives	ISO 14097 – Greenhouse Gas Testing	SAE ARP 5316 – Handling of Halogenated Dielectrics

TABLE A.3
Cooling Fluids/Thermal Transfer Media

Fluid Type	ASTM Standards	IEEE Standards	ISO Standards	SAE Standards
Water-Glycol Coolants	ASTM D1384 – Corrosion Testing of Engine Coolants; ASTM D3306 – Glycol Base Engine Coolant for Automobiles and Light Trucks	N/A	ISO 22241 – DieselExhaust Fluid Quality (related chemistry)	SAE J1034–Engine Coolant Testing
Dielectric Coolants for EVs	ASTM D7042 – Viscosity by Stabinger Viscometer; ASTM D924 – Dielectric Constant	IEEE 1276–Thermal Management in Power Electronics	ISO 12937–Thermal Conductivity of Fluids	SAE J2840 – ElectricVehicle Coolant Performance
Nanofluids	ASTM E2456 – Nanotechnology Terminology; ASTM E3129 – Characterization of Nanoparticles in Colloidal Suspension	N/A	ISO/TR 19716-Nanofluid Terminology and Measurement	N/A
Hydrogen-Cooled Generators	ASTM D3612 – Gas-in-Oil Analysis; ASTM D4057 – Manual Sampling	IEEE C57.106 – InsulatingFluid Maintenance	ISO 14687 – HydrogenFuel Quality	SAE J2579 – HydrogenFuel Systems

TABLE A.4
Condition Monitoring and Diagnostics

Test Category	ASTM Standards	IEEE Standards	ISO Standards	SAE Standards
Oil Analysis	ASTM D445 – Kinematic Viscosity; ASTM D664 – Acid Number; ASTM D4378 – Turbine Oil Monitoring	IEEE 943 – Guide for Field Testing of Fluid Insulation	ISO 2049 – Lubricants-Viscosity Classification	SAE J300 – Engine Oil Viscosity Grades
Dielectric Fluid Testing	ASTM D1816 / D877 – Breakdown Voltage; ASTM D924 – Dielectric Constant	IEEE C57.104 – Interpretation of Dissolved Gas Analysis	ISO 9104 – Electrical Insulating Liquids – Dissolved Gas Analysis	N/A
Coolant Testing	ASTM D1121 – Reserve Alkalinity; ASTM D5827 – Thermal Conductivity	N/A	ISO 22241 (if urea present in cooling systems)	SAE J1034-Coolant Performance Tests

FLUID COMPATIBILITY CHART (ELASTOMERS, METALS, COATINGS)

TABLE A.5
Elastomers vs Fluids

Elastomer	Mineral Oil (Transformer/Turbine)	PAO/Synthetic Hydrocarbon	Natural Ester (Vegetable)	Synthetic Ester (POE/Phosphate Ester*)	Silicone Dielectric	Water-Glycol (EG/PG)	DI Water	Fluoroketone (FK)
NBR (Buna-N)	A	A	B	C	C	C	C	B
HNBR	A	A	B	C	C	C	C	B
FKM (Viton)	A	A	B	C to D*	A	B	B	A
FFKM (Perfluoroelastomer)	A	A	A	A	A	A	A	A
EPDM	C	C	A	A*	C	A	A	B
Aflas (TFE-P)	A	A	A	B	A	A	A	A
VMQ (Silicone rubber)	B	B	B	C	A	C	C	B
FVMQ	B	B	B	C	A	C	C	B
PU (polyurethane)	C	C	C	D	C	C	C	B
PTFE	A	A	A	A	A	A	A	A
PEEK	A	A	A	A	A	A	A	A

*Phosphate esters: very aggressive to many elastomers. EPDM is usually preferred; FKM often swells and may crack. Verify with exact formulation.

TABLE A.6
Metals vs Fluids

Metal/Alloy	Mineral oil	PAO	Natural Ester	Synthetic ester (POE/ Phosphate Ester*)	Silicone	Water-Glycol	DI Water	Fluoroketone
Carbon steel	A (with inhibitor)	A	A	B*	A	B (pH control)	B to C	A
Stainless 304/316	A	A	A	A	A	A	A	A
Copper	B (sulfur/ DBDS risks)	A	B (catalyzes oxidation)	B*	A	C (corrosion if inhibitors depleted)	C	A
Brass/Bronze	B	A	B	C* (zinc leaching)	A	C	C	A
Aluminum	A	A	A	B	A	C (pitting if inhibitor loss)	C (pitting)	A
Zinc Plated	B	B	B	D*	B	D	D	A
Galvanized Steel	B	B	B	D*	B	D	D	A
Magnesium	B	B	B	C	B	D	D	A
Titanium	A	A	A	A	A	A	A	A
Nickel/Ni-P	A	A	A	A	A	A	A	A

*Phosphate esters attack zinc, cadmium, and many yellow metals. Use stainless or nickel alloys. For mineral oils: watch copper-catalyzed oxidation; apply passivators and meet ASTM D130 limits.

TABLE A.7
Coatings, Plastics, Linings vs Fluids

Material/Coating	Mineral Oil	PAO	Natural Ester	Synthetic Ester (POE/Phosphate Ester)	Silicone	Water-Glycol	DI Water	Fluoroketone
Epoxy (Amine-cured)	A	A	A	A	A	A	A	A
Phenolic/Novolac	A	A	A	A	A	A	A	A
Polyurethane Coating	B	B	B	C	B	B	B	B
Alkyd Enamel	A	A	B	D	A	B	B	A
PTFE/PFA Lining	A	A	A	A	A	A	A	A
PVDF	A	A	A	A	A	A	A	A
Nylon (PA6/PA66)	B	B	C (hydrolysis risk)	C	B	C	C	B
Acetal (POM)	A	A	B	B	A	B	B	A
Polycarbonate	B	B	C	D	B	C	C	B
ABS	B	B	C	D	B	C	C	B
HDPE/PP	A	A	A	B	A	A	A	A
Anodized Aluminum	A	A	A	B	A	B	C	A
Zinc-rich Primers	B	B	B	D	B	D	D	A

Critical Application Notes

- Water-glycol and DI water: maintain inhibitor package, pH 8 to 9.5, low chloride. Aluminum and yellow metals need nitrite-free, OAT-style inhibitors or manufacturer-approved packages.
- Natural and synthetic esters: monitor moisture; hydrolysis produces acids that attack zinc and copper. Use stainless, nickel, or passivated copper; specify paint systems resistant to esters.
- Phosphate esters: specify EPDM or FFKM seals, stainless or nickel alloys, epoxy or phenolic linings; avoid zinc, cadmium, galvanized, and common alkyd paints.
- Silicone fluids: generally benign to metals and coatings; some silicone rubbers swell in silicone liquids. Verify with ASTM D471 swell and hardness change.
- Fluoroketones: typically compatible with metals, PTFE, FKM, FFKM, PEEK, and most high-performance plastics. Check vendor datasheet for specific elastomer swell.

Validation Test References

- Elastomers: ASTM D471 or ISO 1817 immersion; report volume change, mass change, Shore hardness, tensile retention.
- Metals: ASTM G31 immersion corrosion; ASTM D130 copper strip for oils; ASTM D665 rust (A and B).
- Polymers/coatings: ASTM D543 chemical resistance; ASTM D1308 coating resistance.
- Coolants: ASTM D1384 glassware corrosion; ASTM D2570 simulated service; pH and reserve alkalinity by ASTM D1287/D1121.

TABLE A.8
Oil and Coolant Test Method Summary Table

Test Method	Standard (ASTM / ISO / IEC)	Property Measured	Typical Fluid(s)	Purpose/Application
Kinematic Viscosity	ASTM D445 / ISO 3104	Flow resistance at 40 °C and 100 °C (cSt)	Oils, dielectric coolants	Detects viscosity shifts from oxidation, contamination, or shearing
Viscosity Index	ASTM D2270 / ISO 2909	Rate of viscosity change with temperature	Oils	Monitors base stock quality, multigrade capability
Acid Number (TAN)	ASTM D664 / ISO 6618	mg KOH/g to neutralize acids	Oils, esters	Tracks oxidation, additive depletion, and acid formation
Base Number (TBN)	ASTM D2896 / D4739 / ISO 3771	mg KOH/g to neutralize acids	Engine oils	Indicates reserve alkalinity; monitors detergent/dispersant health
Water Content	ASTM D6304 (Karl Fischer) / ISO 12937	% or ppm water	Oils, esters, coolants	Detects ingress that accelerates corrosion, oxidation, and dielectric failure
Dielectric Breakdown Voltage	ASTM D1816 / D877 / IEC 60156	kV withstand without arcing	Insulating oils, esters, dielectric coolants	Assesses insulating capability for transformers and HV equipment
Dissolved Gas Analysis (DGA)	ASTM D3612 / IEC 60567	ppm gases (H_2, CH_4, C_2H_2, etc.)	Transformer oils, esters	Identifies electrical/thermal faults in transformers
Flash Point (COC/ PMCC)	ASTM D92 / D93 / ISO 2592	°C ignition temp	Oils, esters, coolants	Assesses volatility and fire hazard
Pour Point	ASTM D97 / ISO 3016	Lowest flow temperature	Oils	Confirms low-temp pumpability
Foaming Tendency	ASTM D892 / ISO 6247	Foam vol/stability	Oils	Evaluates aeration risk in circulation systems
Air Release	ASTM D3427 / ISO 9120	Time to release air	Oils	Prevents pump cavitation, improves bearing film stability

(*continued*)

TABLE A.8 (Continued)
Comparative GHG Impact Table and Low-GWP Substitutions for Thermal and Insulation Fluids

Test Method	Standard (ASTM / ISO / IEC)	Property Measured	Typical Fluid(s)	Purpose/Application
Oxidation Stability (TOST)	ASTM D943 / ISO 4263	Hours to acid/viscosity limits	Turbine oils, esters	Predicts service life under oxidation stress
Copper Strip Corrosion	ASTM D130 / ISO 2160	Tarnish rating (1–4)	Oils, esters, coolants	Detects corrosive sulfur or additives harmful to copper alloys
Rust Prevention	ASTM D665 / ISO 7120	Rust rating in water/salt	Oils	Confirms water contamination control
pH	ASTM D1287 / ISO 10523	pH units	Coolants, water-glycol	Tracks acid/alkaline shift affecting corrosion protection
Reserve Alkalinity	ASTM D1121	mL acid to neutralize	Coolants	Measures remaining corrosion inhibitor buffer
Corrosion (Glassware Test)	ASTM D1384	Metal weight loss	Coolants	Assesses multi-metal corrosion protection
Simulated Service Corrosion	ASTM D2570	Metal weight loss in hot loop	Coolants	Confirms protection in circulating systems
Thermal Conductivity	ASTM E1952 / ISO 22007	W/m·K	Coolants, dielectric liquids	Measures heat transfer capability
Specific Heat Capacity	ASTM E1269 / ISO 11357-4	J/g·K	Coolants, oils	Evaluates heat storage capacity
Density	ASTM D4052 / ISO 12185	g/mL	Oils, coolants	Checks formulation consistency
Particle Count	ISO 4406 / ASTM D7596	Particles/mL by size	Oils, coolants	Monitors cleanliness for wear prevention
Elemental Analysis (Wear Metals, Additives)	ASTM D5185 (ICP) / ISO 21079	ppm elements	Oils, coolants	Tracks wear, contamination, additive levels
FTIR Analysis	ASTM E2412	IR spectrum	Oils	Detects oxidation, nitration, glycol contamination

TRANSFORMER GAS SIGNATURE QUICK REFERENCE

(ppm levels and ratios should always be interpreted with load history, oil type, and sampling integrity in mind.)

TABLE A.9
Key Gases and Their Primary Sources

Gas	Symbol	Primary Fault Indications	Common Generation Mechanisms
Hydrogen	H_2	Partial discharges (PD), corona, some thermal faults	Ionization of oil or cellulose, stray discharges
Methane	CH_4	Low-temp thermal faults (<150 °C)	Slow oil decomposition in localized overheating
Ethane	C_2H_6	Low-to-medium temp thermal faults (150–300 °C)	Oil cracking without significant arcing
Ethylene	C_2H_4	Medium-to-high temp thermal faults (300–700 °C)	Pyrolysis of oil at hot spots
Acetylene	C_2H_2	High-energy arcing (>700 °C)	Arc discharge, severe breakdown
Carbon Monoxide	CO	Cellulose insulation overheating	Thermal degradation of paper and pressboard
Carbon Dioxide	CO_2	General cellulose degradation	Low-energy overheating or long-term aging
Oxygen	O_2	Leaks in sealed units, oxidation	Ambient ingress, oil oxidation
Nitrogen	N_2	Leaks, blanket gas, dissolved air	Normal component of air ingress

TABLE A.10
Typical Gas Signatures by Fault Type

(Based on relative gas concentration patterns; ppm values vary with transformer type/size.)

Fault Type	Signature Gas Pattern	Notes
Partial Discharge (PD)	H_2 ↑↑, CH_4 ↑, others low	Very high H_2, minor CH_4, negligible C_2 gases
Low-Temp Overheating (<150 °C)	CH_4 ↑, C_2H_6 ↑, little C_2H_4, no C_2H_2	Often oil decomposition only
Medium-Temp Overheating (150–300 °C)	CH_4 ↑, C_2H_6 ↑, C_2H_4 ↑, C_2H_2 ~0	More ethylene production
High-Temp Overheating (300–700 °C)	C_2H_4 ↑↑, CH_4 moderate, C_2H_6 low	Copper discoloration likely
Arcing (>700 °C)	C_2H_2 ↑↑, H_2 ↑↑, C_2H_4 ↑	Acetylene is strongest arc indicator
Cellulose Overheating	CO ↑, CO_2 ↑↑	CO_2/CO ratio <3 may indicate active fault
Thermal Fault + PD	H_2 ↑↑, C_2H_4 ↑, C_2H_2 moderate	Mixed signature, investigate both mechanisms

TABLE A.11
Useful Ratios (Duval/Rogers Methods)

Ratio	Formula	Diagnostic Clues
CH_4/H_2	$CH_4 \div H_2$	>1 suggests thermal over PD
C_2H_2/C_2H_4	$C_2H_2 \div C_2H_4$	>1 favors arcing; <0.1 favors overheating
C_2H_4/C_2H_6	$C_2H_4 \div C_2H_6$	>3 indicates high temp faults; <1 low temp
CO_2/CO	$CO_2 \div CO$	<3 suggests active paper insulation fault

Standards References

- **ASTM D3612** – Standard Test Method for Analysis of Gases Dissolved in Electrical Insulating Oil
- **IEEE C57.104** – Guide for the Interpretation of Gases Generated in Oil-Immersed Transformers
- **IEC 60599** – Interpretation of DGA in Oil-Immersed Transformers
- **Duval Triangle & Pentagons** – Empirical graphical method for fault classification

VENDOR PRODUCT INDEX BY APPLICATION

TABLE A.12
Lubricants for Power Generation and Electrified Systems

Application	Product Type	Example Vendor(s)	Example Product(s)	Notes
Turbine lubrication (steam, gas, combined-cycle)	ISO VG 32/46/68 turbine oils	Mobil, Shell, ExxonMobil, Chevron	Mobil DTE 732M, Shell Turbo T 46, Chevron GST Advantage	Meets ASTM D4304
Gearboxes in wind or hydro units	Synthetic PAO gear oils	Klüber, Fuchs, Castrol	Klübersynth GH 6-320, Fuchs Renolin Unisyn CLP 320	ISO 12925-1 CLP
EV driveline / traction motors	Dielectric lubricants	M&I Materials, Shell, Infineum	MIDEL eN 1204, Shell E-Fluids	Low permittivity, high breakdown voltage
Transformer tap changers	Mineral or ester-based lubricants	Shell, NYNAS, Cargill	Shell Diala S4 ZX-I, NYNAS Nytro 10XN	Use IEC 60296 fluid compatibility

TABLE A.13
Insulating Fluids

Application	Product Type	Example Vendor(s)	Example Product(s)	Notes
Mineral insulating oil	Inhibited/uninhibited mineral oils	NYNAS, Ergon, Apar	Nytro Libra, HyVolt II, APAR Transformer Oil	ASTM D3487
Natural ester insulating oil	Biodegradable vegetable-based	Cargill, M&I Materials	FR3, MIDEL eN 1204	ASTM D6871
Synthetic ester insulating oil	POE-based dielectric fluids	M&I Materials	MIDEL 7131	IEC 61099
Silicone insulating fluids	Polydimethylsiloxane	Dow, Elkem	Dow Corning 561, Elkem Syltherm 800	ASTM D2225

TABLE A.14
Cooling Fluids and Thermal Transfer Media

Application	Product Type	Example Vendor(s)	Example Product(s)	Notes
Water-glycol generator coolants	Ethylene/ propylene glycol-based	Dow, Arteco, BASF	Dow DOWTHERM SR-1, Arteco Freecor EV, BASF Glysantin	ASTM D1384 corrosion limits
Dielectric EV battery coolants	Synthetic hydrocarbons / FK fluids	3M, Solvay, Chemours	3M Novec 649, Solvay Galden HT, Chemours Opteon	Low permittivity, high flash point
Nanofluids for power electronics	Al_2O_3, TiO_2, CNT dispersions	Applied Nanotech, NanoAmor	Al_2O_3 nanofluid concentrate	Verify thermal conductivity boost
Hydrogen-cooled generator loop	Deionized water	Veolia, Suez	Veolia DI water supply	IEC 60507 water purity

TABLE A.15
Condition Monitoring and Diagnostics

Application	Product Type	Example Vendor(s)	Example Product(s)	Notes
Oil analysis kits	Field sampling kits	Trico, TestOil, Bureau Veritas	Trico Oil Safe Lab Kit, TestOil Sample Kit	ASTM D4057
DGA gas-in-oil analysis	On-line monitors	Vaisala, Kelman, Qualitrol	Vaisala Optimus, Kelman DGA 900, Qualitrol DGA 150	IEEE C57.104
Coolant chemical analysis	Corrosion inhibitor test strips	Hach, LaMotte	Hach Test Strips, LaMotte CoolTrak	ASTM D1121
Particle counting	Portable particle counters	Pall, Parker, Hydac	Pall PCM500, Parker icount, Hydac FCU 2210	ISO 4406 cleanliness codes

TABLE A.16
Seals, Elastomers and Gaskets

Application	Material	Example Vendor(s)	Example Product(s)	Notes
High-temp mineral oil service	FKM, FFKM	Parker, SKF, Greene Tweed	Parker V8909 FKM, Chemraz FFKM	ASTM D471 compatibility
Phosphate ester fluid service	EPDM	Precision Polymer Engineering, Trelleborg	Perlast ICE G75X, Trelleborg E75 EPDM	Avoid nitrile and FKM

TABLE A.17
Coatings, Linings and Paints

Application	Material	Example Vendor(s)	Example Product(s)	Notes
Tank lining for ester fluids	Phenolic epoxy	Sherwin-Williams, Hempel	Sherwin Macropoxy 646, Hempel PhenGuard	ASTM D543 chemical resistance
Exterior transformer coating	Polyurethane enamel	PPG, AkzoNobel	PPG SigmaCover, Akzo Interthane	ISO 12944 corrosion protection

TABLE A.18
Filtration and Purification

Application	Product Type	Example Vendor(s)	Example Product(s)	Notes
Transformer oil purification	Vacuum dehydrators	GlobeCore, Enervac, Pall	GlobeCore CMM series, Enervac EHV, Pall HLP	Removes water, gases, particulates
Coolant filtration	Side-stream filter skids	Eaton, Pentair	Eaton BECO, Pentair Cuno	ISO 16889 multi-pass

Glossary of Key Terms and Acronyms

A

- **Acid Number (AN or TAN)** – *Total Acid Number*, measured in mg KOH/g, indicating the acidity of a lubricant or insulating fluid (ASTM D664).
- **Air Release Value** – Time for entrained air to separate from oil under standard test conditions (ASTM D3427).
- **ASTM** – *American Society for Testing and Materials*, an international standards organization publishing test methods, specifications, and guides.

B

- **Base Number (BN or TBN)** – *Total Base Number*, measured in mg KOH/g, representing reserve alkalinity in engine oils (ASTM D2896/D4739).
- **Breakdown Voltage (BDV)** – Voltage at which dielectric fluid fails and allows electrical arcing (ASTM D1816/D877).
- **BS&W** – *Bottom Sediment and Water*, contamination found in oil storage or circulation systems.

C

- **C_2H_2 / C_2H_4 / C_2H_6 / CH_4** – Key hydrocarbon gases measured in transformer oil DGA: acetylene, ethylene, ethane, and methane.
- **CO / CO_2** – Carbon monoxide and carbon dioxide gases; key indicators of cellulose degradation in transformers.
- **Coolant Reserve Alkalinity (RA)** – Measure of buffering capacity against acid formation in coolants (ASTM D1121).
- **Copper Strip Corrosion Test** – Test for corrosive sulfur in fluids (ASTM D130).

D

- **DGA** – *Dissolved Gas Analysis*, laboratory or online measurement of gases dissolved in transformer insulating oil to detect faults.
- **DI Water** – *Deionized water*, used in high-purity cooling loops and hydrogen-cooled generator systems.
- **Dielectric Constant (Permittivity)** – Ratio of a material's electrical capacitance to vacuum; lower values desired for electrical insulation fluids.

E

- **EPDM** – *Ethylene Propylene Diene Monomer*, an elastomer often compatible with phosphate ester fluids but not with mineral oils.
- **Ester Fluids** – Natural or synthetic ester-based insulating or cooling fluids, known for biodegradability and high fire point.
- **Evaporation Loss** – Weight loss from oil under heat exposure, measuring volatility (ASTM D972).

F

- **FFKM** – *Perfluoroelastomer*, premium chemical-resistant elastomer used in severe service sealing.
- **Flash Point** – Lowest temperature at which a fluid generates enough vapor to ignite (ASTM D92/D93).
- **Fluid Compatibility** – Assessment of fluid's effect on elastomers, metals, and coatings under service conditions.

G

- **Gas-in-Oil Monitor** – Online DGA device providing continuous transformer health data.
- **Glycol Coolants** – Ethylene glycol (EG) or propylene glycol (PG) water solutions for heat transfer with corrosion inhibitors.

H

- **HNBR** – *Hydrogenated Nitrile Butadiene Rubber*, elastomer with improved heat and oil resistance over standard NBR.
- **Hydrogen (H_2) Signature** – Gas signature indicating partial discharges or early-stage faults in transformer oil.

I

- **IEC** – *International Electrotechnical Commission*, global standards body for electrical equipment and systems.
- **IEEE** – *Institute of Electrical and Electronics Engineers*, publishes transformer and power system guides (e.g., C57.104 for DGA).
- **Inhibited Oil** – Mineral oil with oxidation inhibitors to extend service life.

K

- **Karl Fischer Titration** – Analytical method to measure water content in oils and coolants (ASTM D6304).

L

- **Load Tap Changer (LTC)** – Device allowing voltage adjustment under load; often has separate oil supply from main transformer tank.
- **Lubricity** – Ability of a fluid to reduce wear and friction between surfaces.

M

- **Metal Passivator** – Additive that forms protective films on metals like copper and brass to prevent catalytic oxidation.
- **MIDEL** – Brand of synthetic and natural ester insulating fluids.
- **Mineral Oil** – Petroleum-derived insulating or lubricating fluid meeting ASTM D3487 specifications for electrical use.

N

- **NBR** – *Nitrile Butadiene Rubber*, common seal elastomer for mineral oils; limited in ester or phosphate ester service.
- **Natural Ester Fluid** – Biodegradable vegetable oil-based insulating fluid (e.g., ASTM D6871).

O

- **Oil Sampling Valve** – Fitting used to collect representative oil samples without introducing contaminants.
- **Oxidation Stability** – Resistance of fluid to degradation by oxygen and heat over time (ASTM D943 for turbine oils).

P

- **PAO** – *Polyalphaolefin*, synthetic hydrocarbon base stock for lubricants with excellent stability and low pour points.
- **Partial Discharge (PD)** – Localized electrical discharge not bridging conductors; detected in DGA as high H_2 with low C_2 gases.
- **pH** – Measure of coolant acidity/alkalinity, critical for corrosion prevention.

R

- **Reserve Alkalinity (RA)** – Coolant's buffering capacity against acid formation (ASTM D1121).
- **Rogers Ratio** – DGA fault diagnosis method based on gas ratios (IEEE C57.104).

S

- **SAE** – *Society of Automotive Engineers*, develops standards for lubricants, coolants, and materials.
- **Silicone Fluid** – High-temperature dielectric liquid with stable permittivity and fire resistance.
- **Sludge** – Oxidation byproduct in oils that can clog cooling ducts and reduce heat transfer.

T

- **TAN** – *Total Acid Number*, see Acid Number.
- **TOST** – *Turbine Oil Stability Test*, accelerated oxidation test (ASTM D943).
- **Transformer Oil** – Mineral, ester, or silicone-based insulating liquid used in power transformers.

V

- **Viscosity Index (VI)** – Measure of viscosity change with temperature; higher VI means less change across temperature range.
- **VMQ** – *Silicone Rubber* elastomer used in some sealing applications; may swell in silicone fluids.

W

- **Water Content** – Moisture present in fluid; critical for dielectric strength and corrosion control.
- **Water-Glycol Coolant** – Heat transfer fluid combining water and glycol with inhibitors to protect metals.

Index

For Product Safety Concerns and Information please contact our EU
representative GPSR@taylorandfrancis.com
Taylor & Francis Verlag GmbH, Kaufingerstraße 24, 80331 München, Germany

www.ingramcontent.com/pod-product-compliance
Ingram Content Group UK Ltd.
Pitfield, Milton Keynes, MK11 3LW, UK
UKHW022214190926
475702UK00003B/343

* 9 7 8 1 0 4 1 3 0 5 7 2 9 *